Northern Ecology and Resource Management

Northern Ecology and Resource Management

Memorial Essays Honouring Don Gill

edited by

Rod Olson
Ross Hastings
Frank Geddes

 The University of Alberta Press

First published by
The University of Alberta Press
Athabasca Hall
Edmonton, Alberta
Canada T6G 2E8

Copyright © The University of Alberta Press 1984

ISBN 0-88864-047-1

Canadian Cataloguing in Publication Data

Main entry under title:
Northern ecology and resource management

Includes bibliographical references
ISBN 0-88864-047-1

1. Ecology - Canada, Northern. 2. Ecology - Arctic regions. 3. Natural resources - Canada, Northern - Management. 4. Natural resources - Arctic regions - Management.
I. Olson, Rod. II. Hastings, Ross, 1956–
III. Geddes, Frank, 1955–
QH106.2.N55N67 1984   574.5′09719
C84-091033-9

All rights reserved.
No part of this publication may be reproduced, stored in a retrieval system, or transmitted in any form or by any means, electronic, mechanical, photocopying, recording, or otherwise, without prior permission of the copyright owner.

Typesetting by The Typeworks, Vancouver, British Columbia
Printed by Hignell Printing Limited, Winnipeg, Manitoba, Canada

| *Contents*

Don Gill (1934–1979)   *xiii*

Introduction   *xvii*

Abiotic Components   1

Some Considerations of Soil Development in Northwestern Canada and Some Ecological Relationships   W. W. PETTAPIECE   3
    Introduction   3
        Soil-Forming Processes   3
        What is a Zonal Soil in the North?   6
    Classification Implications   7
    Soil Climate as an Ecological Parameter   8
    Conclusions   14
    Acknowledgments   15
    References   15

Characteristics of Soil Temperature Regimes in the Inuvik Area   C. TARNOCAI   19
    Abstract   19
    Introduction   20
    Area and Climate   20
    Materials and Methods   22
        Location of Sites   22
        Instrumentation   23
        Analysis of Data   23

Results 26
  Mean Annual Soil Temperature 26
  Mean Summer Soil Temperature 27
  Minimum Soil Temperatures 31
  Maximum Soil Temperatures 31
  Frost-Free Days 32
  Days Above 5 C 33
Discussion 33
Summary 36
Acknowledgments 37
References 37

Deflation Measurements of Hietatievat, Finnish Lapland, 1974–77
MATTI SEPPÄLÄ 39
  Introduction 39
  The Study Region 39
  Climate of the Hietatievat Region 40
  Methods of Study 43
  Deflation Measurements 44
  Discussion 47
  Acknowledgments 48
  References 48

Snow and Living Things   WILLIAM O. PRUITT, JR.   51
  Introduction 51
  Taiga Snow Cover 52
    Cultural Effects 61
  Tundra Snow Cover 63
    Cultural Effects 70
  Snow Ecology Models 70
  Zones of Snow Influence and Evolution 72
  Conclusion 73
  References 73

Animal Communities 79

Population Dynamics of the Pine Marten (*Martes americana*) in the Yukon Territory W. R. ARCHIBALD and R. H. JESSUP 81
  Abstract 81
  Introduction 81
  Study Area 82
  Methods and Materials 83
    Study Area Design, Capturing, and Marking 83
    Relocations 84
    Carcass Analysis 84
  Results and Discussion 86
    Density 86

Home Range 88
Dispersal 91
Age and Sex Structure of a Harvested Sample 92
Reproduction 93
Management Application 95
Acknowledgments 96
References 96

Population Dynamics and Horn Growth Characteristics of Dall Sheep (*Ovis dalli*) and Their Relevance to Management
MANFRED HOEFS 99
Abstract 99
Introduction 99
Population Dynamics 101
Model of the Population Structure 102
Horn Growth Characteristics 105
Management Implications 108
On the Compensatory or Additive Nature of Sheep Hunting 111
Conservation Concerns 113
References 114

Winter Range Ecology of Caribou (*Rangifer tarandus*)
D. E. RUSSELL and A. M. MARTELL 117
Abstract 117
Introduction 118
Snow 119
Activity 122
Diet 125
Nutritional Physiology 131
Energetics 134
Summary 137
Acknowledgments 137
References 138

Circumpolar Distribution and Habitat Requirements of Moose (*Alces alces*) E. S. TELFER 145
Introduction 145
Distribution 147
  Pleistocene 147
  Historic and Modern Distribution 150
  Factors Limiting Distribution 151
    LANDFORMS 152
    CLIMATE 153
    VEGETATION 155
    DISEASE 155
    HUNTING 157
    PREDATION 157

Population Densities   157
Habitat Use   158
   Topography   161
   Vegetation   162
   Moose Movements and Habitat   164
      SPRING RANGE   165
      SUMMER-FALL RANGE   165
      EARLY WINTER HABITAT   166
      LATE WINTER HABITAT   167
Habitat Organization   170
   Boreal Forest   171
   Mixed Forest Regions   172
   Alluvial Habitat   173
   Tundra and Subalpine Shrub   173
   Stream Valley Habitats   174
Mechanisms Controlling Habitat Selection   174
   Energy Relationships   174
Summary   177
References   178

Population Growth in an Introduced Herd of Wood Bison (*Bison bison athabascae*)   GEORGE W. CALEF   183
   Abstract   183
   Introduction   184
   The Study Area   185
   Methods   186
      Estimation of Calf Production   187
   Results and Discussion   187
      Population Growth Rate   187
      Causes of the Irruption of the Wood Bison Population   192
      Comparison with Other Ungulate Population Irruptions   195
      The Future of the Wood Bison Population   195
   Afterword   196
   Acknowledgments   198
   References   198

Polar Bear (*Ursus maritimus*) Ecology and Environmental Considerations in the Canadian High Arctic   IAN STIRLING, WENDY CALVERT, and DENNIS ANDRIASHEK   201
   Introduction   201
   Life History of the Polar Bear   203
   Materials and Methods   205
   Results and Discussion   205
      Distribution and Movements   205
      Maternity Denning Areas   210
      Reproductive Parameters   211
      Age Structure and Mortality Rates   213

Estimates of Population Size   213
  Inuit Hunting Patterns and Utilization of Polar Bears   214
Vulnerability to Environmental Disruptions   216
  Disruption of Spring Feeding and Breeding Areas   217
  Disturbance of Maternity Denning Areas   218
  Disturbance of Summer Feeding and Refuge Areas   218
  Man-Bear Conflicts   219
  Changes in Inuit Hunting Patterns   219
  Other Considerations   219
Acknowledgments   220
References   220

Plant Communities   223

Lichen Woodland in Northern Canada   J. STAN ROWE   225
  Introduction   225
  Adaptations of Terricolous Lichens   227
  Woodland Structure Imposed by Substrate   229
  Lichen Woodland on Deep Drift   230
  The Soil Drought Hypothesis   232
  Conclusions   234
  References   235

Tundra Plant Communities of the Mackenzie Mountains, Northwest Territories; Floristic Characteristics of Long-Term Surface Disturbances   G. P. KERSHAW   239
  Introduction and Overview of the CANOL Project   239
  Results and Discussion   243
    Erect Deciduous Shrub Tundra   246
      HEDYSARUM ALPINUM–MOSS COMMUNITY   247
      SALIX RETICULATA–SALIX LANATA–MOSS COMMUNITY   280
      BETULA GLANDULOSA–CLADONIA STELLARIS–MOSS COMMUNITY   281
    Decumbent Shrub Tundra   282
      SALIX BARRATTIANA–MOSS COMMUNITY   283
      SALIX POLARIS–DACTYLINA BERINGICA COMMUNITY   284
      DRYAS INTEGRIFOLIA–CAREX SPP. COMMUNITY   284
    Sedge Meadow Tundra   285
      CAREX MEMBRANACEA–SPHAGNUM SPP. COMMUNITY   286
      MISCELLANEOUS MOSSES–CAREX PODOCARPA COMMUNITY   286
      CAREX SPP.–KOBRESIA SIMPLICIUSCULA COMMUNITY   287
    Lichen Heath Tundra   288
      DRYAS INTEGRIFOLIA–CASSIOPE TETRAGONA COMMUNITY   288
    Fruticose Lichen Tundra   290
      CLADONIA STELLARIS–ALECTORIA OCHROLEUCA COMMUNITY   290
    Cushion Plant Tundra   292

DRYAS INTEGRIFOLIA–CETRARIA TILESII COMMUNITY 292
DRYAS INTEGRIFOLIA–RHIZOCARPON UMBILICATUM COMMUNITY 293
SAXICOLOUS LICHENS–LECANORA EPINYRON COMMUNITY 294
Crustose Lichen Tundra 295
RHIZOCARPON INARENSE–UMBILICARIA PROBOSCIDEA COMMUNITY 296
Summary 297
Conclusions 302
Acknowledgments 305
References 306

Implications of Upstream Impoundment on the Natural Ecology and Environment of the Slave River Delta, Northwest Territories M. C. ENGLISH 311

Introduction 311
Origins of the Slave River Delta 312
The Active Delta 313
Processes 317
Cleavage Bar Development 319
Botanical Development 322
    Equisetum Assemblage 323
    Carex Assemblage 325
    Salix-Equisetum Assemblage 326
    Salix Assemblage 327
Autogenic Succession 327
    Alnus-Salix Assemblage 328
    Alnus Assemblage 328
    Populus Assemblage 330
    Decadent Populus Assemblage 330
    Picea Assemblage 331
Potential Environmental and Ecological Impacts of River Impoundment on the Slave Delta 331
    Short-Term Implications 332
    Long-Term Environmental and Ecological Implications 334
Acknowledgments 336
References 336

Land Use 341

Planning for Land Use in the Northwest Territories NORMAN M. SIMMONS, JOHN DONIHEE, and HUGH MONAGHAN 343

Abstract 343
Introduction 344
Synopsis of Development 345
Institutions and Problems 346

Principles   347
Tools for Implementation   349
A Practical Solution   351
  Proposed Policy and Structures for Northern Land-Use Planning   351
Discussion   356
Appendix — Abstract — Northern Land-Use Planning   358
Acknowledgments   362
References   363

Some Terrain and Land-Use Problems Associated with Exploratory Wellsites, Northern Yukon Territory   H. M. FRENCH   365
Introduction   365
Background   366
Regional Setting   366
Wellsite Terrain Disturbances   371
  Pre–Land-Use Operations   371
  Post–Land-Use Operations   374
  Sump-Related Problems   380
Discussion   382
Conclusions   383
Acknowledgments   383
References   384

Energy Development, Tourism, and Nature Conservation in Iceland   EDGAR L. JACKSON   387
Introduction   387
The Context of Resource Exploitation   388
Energy Resources   389
Recreation and Tourism   393
Resource Use Conflicts and Nature Conservation   396
  Energy-Related Conflicts, Impacts, and Compromises   396
  Recreation-Related Conflicts, Impacts, and Compromises   398
  Nature Conservation   399
Conclusions   400
Acknowledgments   401
References   401

Aklavik, Northwest Territories: "The Town that did not Die"   WILLIAM C. WONDERS and HEATHER BROWN   405
Introduction   405
Background   405
The Impact of Inuvik on Aklavik during the Construction Period, 1955 to 1961   408
The Characteristics of Aklavik-to-Inuvik Migrants, 1955 to 1961   413
Movements between Aklavik and Inuvik and the Characteristics of

Migrants, 1962 to 1974   415
Factors Favouring Migration to Inuvik   418
Factors Favouring Migration to Aklavik   419
Summary   420
Acknowledgments   422
Notes   422
References   423

Acknowledgments   425

About the Contributors   427

About the Editors   437

Don Gill
(1934-1979)

The late Professor Don Gill, Department of Geography, the University of Alberta, was born at Amasa, Michigan on 10 August 1934. Tragically, on Saturday, 28 July 1979, while en route with students to the Northwest Territories to carry out field work, the auto Don was driving collided with another vehicle just outside Peace River, Alberta. Don died before an ambulance could arrive at the scene. Fortunately, the other eleven involved in the accident sustained no serious injuries. It is in tribute to Professor Don Gill and his accomplishments in his brief lifespan of 45 years that *Northern Ecology and Resource Management* is dedicated.

Don is perhaps best remembered by his friends, colleagues, and students for his zest of life, his love of the outdoors, and the many facets to his character. Born into a family of eleven children, he reminisced frequently about his early youth in the forests of northern Michigan and how his grandfather helped to instil a love of nature and the outdoors by taking him out of school to go hunting, often without parental knowledge or acquiescence. His later life revealed a rare drive and determination to succeed at whatever he did. From 1950 to 1952, while still in his teens and with high school incomplete, he worked as a hunting and fishing guide in northern Michigan. At the age of 18 he enlisted in the United States Marine Corps, in which he served (1952-53) as a Marine Ranger in Korea. He rarely spoke of the war, but during his service he completed his high school education. In like manner, while working with the Marquette City Police, he completed his B.Sc. (with high honour) in 1961 at Northern Michigan University, Marquette, Michigan. He then moved to California, where he was science teacher, for the seventh and eighth grades, at Hawthorne Intermediate School, Hawthorne, California. From 1963 to 1965 he did graduate work at the University of California at Los Angeles. Even in an urban environment, Don's interest was in the outdoors. Who else would write an M.Sc. thesis on "Coyote and urban man — An analysis of the relationship between coyote and man in Los Angeles"?

I first met Don Gill in September 1965, when he came to the University of British Columbia for a Ph.D. program in northern studies. Physically, Don was imposing, standing 188 cm (6'2"), weighing 100 kg (220 lbs) or more, with a curly mop of hair and a boyish grin. Don was determined to undertake field work and write his thesis on some aspect of the boreal forest while I, who had field support for work in the tundra just north of the Mackenzie Delta, tried to persuade him to work there. We finally compromised — Don would work in the forest near the limit of trees in the Mackenzie Delta, but within daily view of the tundra of the nearby Caribou Hills. Don chose a field site about 50 km northwest of Inuvik and only a few kilometres from the tiny settlement of Reindeer Station. Travel was by freight canoe. Don was a 100 percent field man. His field camps, instrument layouts, and work schedule were models of efficiency. There was a place for everything and everything had its place — mooring for the canoe, a stand for the washbasin, a place for the axe. But Don did have his lapses, especially after shopping trips to Inuvik, when there were boisterous parties that must have been something to behold, according to the accounts that reached me. At such times Don regaled his audience, which often included assistants, a passing friend, or a nearby trapper, with embellished tales of his past and of antics he had masterminded.

Don accomplished in two seasons of field work what equally competent but less energetic graduate students would require three or four field seasons to achieve. Don was very fortunate in being able to draw upon the teaching, advice, and friendship of Vladimir Krajina, an ecologist of international stature at the University of British Columbia, in his ecological studies in the Mackenzie Delta. Although they were of very different personalities, Krajina was the perfect complement to Don's intellectual needs. Don was forever observant. He was interested in river break-up and freeze-up, the rate of ground thaw, the depth of the snow cover, sedimentation on a river bank, and the growth of a mushroom. His thesis, despite the efforts of his committee to compress its size, totalled nearly 700 pages and contained a wealth of information and research on the "Vegetation and environment in the Mackenzie River Delta, Northwest Territories—A study in subarctic ecology." He completed his thesis in 1971, three years after his appointment as an assistant professor at the University of Alberta.

Our paths crossed only occasionally at society meetings or on field trips after Don's thesis was finished. I have been fortunate in being able to draw liberally from the writings and work of others (Kershaw 1980).

From 1971 on Don was active in the North, conducting field studies, teaching winter and summer school courses, consulting for private industry and government, and leading or participating in professional field trips. He travelled widely in Canada, Alaska, Norway, Sweden, Finland, Iceland, and the British Isles, and attended the Second International Conference on Permafrost at Yakutsk, USSR.

Don made acquaintances with great rapidity and lived each day to the fullest. He was a prolific writer and competent teacher, which earned him a promotion to full professor six years after completing his Ph.D. From 1973 to 1976 he was also Director, Boreal Institute for Northern Studies, the University of Alberta, an independent institute supporting graduate student research in the North.

Don was a popular speaker, frequently invited to lecture to various groups and to present papers at symposia and conferences. His early writings and lectures dealt mainly with the Mackenzie Delta and topics stemming from his field work and thesis. He then became interested in the effects of damming northern rivers, a topic he was given to study for his Ph.D. pre-thesis defence. Although Don continued to maintain a keen interest in the Mackenzie Delta, his area of field work shifted to the Yukon Territory and the upper Mackenzie River Valley. His sudden death has deprived us of much unpublished research. At the time of his death he was about to take a year's leave of absence to conduct research on the Mackenzie River Basin for the Northwest Territories and federal governments and on the Liard River Basin for the British Columbia Hydro and Power Authority.

Northern Ecology and Resource Management is a tribute to the man and his work. To his family, to his son "John Mack" and his daughter Moline, we extend our deepest sympathy. Perhaps there is no more fitting tribute to a professor than an epitaph from a student: "Don's status as a northern ecologist/biogeographer was an inspiration to his students and gave him the international reputation he worked for and justly deserved. His dedication to a scientific understanding of the complete physical environment was admirable by any standards. He attempted to transfer much of his enthusiasm for biogeographical research to his students, hoping to instill in each of them an eagerness for independent ecological research and the rewards which it offered. It is a great loss to the science of northern ecology that he was deprived of his life while at the peak of his academic career and with a creative and productive future ahead of him."

Don Gill was "'one bad-ass dude' to borrow a phrase he often used" (Kershaw 1980).

J. Ross MacKay
Professor Emeritus
University of British Columbia

Reference

Kershaw, P. 1980. Obituary: Don Allyn Gill. Arctic 33(1): 215–219.

Introduction

Professor Don Gill, Department of Geography, the University of Alberta, passed away in July 1979 while en route to Canada's Northwest Territories. At the time of his death he was about to begin a major research project on river basin ecosystems. The accident was a tragic loss to the northern scientific community.

This volume was conceived as a tribute to the late Dr. Gill in recognition of his contribution to northern studies. It consists of invited papers from scientists who either worked with Professor Gill or who conducted research in topics closely related to his interests.

The theme of the book is the "ecology and resource management of northern regions." Most of the papers deal with issues that are directly applicable to northern Canada, but they are relevant to other high-latitude countries because the vulnerable characteristics of northern ecology and resource management are not unique to Canada.

There are many interpretations of what "the North" is. Latitude, position of summer isotherms, limits of permafrost, topographic expression, and tree line are all criteria that have been used to delineate northern regions. In Canada, the North is often politically defined as that vast expanse of land lying north of 60°N latitude. The rigorous climate and sensitive ecology of that region, however, often extend further south into boreal forest.

The North is experiencing unprecedented activity as the search for nonrenewable natural resources escalates in this region. The attractiveness of industrial growth has increased with technological advances that make such proposals more feasible. Throughout much of Canada's north, resource development has been almost wholly restricted to fur harvesting. Up to the Second World War the native inhabitants of Canada's north lived in much the same life style that their ancestors had for centuries. But the traditional ways and means of the northern people have now been subjected to many changes; snowmobiles have replaced dogsleds, rifles have replaced harpoons and snares. Today there are roads, railways, towns, and large industrial developments scattered across the region.

In 1977 Mr. Justice Thomas Berger, in his inquiry into northern development, stated that "to most Canadians the North is the immense hinterland of Canada that lies beyond the narrow southern strip of our country in which we live and work." We concur with his definition. At the same time we look upon the North as our last frontier. It is an arena that will present many challenges to future development. These challenges can best be met if we have an adequate appreciation of the complex nature of northern systems.

The North is playing an increasingly important role in the evolution of Canada. But uncontrolled activity in this sensitive area can lead to environmental and social problems. It is one of the world's last frontiers, and as such it is essential to understand the impact of development on the northern ecosystems and the appropriateness of mitigative measures. This book seeks to present information that will enable us to proceed in a manner that does not seriously compromise the fragile and unique northern ecosystem.

Abiotic Components

Some Considerations of Soil Development in Northwestern Canada and Some Ecological Relationships

W. W. Pettapiece

Introduction

Features of northern soils have been noted in conjunction with other investigations for some time, but it is only in the last two or three decades that specific pedological investigations have been conducted. Much of the early work on permafrost soils was conducted by Tedrow and his co-workers on the Alaskan North Slope and the Arctic Islands (Tedrow and Cantlon 1958, Tedrow et al. 1958) — areas beyond the tree line. In northwestern Canada exploratory surveys along the Mackenzie River noted the presence of soil with permafrost and commented on some of their unusual features (Leahey 1947, Nowosad and Leahey 1960, Day and Rice 1964). But in the early 1970s, in response to the large energy projects, pedological work was greatly accelerated in northwestern Canada (Zoltai and Pettapiece 1973, Tarnocai 1973). Most of this work was in the Subarctic — a hitherto much-neglected area in Canada. This flush of activity gave a greatly expanded data base for soils and soil development in the North. It resulted in a better understanding of pedogenic processes in the permafrost regions and the development of new classification concepts. This paper will attempt to provide clarification of these recent developments, of their relationship to classical concepts, and of some misconceptions in the literature.

Soil-Forming Processes

There are several ways to look at pedogenic processes. One approach is to consider individual processes of addition, removal, or reorganization of constituents. A more common approach is to consider broad, overall processes such as sod formation, podzolization, or calcification. The latter are groups of processes representing a general balance or trend, with the end result being on a particular genetic pathway. Each of these groups may include all the processes of addition, removal, or reorganization under the influence of all the controlling factors such as climate, vegetation, topography, and time. The terms, such as podzolization, represent a particular set of relationships under specific controlling conditions.

Early pedological work was conducted in the more temperate regions of the world, mainly in response to agrarian concerns. As soils investigations were pushed past the agricultural fringes the original concepts of soil development were carried with them, often with some misrepresentation. The term "podzolization," for example, was originally used for any process leading to the formation of an ashy, leached layer in a forest environment. More detailed studies indicated that there were two distinct processes that could result in a net leaching regime and the development of the ashy, near-surface horizon. One involved very acidic conditions, strong chemical weathering, and the release of iron and aluminum. It resulted in strongly coloured reddish subsoils containing a high percentage of amorphous materials. This process is still referred to as "podzolization." The second process is mainly a physical entrainment and movement of the finer soil constituents. It results in clayey subsoils and is called "lessivage." In Canada the current definition of a Podzol depends on the presence of amorphous Fe and Al (and organic matter) rather than on the presence or absence of an ashy Ae horizon. Those soils formed by the lessivage process are called Luvisols. There is consensus that the classical processes of podzolization and lessivage, characteristic of the boreal forests, diminish in intensity in the Subarctic and Arctic (Tedrow and Brown 1962, Retzer 1965, Karavayeva et al. 1965). This attenuation is due to factors such as lower temperatures and concomitant decreases in biological activity, chemical weathering, and leaching. The resultant soil profiles are less strongly differentiated. Gleization, a process associated with wetness, becomes more widespread as one moves into the tundra situation. Precipitation is low, but the cool, short summers with low evapotranspiration rates and the presence of near-surface permafrost often lead to excess moisture in the soil. The role of gleying processes in the classification of northern soils is, however, quite controversial (Tedrow and Cantlon 1958, Ugolini 1966, Liverovskii 1964).

Additional processes that are commonly applied to soil formation involve the distribution of organic material. One process is sod formation,

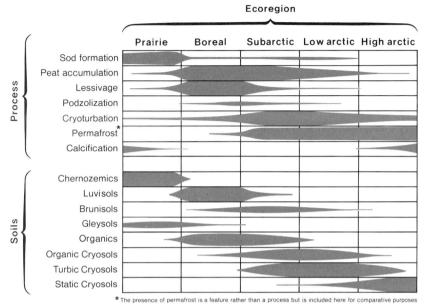

Fig. 1. Pedological processes and soils associated with various ecological regions in northwestern Canada.

the development of organic-mineral horizons associated with grasses and herbs. Another is the accumulation of peat associated with wetland vegetation.

Frost action can become quite disruptive and a definite factor in soil formation. The term "cryoturbation" is used to include all identifiable freeze-thaw processes.

The interrelationships of these processes are complex, but when they are correlated to major ecological regions (fig. 1) there appear to be some definite trends. Sod formation is mainly a prairie phenomenon. It practically disappears in the boreal region, but increases again in the Subarctic–Low Arctic with the increase in graminoid species, herbs, and ericaceous shrubs. Peat accumulation is low in the prairie ecoregion because of rapid mineralization, but increases rapidly in the boreal. It decreases again beyond the tree line, concomitant with diminishing biological activity, to a very low level in the High Arctic.

Lessivage is primarily a boreal process, but it is also evident well into the Subarctic. Podzolization is weak in the subhumid west and northwest, reaching a maximum in the boreal-subarctic region. It is only apparent in coarse-textured, excessively drained materials. Cryoturbation appears to reach a maximum in the Subarctic and Low Arctic. However, it still is a factor even into the prairie region. This is in contrast to the presence of permafrost, which is a Subarctic-Arctic phenomenon. Calcification, the accumulation of calcium carbonate in the solum (or at the surface) is a feature restricted to areas with pronounced dry periods—the Prairie and High Arctic regions.

The Subarctic, which typifies much of northwestern Canada, is characterized by peat accumulations in the wetter sites and lessening of the processes of lessivage and podzolization. Cryoturbation becomes a major factor on all medium and fine textured soils, and permafrost is nearly continuous.

The above discussion has alluded to several other topics that need to be considered. Among them is the problem of defining "zonal" soils in the North.

What is a Zonal Soil in the North?

There has been a good deal of controversy over this question. A zonal soil should be defined as a soil that is in equilibrium with and reflects the influences of the regional climate and vegetation in a major ecological zone. This concept was first espoused by Russian scientists; a discussion by Liverovskii (1964) explains the original intent of the concept and its application to the North. He followed the approach of Vysotskii, who introduced the term "height-of-land" in 1909. This meant a relatively flat water divide, a gently convex surface that received moisture only from atmospheric precipitation. Vysotskii suggested that only soils in these positions could be considered zonal. When the concept was translated from the Russian, height-of-land soils became "well-drained" soils, which indeed was the case for most regions of the world. However, when researchers began to study the soils of the North they used the concept of "well-drained" without regard to the original intent of a drainage divide, or as to how well it represented the zone. The predictable result was that only very coarse-textured or steeply sloping soils, which were not influenced by permafrost, were considered zonal. This idea was first stated by Gorodkov in Russia (1932 referenced in Liverovskii 1964) and carried forward by Tedrow and Cantlon (1958). However, many pedologists could not accept this definition of a zonal soil because it did not represent the normal or usual situation.

Another important aspect of the original definition was that zonal soils are formed from materials of intermediate moisture-retention capacity— medium-textured soils (Liverovskii 1964). This was stressed by several writers (Karavayeva et al. 1965, Ugolini 1966, Pettapiece 1975).

A complicating factor in the zonal argument is the role of permafrost and drainage. Nearly all medium- and fine-textured soils have permafrost close enough to the surface to affect biological activity. In other words, permafrost is part of the pedological realm (Pedosphere). Because permafrost occurs in the medium-textured soils over all landscape elements and supports the zonal vegetation, Liverovskii (1964), Karavayeva et al. (1965), Ugolini (1966), and others argued that permafrost should therefore be considered a characteristic of the zonal soil. It follows that "overmoistening" and gley features in the lower profile should also become zonal features (Karavayeva et al. 1965) and do not relegate these soils to the poorly drained, azonal category. Canadian pedologists (Canada Soil Survey Committee 1973, Pettapiece 1975) have generally followed this argument.

If permafrost is accepted as a zonal phenomenon, should cryogenic processes also be considered a zonal soil-forming factor? It is inescapable that they must. Ugolini (1966) and Karavayeva et al. (1965) both stressed this point. They indicated that the uniqueness of arctic soils was in part due to disturbances caused by cryogenic processes. Surface features caused by cryogenic phenomena have been documented by Washburn (1956), and their relationship to the biotic environment has long been recognized (Hopkins and Sigafoos 1951). In fact, the interrelationships of permafrost itself and of various parts of the environment, particularly organic areas, have been studied by many people (Brown, R. J. E. 1969, Zoltai and Tarnocai 1974). Permafrost and cryogenic processes are zonal characteristics. They are inexorably involved in the ecology of the North and must be recognized as factors in the ecosystem.

However, not all soils in the North are cryogenic. The arctic brown (Tedrow et al. 1958), podzolic and brown non-differentiated of Russia (Sokolov and Sokolova 1962), and Canadian brunisols are examples. Karavayeva et al. (1965) suggested that they could be considered zonal even though these soils are genetically independent of the cryogenic types. However, one would then have to qualify their occurrence by such statements as "on coarse-textured, freely drained materials...."

Classification Implications

Cryoturbation imparts many unique features to northern soils. It also affects the approach to their classification. For illustrative purposes the strong microrelief of earth hummock formations (Washburn 1956) can be discussed. Hummocks are common in the Subarctic and Arctic of northwestern Canada (Zoltai and Pettapiece 1973, Zoltai and Tarnocai 1974) and have received a good deal of study (Karavayeva et al. 1965, Pettapiece 1974, Tarnocai and Zoltai 1978).

Individual hummocks are commonly about 50 cm high and 100 to 150 cm in diameter (Tarnocai and Zoltai 1978). They have a marked effect on the distribution of vegetation. Shrubs (trees) and lichens are found on the raised portion and sphagnum mosses in the depressions (Sigafoos 1952, Zoltai and Tarnocai 1974). Cross-sections of the soil body show many contrasting features, including subsurface organic accumulations. Several workers have commented on this organic distribution (MacKay et al. 1961, MacNamara and Tedrow 1966, Brown, J. 1969). The consensus is that this distribution is a pedogenic feature caused by cryoturbation. Ugolini (1966) suggested that if cryogenic processes are soil-forming factors, then all cryopedogenic soils might be considered to have formed from a single process, and the variable soil complex should be classed as a single unit. Karavayeva et al. (1965) and Pettapiece (1975) also stressed this approach.

Early classification systems could not accommodate in detail the extreme lateral variations in soil features caused by cryogenic processes. However, the introduction of the pedon concept (Soil Survey Staff 1960, Knox 1965), which considers a three-dimensional body and recognizes cyclic variation within the unit definition, was particularly useful for the classification of cryopedogenic soils. It became the basis for the Canadian Soil Classification (Canada Soil Survey Committee 1973, 1978) and for the introduction of a new soil order. This order, the Cryosolic order, was based primarily on the presence of permafrost and its effect on soils and vegetation.

The Subarctic is characterized by Turbic Cryosols, Organic Cryosols, and Organics, with lesser amounts of Luvisols and Brunisols. However, the proportions can vary depending on surface material and landform. A more detailed discussion of this can be found in Pettapiece et al. (1978).

Soil Climate as an Ecological Parameter

The climatic elements of temperature and moisture are the driving and controlling forces in nature. They are the basis of ecological zonation. But there are many instances of apparent anomalies when only one part of the ecosystem is considered by itself. Some of the most obvious anomalies deal with vegetation. For example, organic soils or sandy soils do not usually support "zonal" vegetation. Aspect can also have a marked effect on the vegetation assemblage. One can identify the specific causes for a change in each of the above cases—the organics are wetter, the sands are drier, the south slope is warmer, the north slope cooler. But it would be useful to have a single overall concept within which to discuss these variances. Since all are climatic parameters, a convenient concept here is that of soil climate.

The above examples could occur within 1 km or even several hundred metres of each other. Obviously, the regional climate is the same and the above-ground portions of the plants are subject to a similar climatic regime. It is the soil condition that is different. Let us consider specific examples from the Subarctic around Inuvik (Pettapiece et al. 1978). The normal "height-of-land" or zonal situation with medium-textured materials supports a black spruce–lichen forest on cryogenic (earth hummock) soils (fig. 2a). Permafrost is generally less than 70 cm from the surface. The organic soils in this area (peat plateaus) support mainly a moss-lichen cover with some shrubs along the polygonal cracks, but no trees (fig. 2b). This is a vegetation assemblage usually associated with an arctic environment. Investigation of the soils indicated permafrost at 20 to 30 cm. This indicates a very cold soil climate with a shallow rooting zone, which is typical of the Arctic. On the other hand, a nearby sandy glaciofluvial deposit has an active layer of over 100 cm. The soil, an Eluviated Dystric Brunisol (fig. 2c), is similar to those on sands in the boreal region. This warmer, drier, noncryogenic soil should and does support a more southerly vegetation. Just a few miles further south a similar soil site has an open white spruce–birch forest with some juniper and lichen. Therefore, within one regional climate in the Subarctic one can encounter ecosites with soil-vegetation relationships that are also typical of arctic and boreal regions. These apparent anomalies are justifiable and can be logically explained in terms of soil-site characteristics, and more specifically, soil climate (Pettapiece and Zoltai 1974, Pettapiece et al. 1978).

Another phenomenon of the Subarctic that can be explained in terms of soil climate is the vegetation patterns resulting from fire history (Pettapiece 1974). The first generation of trees on recent burns are much larger than those found in adjacent areas that have not been burned for several hundred years (fig. 3). The explanation lies in the underground environment. The old forest has a buildup of moss and lichen, which insulates the soil and allows for the aggradation of the permafrost table. The result is a progressively colder, shallower rooting zone until equilibrium is established— a harsh soil climate for the trees to operate in. After a fire, which destroys

Fig. 2. Ecosite characteristics in the Subarctic.
(a) Typical black spruce–lichen forest on moderately fine-textured earth hummock soils (Orthic Turbic Cryosol). Note the leaning ("drunken") trees.

Fig. 2. (b) Arctic vegetation associated with a cold organic soil and a very shallow active layer (Organic Cryosol).

Fig. 2. (c) Boreal vegetation on a sandy noncryogenic soil (Eluviated Dystric Brunisol).

or reduces the surface organic layer, the permafrost table recedes and the soil becomes drier and warmer—a more desirable rooting environment for growth. In this case there was a change in soil climate on the same site, with the same regional climate.

Fire, however, does not always lead to improved growing conditions. Near the arctic tree line, under conditions of restricted drainage and poor soil structure, fire may lead to oversaturation and a tundra condition (Pettapiece and Zoltai 1974). In the same area (fig. 4) natural exposures such as stream banks can often have improved drainage, warmer, drier soils, and abrupt changes in plant growth. It is in these privileged locations that the extreme northerly range extensions of many species are found (Drew and Shanks 1965).

A special case for soil microclimate considerations arises where cryogenic activity results in changes in microrelief (fig. 5). For example, the top of an earth hummock may have a relatively warm, very dry climate with permafrost at about 70 cm. The adjacent depression, only 50 cm away, may be cold, water-saturated, and have ice at 15 to 20 cm below the surface. Detailed ecological studies must recognize this effect on features such as plant distribution and depth of the active layer. It may appear anom-

Fig. 3. Ecosite variations associated with burn history.
(a) Fire patterns in the Subarctic.

Fig. 3. (b) An old stand with a thick insulating mat and a very shallow rooting zone.

Fig. 3. (c) First-generation forest responding to a deeper, warmer soil environment.

Fig. 4. Ecosite variations associated with different landscape positions. Note the "fire-induced" tundra of the foreground and the thin band of white spruce associated with the better-drained soils along the stream.

Fig. 5. An earth hummock soil landscape (Orthic Turbic Cryosol) illustrating the relationship of vegetation to microsites. The hummock tops are covered by lichens while the depressions support mosses. Shrubs tend to be concentrated on the sides of the hummock.

alous that a soil can be dry when ice is only 50 cm away, but Black (1977) found that a major contributor to black spruce seedling mortality was moisture stress on the hummock crests.

The Subarctic, like other transition ecotones such as the subalpine and parkland areas, is particularly sensitive to small changes in any component of the environment. Soil is often considered to be a passive component of the ecological system, but in these areas it can become a controlling factor. Any interpretations of the ecology of the area should recognize the interrelationships of all parts of the system to assure correct conclusions.

Conclusions

This paper set out to discuss or at least expose some of the pedological peculiarities of northwestern Canada, in order to provide a basis of understanding for those people who might work in the North but who have had no formal soils training. It is an attempt to put into perspective some of the concepts and arguments that are in the literature. Particular attention was directed to:

(a) soil-forming processes—cryogenic processes should be considered a pedogenic factor.
(b) the concept of a zonal soil—"height-of-land" means a position that does not receive excess water. It is not identical to "well-drained." There are zonal (cryopedogenic) processes and a zonal soil characteristic of permafrost areas.
(c) soil classification—using the pedon concept, all the elements of cryogenic forms can be considered and classed as one unit.
(d) soil climate—this is a very convenient and useful concept for explaining many variations in soil, vegetation, and ecosite properties.

Acknowledgments

The field investigations upon which much of the paper is based were carried out under the Environmental-Social Program, Northern Pipelines, of the Task Force on Northern Oil Development, Government of Canada. Dr. O. L. Hughes co-ordinated the field support and along with S. C. Zoltai provided valuable breadth of experience and discussion. A special acknowledgment to Don Gill for his enthusiasm for the North, his inclusion of soils in the interdisciplinary course on northern studies, and his personal co-operation and friendship.

References

Black, R. A. 1977. Reproductive biology of black spruce at treeline, Inuvik, N.W.T. Ph.D. thesis. Univ. Alberta.
Brown, J. 1969. Soil properties developed on the complex tundra relief of northern Alaska. Biul. Perygl. 18: 153–167.
Brown, R. J. E. 1969. Permafrost as an ecological factor in the Subarctic: Ecology of the Subarctic regions. Proc. Helsinki Symp.: 129–140. UNESCO, Paris.
Canada Soil Survey Committee. 1973. Tentative classification system for Cryosolic soils. Proc. 9th Meeting, Saskatoon, Soil Res. Inst., Central Exp. Farm. Ottawa.
———. 1978. The Canadian system of soil classification. Can. Dep. Agr. Pub. 1646. Ottawa. 164 pp.
Day, J. H., and H. M. Rice. 1964. The characteristics of some permafrost soils in the Mackenzie Valley, N.W.T. Arctic 17(4): 222–236.
Drew, J. V., and R. E. Shanks. 1965. Landscape relationships of soils and vegeta-

tion in the forest-tundra ecotone, upper Firth River Valley, Alaska-Canada. Ecol. Mon. 35(3): 285–306.
Hopkins, D. M., and R. S. Sigafoos. 1951. Frost action and vegetation patterns on Seward Peninsula, Alaska. Geol. Surv. Bull. 954-C: 50–101.
Karavayeva, N. A., I. A. Sokolov, T. A. Sokolova, and V. O. Targul'yan. 1965. Peculiarities of soil formation in the tundra-taiga frozen regions of eastern Siberia and the Far East. Soviet Soil Sci. (Pochv. transl.) 7: 756–766.
Knox, E. G. 1965. Soil individuals and soil classification. Proc. Soil Sci. Soc. Amer. 29: 79–84.
Leahey, A. 1947. Characteristics of soils adjacent to the MacKenzie River in the Northwest Territories of Canada. Proc. Soil Sci. Soc. Amer. 12: 458–461.
Liverovskii, Yu. A. 1964. Soils of the far north and problems of their future study. Problems of the North 8: 165–179.
MacKay, J. R., W. H. Mathews, and R. S. MacNeish. 1961. Geology of the Angigstciak archaeological site, Yukon Territory. Arctic 14(1): 25–52.
MacNamara, E. E., and J. C. F. Tedrow. 1966. An arctic equivalent of the Grumisol. Arctic 19(2): 145–152.
Nowosad, F. S., and A. Leahey. 1960. Soils of the Arctic and Subarctic regions of Canada. Agr. Inst. Rev. (March-April) 48: 48–50.
Pettapiece, W. W. 1974. A hummocky permafrost soil from the Subarctic of northwestern Canada and some influences of fire. Can. J. Soil Sci. 54(4): 343–355.
―――. 1975. Soils of the Subarctic in the lower Mackenzie Basin. Arctic 28(1): 35–53.
―――, and S. C. Zoltai. 1974. Soil environments in the western Canadian Subarctic. In W. C. Mahaney, ed. Quaternary environments symposium. York Univ.-Atkinson College Geogr. Monogr. Ser. No. 5. Toronto.
―――, C. Tarnocai, S. C. Zoltai, and E. T. Oswald. 1978. Soil, permafrost and vegetation relationships in northwestern Canada. Guidebook for Tour 18, Int. Soc. Soil Sci. Edmonton. 165 pp.
Retzer, J. L. 1965. Present soil-forming factors and processes in arctic and alpine regions. Soil Sci. 99: 38–44.
Sigafoos, R. S. 1952. Frost action as a primary physical factor in tundra plant communities. Ecol. 33: 480–487.
Soil Survey Staff, USDA. 1960. Soil classification, a comprehensive system, 7th approximation. U.S. Gov. Printing Office, Washington, D.C.
Sokolov, I. A., and T. A. Sokolova. 1962. Zonal soil groups in permafrost regions. Soviet Soil Sci. 10: 1130–1136.
Tarnocai, C. 1973. Soils of the Mackenzie River area. Environmental-Social Program, Task Force on Northern Oil Development, Govt. of Canada, Report 73-26. 136 pp.
―――, and S. C. Zoltai. 1978. Earth hummocks of the Canadian Arctic and Subarctic. Arctic and Alpine Research 10(3): 581–594.
Tedrow, J. C. F., and J. Brown. 1962. Soils of the northern Brooks Range, Alaska: Weakening of the soil-forming potential at high arctic altitudes. Soil Sci. 93(4): 254–261.
―――, and J. E. Cantlon. 1958. Concepts of soil formation and classification in arctic regions. Arctic 11: 166–179.

———, J. V. Drew, D. E. Hill, and L. A. Douglas. 1958. Major genetic soils of the arctic slope of Alaska. J. Soil Sci. 9: 33–45.

Ugolini, F. C. 1966. Soils of the Mesters Vig district, northeast Greenland 2, exclusive of the Arctic Brown and Podzol-like soils. Medd. om Grønl. 176(2). 25 pp.

Washburn, A. L. 1956. Classification of patterned ground and review of suggested origins. Bull. Geol. Soc. Amer. 67: 823–866.

Zoltai, S. C., and W. W. Pettapiece. 1973. Studies of vegetation, landform and permafrost in the Mackenzie Valley: Terrain, vegetation and permafrost relationship, in the northern part of the Mackenzie Valley and northern Yukon. Env. Soc. Prog. Northern Pipelines Task Force, Gov. of Can. 98 pp.

———, and W. W. Pettapiece. 1974. Tree distribution on perennially frozen earth hummocks. Arctic and Alpine Res. 6(4): 403–411.

———, and C. Tarnocai. 1974. Soils and vegetation of hummocky terrain. Environmental-Social Committee, Northern Pipelines, Task Force on Northern Oil Dev. Rep. No. 74-5.

Characteristics of Soil Temperature Regimes in the Inuvik Area

C. Tarnocai

Abstract

Soil temperature data obtained during 2½ years of a 5-year study were evaluated to determine the thermal regime of the active layer and the near-surface permafrost and its relationship to soil properties, patterned ground type, vegetation, and snow cover. These soil temperatures were collected at the 2.5, 5, 10, 20, 50, and 100 cm depths on eight of the most common soil types in the Inuvik area. The highest mean annual soil temperature (0.7 C), the highest mean summer soil temperature (10.8 C), the lowest minimum soil temperature (-17.5 C), the highest maximum soil temperature (24 C), and the greatest number of frost-free days (151 days) were all associated with the 2.5-cm depth. On the other hand, the lowest mean annual soil temperature (-3.8 C), the lowest mean summer soil temperature (-1.8 C), the highest minimum soil temperature (-5.2 C), and the lowest maximum soil temperature (-0.4 C) were all associated with the 100-cm depth. Only two of the eight soils had frost-free days at the 100-cm depth. Soil texture and type of soil material had little effect on the soil temperature, but vegetation cover, thickness of the surface organic layer, moisture content, and topographic location had a much greater effect. Mineral soils associated with earth hummocks had various soil temperature regimes, but organic soils associated with polygonal peat plateaus had lower soil temperatures than organic soils associated with peat plateaus. Soils on the Mackenzie Delta were generally found to be cooler than soils on the till uplands.

Introduction

The near-surface temperature is a very important property of soils in the permafrost region. In this area the soil temperatures not only influence the biological processes and most of the chemical and physical processes, but the low soil temperatures also trigger the cryogenic processes. The most common effects of these cryogenic processes are cryoturbation and frost heave. If, however, the soil temperature increases because of surface disturbance, degradation of permafrost occurs, resulting in severe erosion or subsidence. Thus a knowledge of the thermal regime of the soil (active layer and the near-surface permafrost) is very important for land use decisions and for the construction of roads or buildings on permafrost soils. This knowledge is vital for determining the method of land use or construction that will cause the least change in the thermal regime of the soil.

Judge (1973) measured soil temperatures at a number of sites in the Mackenzie Delta area. All of these temperatures were taken in deep bore holes at various depths, beginning several metres below the surface. Soil temperatures were also measured under buildings, roads, and the airport landing strip in the Inuvik area by the National Research Council. Soil temperature studies were carried out in the Mackenzie Delta by Gill (1971) and on hummocky terrain near Inuvik by Mackay and MacKay (1976).

This study, which was initiated in 1978 and is scheduled to be completed in 1983, was set up to monitor the soil temperatures of the most common soils occurring in the Inuvik area. This paper presents characteristics of the thermal regime of the active layer and the near-surface permafrost and their relationship to soil properties, patterned ground type, vegetation, and snow cover, based on data collected between September 1978 and March 1981.

Area and Climate

The study area is located in the Inuvik and Arctic Red River areas in the Northwest Territories (fig. 1). The area encompasses two different terrain types, the Mackenzie Delta and the rolling-to-hilly area of the Caribou and Campbell Lake Hills (Mackay 1963).

The Mackenzie Delta is a maze of channels and lakes, with the dominant soil material being a silt loam-textured alluvium. The rolling-to-hilly terrain of the Caribou Hills, where Inuvik is located, is composed domi-

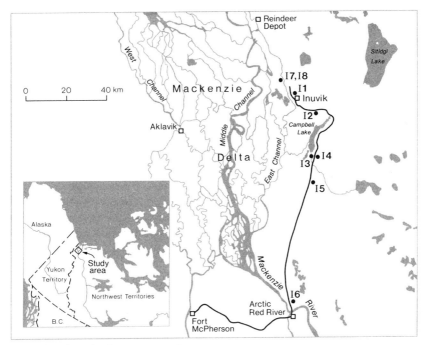

Fig. 1. Location of soil temperature sites.

nantly of glacial till. The Campbell Lake Hills area, south of Inuvik, is an upland where bedrock either occurs as outcrops or lies close to the surface. In this area glacial drift of variable thickness is composed mainly of moderately fine-textured till, with local areas of outwash and ice-contact deposits. Peat deposits are commonly found in depressions.

The vegetation of the Mackenzie Delta portion of the study area presents a contrast to that of the rolling till uplands. A close canopy of white spruce (*Picea glauca*) with a continuous feather moss carpet grows on the least frequently inundated areas of the Mackenzie Delta. This association resembles the forests of the boreal forest region. The vegetation on areas of the Mackenzie Delta that are periodically inundated is dominated by horsetails (*Equisetum* spp.), willows (*Salix* spp.), and alder (*Alnus* spp.). A more detailed description of the vegetation occurring on the Mackenzie Delta can be obtained from Gill (1971).

The rolling till upland area between Inuvik and Arctic Red River supports an open subarctic forest of stunted black spruce (*Picea mariana*) with some willow (*Salix* spp.), and has a lichen (*Cladonia* spp.) and moss ground cover. Black spruce growing on severely cryoturbated soils associated with earth hummocks are usually tilted in all directions by the frost-heaved soil.

The vegetation on peatlands is either dwarf shrubs, lichens, and mosses, or open black spruce, ericaceous shrubs, lichens, and mosses.

The dominant soils on the alluvial deposits in the Mackenzie Delta area are Regosolic Static Cryosols and Gleyed Cumulic Regosols (soil classification system according to Canada Soil Survey Committee 1978), with the latter occurring along the channels, which are periodically inundated. The remainder of the area is dominated by Regosolic Static Cryosols. The till upland area of the Caribou and Campbell Lake Hills is dominated by Turbic Cryosols, with the Orthic and Brunisolic subgroups being the most common. The soils on the coarse-textured ice-contact deposits are Eluviated Dystric Brunisols. Most of the Turbic Cryosols are associated with a thin layer of surface peat.

The study area has a continental climate, but does not have the temperature extremes exhibited further inland in the Mackenzie Valley (as, for example, at Fort Good Hope). The aerial climatic data presented by Burns (1973) indicates that Inuvik has a mean annual temperature of −9.6 C and total annual precipitation of 260 mm, of which about 40% occurs as rainfall. Snowfall averages 174 cm. The coldest month is February, with a mean temperature of −29.2 C (−23.9 C maximum and −35 C minimum). The warmest month is July, with a mean temperature of 13.2 C (19.2 C maximum and 7.4 C minimum). The extreme maximum and minimum temperatures recorded at Inuvik are 31 C and −57 C, respectively.

According to Brown (1956, 1967), the entire study area lies within the continuous permafrost zone. Mackay (1963), however, points out that when the depths of the Mackenzie Delta channels or lakes exceed the thickness of winter ice, the subjacent bottom sediments will remain unfrozen. In addition, Gill (1971) indicates that permafrost-free soil also exists both where large channels have undergone recent shifts and in newly deposited slipoff slopes. These areas, according to Gill (1971), coincide with the *Salix-Equisetum* communities in the Mackenzie Delta.

Materials and Methods

Location of Sites

Soil temperatures are being monitored on eight sites (fig. 1). Site I1 is located approximately 5 km north of Inuvik, sites I2 to I6 are located along the Dempster Highway between Inuvik and Arctic Red River, and sites I7 and I8 are located on the southwest side of Bombardier Channel near its confluence with the East Channel in the Mackenzie Delta, approximately 8 km north of Inuvik.

Sites I1, I3, and I6 are located on undulating morainal terrain, which is associated with earth hummocks. Site I4 is situated on the top of an esker, while sites I2 and I5 are associated with a polygonal peat plateau and peat plateau, respectively. Polygonal peat plateaus and peat plateaus are perennially frozen peat landforms that commonly occur in the Subarctic (Zoltai and Tarnocai 1975). Sites I7 and I8 are both located on a recent fluvial terrace on the Mackenzie Delta.

These sites are situated on the most common soils in the area: Brunisolic and Orthic Turbic Cryosols, associated with earth hummocks on finetextured till (sites I1, I3, and I6); Mesic Organic Cryosol, associated with a polygonal peat plateau (site I2) and with a peat plateau (site I5); Eluviated Dystric Brunisol, cryic phase, associated with a coarse-textured sandy deposit (site I4); and Regosolic Static Cryosol, associated with loamtextured alluvium (sites I7 and I8).

Some of the properties of these sites, together with the associated soils, landforms, and vegetation, are listed in table 1.

Instrumentation
The thermistor cables were constructed by using Yellow Spring epoxycoated thermistor beads #44033. These thermistor beads were positioned on the cable so as to allow the soil temperatures to be read at the 2.5, 5, 10, 20, 50, and 100-cm depths.

The installation of thermistor cables was done with as little disturbance as possible of the site and the soil. The thermistor cables, fastened to halfround dowelling, were placed in a hole drilled by a 2.7-cm diameter permafrost auger. The half-round wooden dowelling was then placed in the auger hole so that the round part of the dowelling, where the thermistors were fastened, made contact with the auger-hole wall. The contact was further secured by packing soil material in the other half of the auger hole.

A Data Precision Model 245 digital multimeter was used to monitor soil temperatures weekly at all sites except site I6, which was monitored biweekly.

Analysis of Data
Computerized techniques were utilized to streamline data processing. For each location, analysis of data was based on measurements obtained from

Table 1 Description of soil temperature sites

Site No.	Location Lat.N	Long.W	Elev. m (a.s.l.)	Landform	Material	Topography Aspect	Slope %	Sub-group	Soil Texture	Drainage	Thickness of surface organic layer (cm)	Ice content (%)	Active layer (cm)	Patterned ground	Vegetation
11	62°23'	133°44'	25	Undulating morainal blanket	Colluviated till	W	2	Brunisolic turbic cryosol	Silty clay	Imperfect	less than 1	High	98	Earth hummock	Black spruce, ericaceous shrubs, lichen
12	68°19'	133°25'	100	Polygonal peat plateau	Mesic sedge peat	—	0	Mesic organic cryosol	—	Imperfect to poor	+	High	39	Polygonal peat plateau	Dwarf shrubs, lichens, moss
13	68°08'	133°27'	30	Undulating morainal blanket	Till	SW	3	Orthic turbic cryosol	Clay	Moderately well to poor	6*	High	65	Earth hummock	Black spruce, ericaceous shrubs, lichen
14	69°07'	133°26'	40	Ridged glaciofluvial (esker)	Glaciofluvial	—	0	Eluviated dystric brunisol, cryic phase	Sandy loam to sand	—	3	Low	100	—	Black spruce, lichen
15	67°57'	133°28'	75	Peat plateau	Mesic sedge peat	—	0	Mesic organic cryosol	—	Imperfect to poor	+	High	45	—	Dwarf shrubs, lichen, moss

						E								Earth hummock	
16	67°30'	133°46'	46	Undulating morainal blanket	Till		2	Brunisolic turbic cryosol	Loam	Moderately well to imperfect	5*	High	68		Black spruce, ericaceous shrubs, lichen
17	68°25'	133°52'	8	Fluvial terrace	Fluvial	—	0	Regosolic static cryosol	Silt loam	Well	2*	Medium	71	—	Willow, alder
18	68°25'	133°52'	10	Fluvial terrace	Fluvial	—	0	Regosolic static cryosol	Silt loam	Well	10*	Medium	35	—	White spruce, willow, alder

* Thickness of organic layer measured at the top of hummock
+ Organic soil

all six depths. For each depth a measure of temperature fluctuation throughout the year was achieved by mathematically calculating the best-fitting line to the data, i.e., by determining an equation of the form: Temperature = A Function of Date (Mills et al. 1977). After determining this equation the following parameters were derived:
— mean annual soil temperature (MAST)
— mean summer soil temperature (MSST)
— date of spring thaw — 0 C
— date of fall freeze — 0 C
— number of frost-free days
— maximum (XST) and minimum (MST) soil temperature and dates of occurrence
— date in spring and fall when soil temperature rises above and falls below 5 C
— length of season when soil temperature was above 5 C.

Results

Mean Annual Soil Temperature

The highest mean annual soil temperature (MAST), 0.7 C, was found at the 2.5-cm depth in soil I1, while the lowest MAST, -3.8 C, was found at the 100-cm depth in soil I6 (fig. 2, table 2). All MAST values were below 0 C, with the exception of those for soils I1 and I5. For both of these soils the MAST was slightly above 0 C at the 2.5-cm and 5-cm depths. Soils I1 and I5 had the highest, and soils I6, I7, and I8 the lowest, MAST values at all depths.

Soil texture and patterned ground type had little effect on the MAST values in mineral soils. Fine-textured soil I1 had the highest MAST values of all soils at nearly all depths. On the other hand, fine-textured soil I3 and coarse-textured soil I4 had similar, moderate MASTs in the active layer, but soil I3 had a lower MAST value at the 100-cm depth than did soil I4. The coldest MAST values were associated with soils I6, I7, and I8. All of these soils were medium-textured loams and silt loams. Soils I1 (silty clay), I3 (clay), and I6 (loam) were all associated with earth hummocks, but their MAST values differed significantly (fig. 2, table 2), with the highest being soil I1 (-2.1 C at the 50-cm depth and -1.9 C at the 100-cm depth) and the lowest being I6 (-3.4 C at the 50-cm depth and -3.8 C at the 100-cm depth). Soil I4 (sand) had the coarsest texture of all mineral soils monitored, but its MAST values were only slightly higher than those of all finer-textured mineral soils except soil I1.

On peat deposits, however, the MAST values were lower on soil I2, associated with polygonal peat plateaus, than on soil I5, associated with peat plateaus, in spite of the fact that both of these soils were composed dominantly of moderately decomposed sedge peat.

The MAST values of the two soils (I7 and I8) occurring on the Mackenzie Delta were among the lowest measured, but the near-surface portion of the active layer of soil I8 was somewhat cooler than that of soil I7. Both of these soils were silt loam in texture.

Mean Summer Soil Temperature

The highest mean summer soil temperature (MSST), 10.8 C, was associated with the 2.5-cm depth in soil I1, while the lowest MSST, −1.8 C, was associated with the 100-cm depth in soil I6 (fig. 3, table 2). The MSST values were above 0 C in all soils at a depth of 20 cm or less and above 0 C in soils I1, I3, I4, I6, and I7 to a depth of 50 cm.

The highest MSST values were associated with soil I1 (silty clay in texture with earth hummocks). These high MSST values for soil I1 are probably due to its western exposure and the very thin surface organic layer. Soils I3 and I6, which were also fine-textured soils associated with earth hummocks, have lower MSST values. Both of these soils had much thicker surface organic layers than did soil I1 (table 1). At the 20-cm depth, for example, MSST values for these soils were 5.1 C (I1), 2.6 C (I3), and 1.9 C (I6).

The MSST value of the active layer of soil I2 was somewhat higher than that of the active layer of soil I5. These soils were both associated with mesic peat materials, but their vegetation cover differed. Soil I5 had a thick, continuous lichen cover, while for soil I2 this lichen cover was discontinuous, the vegetation cover was not as heavy, and in some areas, dark peat surfaces were exposed. The reverse was true of the MSST values of the near-surface permafrost layer (100-cm depth) of these two soils, however, with soil I2 having a lower MSST (−1.6 C) than soil I5 (−0.8 C).

The MSST values of the active layers of the two Mackenzie Delta soils were also different, although their MSST values at the 2.5-cm depth were very similar (6.1 C and 7.7 C). At lower depths, however, soil I7 had higher MSST values than soil I8 had. Soil I7 was still periodically inundated by the Mackenzie River in late May and June. This may have helped to accelerate the thawing process. It may also have caused the slightly

Table 2 Soil temperature parameters for soils 11-18

Site No.	Depth (cm)	MAST (°C)	MSST (°C)	Number of frost-free days	Date of 0°C SPRING	Date of 0°C FALL	Min. temp. (°C)	Min. temp. DATE	Max. temp. (°C)	Max. temp. DATE	Date of 5°C SPRING	Date of 5°C FALL	Number of days above 5°C
11	2.5	0.7	10.8	136	May 18	Oct 1	-13.0	Mar 16	24.0	Aug 17	May 22	Sep 10	111
	5	0.2	9.4	136	May 18	Oct 1	-13.1	Mar 16	21.9	Aug 17	May 25	Sep 10	108
	10	-0.4	7.2	129	May 25	Oct 1	-12.6	Mar 16	16.8	Jul 23	Jun 5	Sep 7	94
	20	-0.9	5.1	122	Jun 1	Oct 1	-11.9	Mar 16	12.2	Aug 17	Jun 22	Sep 1	71
	50	-2.1	1.6	108	Jun 15	Oct 1	-10.0	Mar 16	6.0	Aug 17	Jul 23	Aug 17	25
	100	-1.9	-1.1	34	Aug 3	Sep 6	-7.6	Mar 23	0.0	—	—	—	0
12	2.5	-1.1	8.6	132	May 18	Sep 27	-17.5	Feb 21	17.5	Aug 2	May 30	Sep 10	103
	5	-0.4	7.7	132	May 18	Sep 27	-16.5	Feb 21	15.4	Aug 17	Jun 7	Sep 10	95
	10	-1.5	6.0	132	May 18	Sep 27	-15.9	Mar 14	15.0	Jul 19	Jun 2	Sep 2	92
	20	-2.5	3.2	98	Jun 21	Sep 27	-15.0	Mar 14	7.6	Jul 26	Jul 14	Aug 15	42
	50	-2.9	-0.7	84	Aug 17	Nov 9	-10.7	Mar 16	0.2	Sep 6	—	—	0
	100	-2.7	-1.6	0	—	—	-8.4	Mar 23	-0.3	—	—	—	0
13	2.5	-0.8	7.8	151	May 3	Oct 1	-13.9	Mar 15	16.9	Jul 26	Jun 19	Sep 7	80
	5	-0.4	7.3	130	May 29	Oct 1	-13.8	Mar 15	13.6	Jul 26	Jun 20	Sep 7	79
	10	-1.1	5.5	116	Jun 7	Oct 1	-13.5	Mar 15	9.7	Jul 26	Jun 30	Aug 3	34
	20	-2.3	2.6	110	Jun 30	Oct 18	-12.7	Mar 15	6.8	Aug 16	Jul 24	Aug 20	27
	50	-2.8	0.2	91	Jul 19	Oct 18	-11.2	Mar 15	2.4	Aug 23	—	—	0
	100	-3.0	-1.4	0	—	—	-9.4	Mar 22	0.4	—	—	—	0
14	2.5	-1.0	6.5	133	May 24	Oct 4	-11.4	Mar 15	15.5	Aug 16	June 18	Sep 2	76
	5	-1.2	5.5	133	May 24	Oct 4	-11.1	Mar 15	13.0	Aug 16	Jun 21	Sep 2	73
	10	-1.5	4.2	127	May 31	Oct 5	-11.0	Mar 15	10.3	Aug 16	Jul 1	Aug 30	60
	20	-1.3	3.5	125	Jun 7	Oct 10	-10.5	Mar 15	8.1	Aug 16	Jul 5	Aug 30	56
	50	-1.7	1.8	140	Jun 21	Nov 8	-9.1	Mar 15	5.3	Jul 26	Jul 26	Aug 16	21
	100	-2.1	-0.7	78	Jul 19	Oct 5	-7.2	Apr 19	1.3	Aug 23	—	—	0

15	2.5	0.5	7.7	141	May 17	Oct 5	-10.8	Mar 15	20.1	Aug 16	Jun 20	Sep 10	81
	5	0.1	6.5	128	May 30	Oct 5	-10.5	Mar 15	17.1	Aug 16	Jun 20	Sep 2	74
	10	-0.4	4.6	127	May 31	Oct 5	-10.1	Mar 15	12.2	Aug 16	Jul 4	Sep 2	60
	20	-1.1	1.8	112	Jun 20	Oct 10	-8.8	Mar 15	6.8	Aug 16	Jul 20	Aug 20	31
	50	-1.4	-0.7	133	Aug 23	Jan 3	-7.2	Mar 15	0.1	—	—	—	0
	100	-1.1	-0.8	0	—	—	-5.2	Mar 30	0.0	—	—	—	0
16	2.5	-1.3	7.0	133	Jun 2	Oct 13	-14.1	Feb 15	16.6	Aug 23	Jun 6	Sep 5	91
	5	-1.6	5.9	133	Jun 2	Oct 13	-13.9	Feb 15	14.8	Aug 23	Jun 6	Sep 5	91
	10	-2.4	3.8	127	Jun 5	Oct 10	-13.7	Feb 15	11.2	Aug 23	Jul 1	Sep 2	63
	20	-2.9	1.9	136	Jun 8	Oct 22	-13.3	Feb 15	6.8	Aug 16	Jul 19	Aug 20	32
	50	-3.4	0.1	103	Jul 1	Oct 12	-12.4	Mar 22	2.8	Aug 16	—	—	0
	100	-3.8	-1.8	0	—	—	-11.4	Mar 22	-0.1	—	—	—	0
17	2.5	-2.2	5.9	130	May 26	Oct 3	-16.3	Mar 16	13.6	Aug 17	Jun 29	Sep 7	70
	5	-2.1	5.2	128	May 28	Oct 3	-16.1	Mar 16	11.8	Aug 17	Jul 5	Sep 7	64
	10	-2.5	3.9	123	Jun 4	Oct 5	-15.6	Mar 16	9.1	Aug 17	Jul 6	Sep 12	68
	20	-3.1	2.4	113	Jun 15	Oct 6	-15.3	Mar 16	7.0	Aug 17	Jul 15	Aug 29	45
	50	-3.5	0.1	88	Jul 9	Oct 5	-13.3	Mar 16	2.9	Aug 17	—	—	0
	100	-3.3	-1.4	0	—	—	-11.1	Mar 23	-0.3	—	—	—	0
18	2.5	-1.7	6.3	134	May 22	Oct 3	-14.2	Mar 16	12.0	Aug 10	Jun 24	Sep 7	75
	5	-2.4	5.1	126	May 30	Oct 3	-13.8	Mar 16	10.1	Aug 10	Jul 6	Sep 7	63
	10	-3.0	2.9	116	Jun 11	Oct 5	-13.0	Mar 16	7.0	Aug 10	Jul 9	Aug 28	50
	20	-3.2	1.1	110	Jun 24	Oct 3	-11.1	Mar 22	4.1	Aug 17	—	—	0
	50	-3.2	-0.6	87	Jul 10	Oct 5	-10.0	Mar 16	0.2	Sep 1	—	—	0
	100	-3.1	-1.2	0	—	—	-8.6	Mar 22	-0.4	—	—	—	0

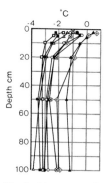
Fig. 2. Mean annual soil temperatures for soils I1 to I8.

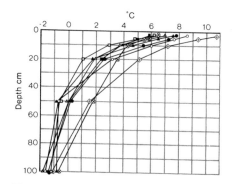
Fig. 3. Mean summer soil temperatures for soils I1 to I8.

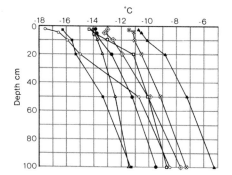
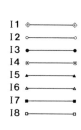
Fig. 4. Minimum soil temperatures for soils I1 to I8.

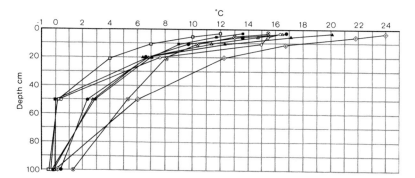
Fig. 5. Maximum soil temperatures for soils I1 to I8.

higher summer soil temperatures as compared with soil I8, which was inundated only during the extreme spring high-water levels in some years.

Minimum Soil Temperatures

The lowest minimum soil temperature (MST), −17.5 C, was measured at the 2.5-cm depth in soil I2, while the highest MST, −5.2 C, was recorded at the 100-cm depth in soil I5 (fig. 4, table 2). This soil (I5) had the highest MST values at all depths of all eight soils studied. Soil I2 had the lowest MST values at the 2.5, 5, and 10-cm depths. At greater depths (20, 50, and 100 cm), soil I7 had the lowest MST values.

For the three hummocky soils, soil I1 had the highest MST and soil I6 had the lowest MST at all depths. Although the near-surface MST values of these three soils were similar, at lower depths (50 and 100 cm) the differences were greater. At the 50-cm depth the temperature difference between soils I1 and I6 was 2.4 C, and at the 100-cm depth, 3.8 C. Soil I6 had the lowest temperatures even though it had the deepest snow cover, 55 cm as compared with 30 cm for I1 and 39 cm for I3.

Organic soils associated with polygonal peat plateaus (I2) have much lower MST values than do organic soils associated with peat plateaus. Although the differences were especially large at near-surface depths, at lower depths (50 and 100 cm) soil I5 was still more than 3 C warmer than soil I2. The snow cover on these soils was similar (28 cm on soil I2 and 21 cm on soil I5).

When the MST values of the two Mackenzie Delta soils (I7 and I8) were compared, it was found that soil I7 had the lower MST at all depths. When the MST occurred, the snow covers were 45 cm (I7) and 31 cm (I8).

Soils I6 and I7 were both associated with the lowest MST, in spite of the fact that they had deeper snow cover than other sites. In the case of soil I6, this difference in snow depth was 25 cm when compared with soil I1 and 16 cm when compared with soil I3, all these soils occurring on hummocky sites. This added snow cover probably did not provide sufficient extra insulation to prevent a greater cooling of soil I6. This was probably also the case with soil I7, which had 14 cm more snow than soil I8. Soil I7 is situated close to the channel and is more exposed to the winds than is soil I8. Thus the snow on soil I7 was more wind-packed and therefore of lower insulating capacity than the soft, low-density snow associated with soil I8.

Maximum Soil Temperatures

The highest maximum soil temperature (XST), 24 C, was measured at the 2.5-cm depth in soil I1, and the lowest XST, −0.4 C, was recorded at the

100-cm depth in soil I8 (fig. 5, table 2). Soil I1 had the highest XST values at all depths except 100 cm, where soil I4 had the highest XST. The lowest XST values were observed in soil I8 at all depths except at 50 cm, where soil I5 was 0.1 C cooler than I8.

For the three soils (I1, I3, and I6) associated with earth hummocks, soil I1 had the highest XST values at depths between 2.5 and 50 cm. At the 100-cm depth these soils had very similar XST values.

Organic soils (I2 and I5) had very similar XST values. Near the surface (10 cm or less), however, soil I2 had the higher XST.

For the two soils located on the Mackenzie Delta (I7 and I8), soil I7 had the higher XST. This was especially noticeable at the 10, 20, and 50-cm depths.

Frost-Free Days

The greatest number of frost-free days (151) occurred in soil I3 at the 2.5-cm depth. Below this depth the greatest numbers of frost-free days were 136 days (soil I1) at the 5-cm depth; 132 days (soil I2) at the 10-cm depth; 136 days (soil I6) at the 20-cm depth; 140 days (soil I4) at the 50-cm depth; and 78 days (soil I4) at the 100-cm depth (table 2).

Of the three soils (I1, I3, and I6) associated with earth hummocks, soil I1 had the greatest number of frost-free days. Soil I1, which thawed to the 100-cm depth for slightly over one month, is the only hummocky soil to thaw to this depth.

The length of the frost-free period (approximately 130 days) in the near-surface layers (2.5 to 10 cm) of the two organic soils was very similar. At greater depths, however, soil I5 had a significantly longer frost-free period than soil I2.

The number of frost-free days in the two delta soils (I7 and I8) were very similar at all depths. In the near-surface layer (2.5 to 10 cm) the number of frost-free days ranged from 116 to 134; at greater depths (20 to 50 cm) it ranged from 87 to 113.

The thawing of the soil surface generally began in May, with rapid thawing in the near-surface soil. The thaw reached the 50-cm depth in early June in soil I6, in mid-June in soils I1 and I4, in late June in soils I2, I3, and I5, and in early July in soils I7 and I8. In the fall the freezing process occurred much more quickly in mineral soils than in organic soils. At the 50-cm depth the soil froze in early October in soils I1, I7, and I8, in mid-October in soils I3 and I6, in early November in soils I2 and I4, and in early January in soil I5 (table 2).

Since these soils froze from both the surface and the permafrost table, the middle portion of the active layer usually froze last. Soils I1, I7, and I8 froze quickly, generally taking less than one week. On the other hand, the freezing process took approximately 1.5 weeks in soil I6, 3 weeks in soil I3, 4 weeks in soils I2 and I4, and 11 weeks in soil I5.

During the period of thawing, especially in the late summer, freeze-back occurred in mineral soils in periods of cool weather. This was especially noticeable in soils I1 and I8 during late August and early September. The magnitude of this freeze-back was 14 cm in soil I1 and 4 cm in soil I8. No freeze-back was observed in any of the organic soils monitored.

Days Above 5 C

The greatest number of days above 5 C was found in soil I1 at depths of 20 cm or less, with 111 days at 2.5 cm, 108 days at 5 cm, 94 days at 10 cm, and 76 days at 20 cm (table 2). The least number of days above 5 C was 70 days for soil I7 at 2.5 cm, 63 days for soil I8 at 5 cm, 34 days for soil I3 at 10 cm, and 27 days for soil I3 at 20 cm. Only soils I1 and I4 had temperatures above 5 C (25 days and 21 days, respectively) at the 50-cm depth, and no soils had temperatures above 5 C at the 100-cm depth.

Discussion

The temperature regimes of Cryosolic soils (soils associated with permafrost) are not necessarily controlled by the factors (e.g., soil texture and type of soil material) that control the temperature regimes of soils without permafrost. In temperate regions organic and fine-textured mineral soils have much lower MAST and MSST values than coarser-textured mineral soils occurring in a similar climatic region (Mills et al. 1977). In this study the organic and mineral Cryosols were found to have similar soil temperature regimes. Soil temperature regimes of the mineral soils, however, were greatly influenced by their topographical location and vegetation cover. Patterned ground types had little influence on the soil temperatures of these subarctic mineral soils.

Earth hummocks are common in the Mackenzie Valley. Soils associated with these earth hummocks have a variety of temperature regimes. The MAST and MSST values of soil I1, located north of Inuvik, were the highest of all eight soils studied. This soil was one of the warmest monitored in the area, probably because of its topographic position on a slight (less than 2%) westerly slope and the very thin surface organic layer. On the other hand, soil I6, which was located 136 km south of Inuvik and which was also associated with earth hummocks, had the lowest values for both the MAST and the MSST. The location of this soil in a depression was partly responsible for the cooler soil temperatures.

In organic Cryosolic soils the patterned ground type correlated much better with soil temperature regimes than it did in the mineral Cryosolic soils. Soils associated with a polygonal peat plateau had a lower MAST than soils associated with a peat plateau in the same area. The MSST of the near-surface permafrost was also lower in the soil associated with the polygonal peat plateau. The MSST values of the active layer of these soils were lower in the soil associated with the peat plateau, where the surface was covered with a thick, continuous lichen cover. They were higher in the soil associated with the polygonal peat plateau, where the vegetation cover was discontinuous and not as heavy as on the peat plateau. Thus it can be seen that the vegetation cover has a significant effect on the MSST values of the near-surface active layer (0 to 20-cm depth) of these organic soils. Thick lichen, moss, and forest cover decrease the MSST. This was also the case in soil I4, which had relatively low MSST values in the near-surface soil due to both the heavy forest canopy and the thick lichen layer.

The moisture content of the soils monitored in this study affects their thermal regime in several ways. The low moisture (ice) content in soil I4 is responsible for deeper active-layer development, since the very low ice content facilitates thawing to a greater depth. On the other hand, the high moisture content in organic soils retards the freezing process in the fall because of the higher latent heat associated with high water content.

The lack of correlation between soil texture and the soil temperature regime of these subarctic soils is probably due to the low rate of evapotranspiration in this region. Evapotranspiration is the main factor in soil temperature differences in the temperate region where coarse-textured soils are generally dryer than fine-textured soils. Evaporation of this higher moisture content in the fine-textured soils produces cooler soil temperatures than in the coarser, and dryer, soils. In cooler regions the rate of evapotranspiration is low, and thus differences in soil temperature due to soil texture are minimal (Clarke Topp personal communication).

The lowest minimum temperatures were observed in those soils located in depressions on soils I2, I6, and I7. The snow cover, which varied from site to site, does not seem to correlate with the lowest minimum temperature. This may indicate either that the snow cover was not deep enough to provide sufficient insulation from the cold air temperatures or that the snow was a high-density type due to wind-packing.

Using the 50-cm depth for the purpose of comparison, the highest number of frost-free days was found in soils I4 and I5 (140 and 133 days, respectively). Soils I1 and I6 occurred in the mid-range of frost-free days (108 and 103 days, respectively), while the lowest number was found in soils I2, I3, I7, and I8 (84, 91, 88, and 87 days, respectively).

The two Mackenzie Delta soils were found to be generally cooler than the other soils, possibly because of their low topographic positions. When these two soils were compared with each other it was found that soil I7, located in the alder-willow zone, was colder in winter and warmer in summer than soil I8, which was located in a relatively higher position under white spruce vegetation. The colder winter soil temperatures of soil I7 result from this soil being more exposed during the winter and being associated with a higher density of wind-packed snow, which provides less insulation than the soft, low-density snow associated with I8. The higher summer soil temperatures in soil I7 may result from the location of this soil in the alder-willow zone; it is subjected to nearly annual spring inundation, which speeds up the thawing process in the soil. Soil I8 is inundated only during exceptionally high spring floods and, based on the small amount of alluvial material deposited by these floods, it appears that the length of this inundation is much shorter than that of site I7. The thickness of the surface organic layer is only 2 cm on soil I7 as compared with 10 cm on soil I8. The thinner organic surface layer on soil I7 allows the soil to warm up much more during the summer than is possible on soil I8, which is insulated with a thicker blanket of organic matter. The alder-willow type of vegetation associated with soil I7 provides less shading, especially in the early part of the summer, than does the white spruce vegetation on soil I8; hence more incoming radiation is able to reach soil I7.

Although the MSST of the rooting zone (0 to 30 cm) of soil I8 was the lowest of the eight soils monitored in this study, its forest productivity is probably the highest. This would suggest that forest growth is controlled more by the nutrient status of the soil and to a lesser extent by the soil temperature. The higher nutritive and pH values result from periodic inundation by the Mackenzie River. A similar phenomenon was also found on disturbed sites in Alaska by Chapin and Shaver (1981).

The active layer of these Cryosolic soils had very little buffering capacity and responded quickly to a change in air temperature. This was due both to the shallow active layer, which was capable of containing only small amounts of stored heat, and to the underlying permafrost, which acted as a heat sink and removed heat from the active layer as long as this layer had a higher temperature. In fact, when the temperature regime of the near-surface permafrost layer (100-cm depth) was compared with the temperature regime of the active layer, it was found that mineral soils 16, 17, and 18 and organic soil 12 had the lowest MAST values at the 100-cm depth. These soils also had the lowest MAST values and the shortest frost-free period in the active layer. This would indicate that the temperature regime of the active layer was controlled not only by the aerial climate and other environmental factors, but also by the temperature of the permafrost layer.

In subarctic Cryosolic soils the occurrence of freeze-back, which takes place during the period of cooler weather long before the surface soil freezes, is neither as rapid nor as common as in Cryosolic soils in the arctic region (Tarnocai 1980). Freeze-back is a thermal process during which soil materials become frozen (having temperatures below 0 C). This freeze-back was especially noticeable in two of the mineral soils (soils 11 and 18). In organic soils, however, no freeze-back was observed.

Results obtained during this study indicate that the main factors affecting soil temperature are topographic location, moisture content, vegetation, and surface organic matter. Depressional topography, high moisture content, dense vegetation cover, and any surface organic matter have a negative effect on soil temperatures. If any of these conditions do not occur, the soil temperature will be increased. For land-use decisions, especially in areas where the soil is affected by these negative conditions, these factors should be considered in order to avoid long-lasting degradation and disturbance of the land. This is especially critical when the soil is associated with high ice content, as is the case with the majority of the soils in the study area. If land-use practices affect these conditions, soil temperatures will increase, resulting in rapid melting of the ice-rich subsoil and serious degradation of the environment.

Summary

Temperatures in eight soils, all associated with permafrost, were monitored over a period of 2½ years. It was found that topographic position, vegetation, and thickness of the surface organic matter had the greatest effect on the temperature regime of these soils. Soil moisture affected not only the rate of both thawing and freezing, but also the depth of the active-layer development. Soil texture and soil material had little effect on soil temperature since the cold, subarctic climate is associated with a low rate of evapotranspiration, and these two soil properties have an effect only when evapotranspiration is high.

Acknowledgments

The Inuvik Scientific Resource Centre has provided the manpower necessary for monitoring the soil and air temperatures, the depth to frost table, and the snow depth. Special thanks are due to D. A. Sherstone, scientist-in-charge of the Centre and to J. D. Ostrick, M. Bosh, and M. McRae, who took the actual readings.

References

Brown, R. J. E. 1956. Permafrost investigations in the Mackenzie Delta. Can. Geogr. 7: 21–26.
———. 1967. Permafrost in Canada. Geol. Surv. Can. Map 1246A.
Burns, B. M. 1973. The climate of the Mackenzie Valley-Beaufort Sea. Vol. I. Env. Can. Climatol. Stud. No. 24. 227 pp.
Canada Soil Survey Committee. 1978. The Canadian system of soil classification. Can. Dep. Agr. Pub. 1646. 164 pp.
Chapin, F. Stuart III, and Gaius R. Shaver. 1981. Changes in soil properties and vegetation following disturbance of Alaskan arctic tundra. J. Appl. Ecol. 18: 605–617.
Gill, D. 1971. Vegetation and environment in the Mackenzie River Delta, Northwest Territories. Ph.D. thesis. Univ. British Columbia, Dep. Geogr. 610 pp.
Judge, A. S. 1973. The thermal regime of the Mackenzie Valley: Observation of the natural state. Environmental-Social Program, Northern Pipelines, Rep. No. 73–38. 177 pp.
Mackay, J. R. 1963. The Mackenzie Delta area, N.W.T. Memoir 8, Geogr. Branch, Mines and Technical Surveys, Ottawa. 202 pp.
———, and D. K. MacKay. 1976. Cryostatic pressures in non-sorted circles (mud hummocks), Inuvik, Northwest Territories. Can. J. Earth Sci. 13: 889–897.
Mills, G. F., C. Tarnocai, and C. F. Shaykewich. 1977. Characteristics and distribution of soil temperature regimes in Manitoba, Canada. Proc. 21st Annu. Manitoba Soil Sci. Meeting: 56–85.
Tarnocai, C. 1980. Summer temperatures of Cryosolic soils in the north-central Keewatin, N.W.T. Can. J. Soil Sci. 60: 311–327.
Zoltai, S. C., and C. Tarnocai. 1975. Perennially frozen peatlands in the western Arctic and Subarctic of Canada. Can. J. Earth Sci. 12: 28–43.

Deflation Measurements on Hietatievat, Finnish Lapland, 1974-77

Matti Seppälä

Introduction

Because of scattered vegetation, strong winds and deflation are typical phenomena in arctic and subarctic environments. Eolian features have often been reported in the geomorphological literature of periglacial regions (e.g., Samuelsson 1926, Black 1951, Fristrup 1952-53, Thorarinsson 1962, Seppälä 1971, 1972, David 1977, Åkerman 1980). However, few quantitative studies of deflation rates and amounts have been carried out in these remote areas. This study is a direct continuation of an earlier study made by the author (Seppälä 1974), which reported the character of deflation during a two-week period in a particular deflation basin. During that study, moving sand was trapped and important data about weather conditions during the sand drift were received. Some information about long-term deposition was then reported.

The questions to which this study tries to give answers are: (1) What is the annual amount of deflation needed to keep the deflation basin uncovered by vegetation in subarctic Lapland? (2) How quickly have the deflation basins been built up at the present deflation rate?

The Study Region

The study region is called Hietatievat; it is situated in eastern Enontekiö in Finnish Lapland (68°27′N, 24°42′E) (fig. 1). Hietatievat is part of a larger system of eskers, which begins in the upper course of the Käkkälö River near the Norwegian border and continues south all the way to Rupivaara fell (Tanner 1915: 262–264). Hietatievat is a broad esker, about 1.5 km wide and 4 km long (fig. 1). The relative height of Hietatievat esker is about 35 m and its summits are about 385 m above sea level (Seppälä 1966: 276).

Parabolic sand dunes have been deposited on the esker and its edges. They are at present partly anchored by vegetation, and are partly open blowouts (Ohlson 1957, Seppälä 1966, 1974) (figs. 1 and 2). The largest sand dunes lie on the eastern side of the esker, running parallel with the esker and rising from 4 to 10 m above the surface of the peat bog at the edge of the dunes (Seppälä 1974).

The forests of the Hietatievat region consist of mountain birch (*Betula pubescens* subsp. *tortuosa*). On the top of the esker there are also single pines (*Pinus silvestris*). The region is not far from the pines' northern limit (Atlas of Finland 1960: map 10 18). Details of the vegetation succession on Hietatievat can be found in Tobolski (1975).

The periglacial character of the climate in the Hietatievat region is evident from the palsa bogs, which have permafrost situated nearby (Ohlson 1964, Seppälä 1976, 1979). On the Hietatievat esker there are also frost wedges filled with eolian sand, which are found in fine silt that was originally deposited in a late glacial lake during deglaciation (Seppälä 1966, 1981*b*). Polygonal frost wedges that are still active can be seen in the old eolian basins between sand dune ridges (Seppälä 1974, 1981*b*).

Climate of the Hietatievat Region

The studied region is in the rain shadow of the Scandinavian mountains, which are located west of the region. The Gulf Stream keeps the climate relatively mild compared with other regions in the world north of the 68th parallel.

The mean annual temperature is about −1 C (Atlas of Finland 1960: map 5). The warmest month is July, with a mean temperature of +14 C; the coldest is February with a mean temperature of −13 C (Kolkki 1959).

The annual rainfall in the area is about 400 mm, of which 55 to 60% falls in snow. The average thickness of the snow cover is 60 cm. The permanent winter snow falls — according to observations made from 1892 to 1941 – on 25 October and disappears from the open places on 20 May (Atlas of Finland 1960).

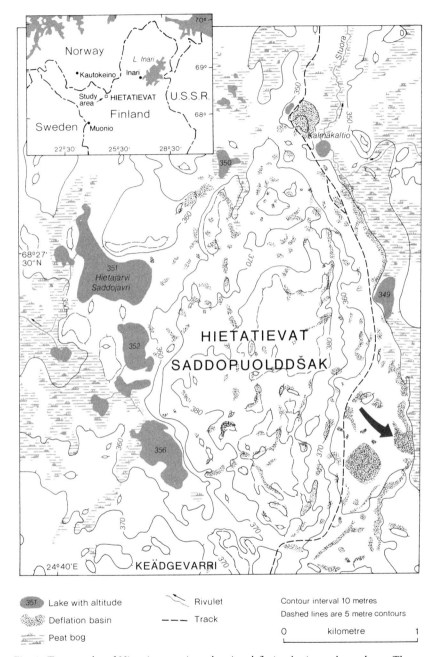

Fig. 1. Topography of Hietatievat region, showing deflation basins and peat bogs. The arrow indicates the studied blowout. Map of the general location of the study region inset in the upper left corner. Redrawn from the topographic map of Finland.

Fig. 2. Aerial photograph from Hietatievat, looking north. White areas indicate deflation basins. Photographed on 30 June 1973. Published with the permission of Topografikunta.

Fig. 3. The studied blowout, facing northwest. Photographed by R. Å. Larsson, 26 June 1973.

Westerly winds are predominant throughout the year. Less than 3% of all winds are strong (Atlas of Finland 1960: 6); this means that, generally, the sand movement is not very intensive (Seppälä 1974).

Methods of Study

On the same blowout (figs. 1 and 3) described in the earlier study by the author (Seppälä 1974), 20 wooden stakes 3 to 4 cm in diameter were driven in at a depth of 50 to 70 cm (fig. 4). Stakes stood at intervals of 20 m in four lines, in a west-east direction (fig. 5). The longest line was 140 m in length. At the beginning of the study the height of the sand surface was marked on each stake with a knife (fig. 4). The top 20 to 30 cm of the stakes were above the sand surface.

The study began on 1 July 1974. The position of the sand surface was observed four times over 3 years: on 26 August 1975, 11 August 1976, 27 August 1976, and 7 July 1977 (table 1). The vertical position of the sand surface at each point was measured to an accuracy of ±0.5 cm.

Although people did not move the stakes, one stake was broken at the sand surface by running reindeer and was replaced by a new one. Since eolian sand is not frost-active material, frost did not lift the stakes.

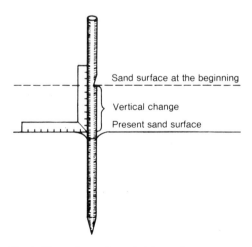

Fig. 4. Measuring stake and right-angle measure.

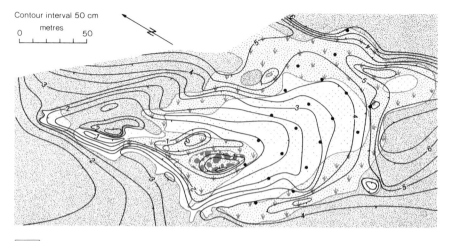

☐ Open sand surface (mainly deflation surface)

▦ Sand surface partly covered by vegetation (present deposition area)

▦ Completely anchored sand (no present aeolian activity or only very little accumulation)

▦ Stone and block surface of the basic esker material lying under the former sand dune

• Observation point; compare with fig. 6

Fig. 5. Topographic map of the studied blowout. (1) Open sand surface (mainly deflation surface). (2) Sand surface partly covered by vegetation (present deposition area). (3) Completely anchored sand (no present eolian activity or only very little accumulation). (4) Stone and block surface of the basic esker material lying under the former sand dune. (5) Observation point; compare with fig. 6. Mapped in the field in June 1973.

Deflation Measurements

During the first year of the study, 1974–75, the maximum deflation measured was 7.5 cm, which occurred at the west side of the basin (fig. 6). The most effective winds seem to have blown at that time, most likely from the northwest, deepening the centre of the blowout. The mean amount of deflation counted during the 20 observations was then 2.5 cm. About 190m³ of sand, which corresponds to 388 tons (counted with the density coefficient of 2.0) (table 1), moved from the measured area of 7,600 m².

During the second year, 1975–76, the deflation rates were much smaller, averaging 1.6 cm, and about 121 m³ of sand (242 tons) were moved. When looking at the deflation isopletes (fig. 6), the southwest and/or northeast winds were the most effective. The deepening of the blowout occurred mainly in its southwestern part.

The third measurement was made only 16 days after the second (11 and 27 August 1976). West winds transported the sand in the basin, but the

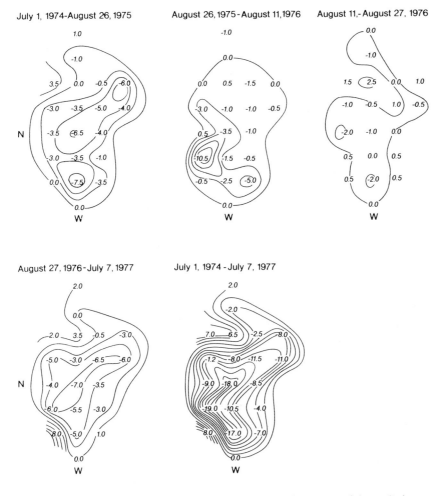

Fig. 6. Isopleth maps of the vertical deflation (−) and accumulation (+) of the studied blowout in different periods of Hietatievat, Finnish Lapland. Interval of points 20 m and of curves 2 cm. For the position of figures indicating the observations see fig. 5.

total amount of sand did not increase during the period. The deepening continued at the western and central parts of the deflation basin (figs. 5 and 6).

From 27 August 1976 to 7 July 1977 the most effective deflating winds blew from the northwest. The basin deepened, on average, 2.1 cm, and the values ranged from −7 cm to +8 cm. The amount of drifted sand totalled 158 m³ (316 tons).

If the whole 3-year period (from 1 July 1974 to 7 July 1977) is seen as a

Table 1 Vertical changes of sand surface (- erosion and + accumulation) and amounts of out-transported sand masses of the studied blowout (figs. 3 and 5) on Hietatievat, Finnish Lapland, during the measuring periods from 1 July 1974 to 7 July 1977 (See also fig. 6. Studied area 7600 m^2).

Period of observation	Range of vertical change of sand surface (cm)	Average change (cm)	Amount of out transported sand (m^3)	(tons)
1 July 1974 to 26 August 1975	-7.5 – +3.5	-2.55	194	388
26 August 1975 to 11 August 1976	-10.5 – +0.5	-1.6	121	242
11 August 1976 to 27 August 1976	-2 – +2.5	0	0	0
27 August 1976 to 7 July 1977	-7 – +8	-2.07	158	316
1 July 1974 to 7 July 1977	-19 – +8	-6.22	473	946

cumulaltive deflation and deposition on each measuring point, we find that at 15 points, deflation ranging vertically from -2 to -19 cm took place; at 4 points, accumulation dominated, ranging from +2 to +8 cm; and at the westernmost point, which was sited as a control point just at the edge of the blowout, the value was always 0 (figs. 5 and 6). The winds in the sector between northwest and southwest seemed to be the most effective in the formation of the blowout. The edge of the blowout may have caused a turbulence, so that the deepening of the basin mainly took place close to the edge. The sand was then transported on the surface to the other side of the blowout.

The total vertical deflation in the 3 years, on average, was 6.2 cm counted from all the measurement points. This value corresponds to about 470 m^3 of sand (more than 946 tons) (table 1). The average values for 1 year are 2 cm, 157 m^3, and 314 tons.

In 1973 some long-term deposition observations were made according to the vegetation at a particular blowout. (Seppälä 1974). The measurements were based on perennial plants (*Carex* spp., *Descampsia flexuosa*, *Juncus trifidus,* and *Solidago virga-aurea*), which are able to grow on wind-blown surfaces that have sand accumulations. When these plants are buried by drift sand, they form a new leaf node above the former one every spring. By measuring the distances between the leaf nodes we can arrive at a relatively clear picture of the annual intensity of accumulation. The details of these measurements have been described by Seppälä (1974). The vertical annual accumulation of drift sand on the edges of the blowouts was 2.1 to 2.2 cm on average, which is remarkably close to the deflation values (2 cm) presented in this study. It seems obvious that the accumulation area with scattered vegetation may be as large as the deflation area (fig. 5). The total area of deposition has not been measured and is therefore unknown.

Discussion

The method used in this study gives a much better view of long-term deflation and sand transportation than does the trapping of moving sand (cf. Seppälä 1974). The stakes do not obstruct sand movement, so there is no artificial accumulation as is the case when traps are used. There is the added advantage that we obtain a complete picture of the sand budget (both accumulations and depletions) over a long period in the whole area of the blowout. By trapping sand we can of course collect detailed information about the weather conditions that form the background of the geomorphic process. In this way we can get an idea of how and when the process works, but the conclusions about the formation of the geomorphic feature are more speculative than those presented here.

Some of the greatest deflation basins in the Hietatievat region are 10 m deep (Seppälä 1974). At the average rate of deflation (2 cm per year), it has taken at least 500 years to form them. Deflation rates differ greatly from place to place, even within the same blowout. The wind digs deep hollows, but in these subarctic conditions the deposition occurs on larger areas around the basins, so large dunes cannot form. When deflation concentrates at certain points, the formation of basins may take only a few hundred years. Redeposition of sand by the sliding of water and slush down the steep sides of deep blowouts may disturb the system, providing more material for the wind to transport and thereby enlarging the blowout.

The last forest fire on Hietatievat took place about 1825 (Seppälä 1966). Seventy kilometres west of Hietatievat there is a place where the deflation and accumulation of dune sand has continued for some 1,300 years after a forest fire (Seppälä 1981a).

The main causes of deflation in the present climatic conditions in Finnish Lapland are forest fires, reindeer, and vehicles, which destroy vegetation on the surfaces of sand dunes, glaciofluvial deposits, and sandy till surfaces. Once deflation has begun, it may cease for any one of five reasons

(Seppälä 1971): (1) the fine sand is carried away, leaving coarser (fig. 5) or finer material underneath (e.g., silt); (2) deflation reaches the level of the ground water (e.g., Ohlson 1957); (3) the ground-water level rises; (4) a blowout becomes so deep that the wind can no longer reach deflating velocities at the bottom of the blowout (deflation continues, however, at the edges of the basin); and (5) the climate becomes wetter and/or warmer, so that conditions favouring plant growth are improved. All these factors encourage the growth of vegetation in the basins.

A 2-cm thick layer of moving sand can, in a year, prevent the growth of vegetation in the blowouts. Seeds are transported away from the basin even when sand movement is minor, leaving the sand surface consistently bare.

Deflation in Finnish Lapland has some economic impact because the best growths of *Cladonia* lichen, on which reindeer feed, are located on the fine-sand areas. If reindeer populations in the region are allowed to increase unchecked, the lichens are depleted and the sand starts to move. It then takes many years to re-establish the lichens. No calculations of the value of lichen growth lost by deflation can be presented. However, they can be produced when deflation patterns have been mapped.

Acknowledgments

Messrs. Kai Lundén, Rolf A. Larsson, Viljo Säntti, and Jukka Rastas helped me during the field work. The study trips were financially supported by the National Council for Natural Sciences in Finland.

References

Åkerman, Jonas. 1980. Studies on periglacial geomorphology in West Spitsbergen. Meddelanden från Lunds Univ. Geogr. Inst., Avhandlingar 89. 297 pp.
Atlas of Finland. 1960. Geogr. Soc. Finland, Helsinki.
Black, Robert F. 1951. Eolian deposits of Alaska. Arctic 4: 89–111.
David, P. P. 1977. Sand dune occurrences of Canada. National Parks Branch, Can. Dep. Indian and Northern Affairs, Contract 74–230, Rep. 183 pp.
Fristrup, Børge. 1952–53. Wind erosion within the Arctic deserts. Geogr. Tidsskrift 52: 51–65.

Kolkki, Osmo. 1959. Temperaturkarten und tabellen von Finnland für den zeitraum 1921–50. Beil. zum Meteorol. Jahrb. Finnl. 50: 1.

Ohlson, Birger. 1957. Om flygsandfälten på Hietatievat i östra Enontekiö. Summary: On the drift-sand formation at the Hietatievat in eastern Enontekiö. Terra 69: 129–137.

———. 1964. Frostaktivität, verwitterung und bodenbildung in den fjeldgegenden von Enontekiö, Finnisch-Lapland. Fennia 89: 3. 180 pp.

Samuelsson, Carl. 1926. Studien über die wirkungen des windes in den kalten und gemässigten Erdteilen. Bull. Geol. Inst. Univ. Uppsala 20: 55–230.

Seppälä, Matti. 1966. Recent ice-wedge polygons in eastern Enontekiö, northernmost Finland. Pub. Inst. Geogr. Univ. Turkuensis 42: 274–287.

———. 1971. Evolution of eolian relief of the Kaamasjoki-Kiellajoki river basin in Finnish Lapland. Fennia 104. 88 pp.

———. 1972. Location, morphology and orientation of inland dunes in northern Sweden. Geogr. Ann. 54A: 85–104.

———. 1974. Some quantitative measurements of the present-day deflation on Hietatievat, Finnish Lapland. Abhandlungen der Akademie der Wissenschaften in Göttingen, Mathematisch-Physikalische Klasse, III. Folge 29: 208–220.

———. 1976. Seasonal thawing of a palsa at Enontekiö, Finnish Lapland, in 1974. Biul. Perygl. 26: 17–24.

———. 1979. Recent palsa studies in Finland. Acta Univ. Ouluensis A82: 81–87.

———. 1981*a*. Forest fires as activator of geomorphic processes in Kuttanen eskerdune region, northernmost Finland. Fennia 159: 221–228.

———. 1981*b*. Present-day periglacial phenomena in northern Finland. Biul. Perygl. [In press.]

Tanner, V. 1915. Studier öfver kvartärsystemet i Fennoscandias nordliga delar III. Résumé: Etudes sur le système quarternaire dans les parties septentrionales de la Fennoscandie. Fennia 36: 1. 815 pp.

Thorarinsson, Sigurdur. 1962. L'érosion éolienne en Islande a la lumière des études téprochronologiques. Rev. Géomorph. Dynam. 13: 107–124.

Tobolski, K. 1975. Succession of vegetation on drifting sands of Finnish Lapland dunes. Quest. Geogr. 2: 157–168.

Snow and Living Things

William O. Pruitt, Jr.

Introduction

In order to study and appreciate the effects of snow on living things, we must be prepared to go beyond the limitations of the English language, because English, having evolved in a misty, maritime climate, is notably deficient in snow terms. Table 1 gives a sample of terms useful in snow ecology. In previous works I defined and discussed a number of terms useful for understanding the ecology of snow (Pruitt 1978, 1979).

The characteristics of a snow cover that are important to living things are duration, thickness, hardness, and density. The latter two characteristics are influenced primarily by wind and the occurrence of winter thaws or freeze-thaw cycles. Thus there are four combinations that agree with four major geographic types of snow cover: (1) steppes and coastal regions with freeze-thaw and wind, (2) tundra with wind but no freeze-thaw, (3) inland southern and maritime regions with freeze-thaw but no wind, and (4) taiga with no freeze-thaw and no wind. For some organisms (i.e., mammals) high mountains constitute a separate ecological realm (Shvarts 1963). Because of variations in height and latitude, however, and because wind has such an influence on the snow cover at high elevations, it seems best to consider the snow cover of high mountains as special cases of categories (1) and (2). All snow on the ground, no matter where it is located, is subject to metamorphosis or changes that modify the crystals and affect the internal physical properties of the snow cover. This paper is concerned particularly with types (2) and (4).

Taiga Snow Cover

In the taiga or northern coniferous forest (Pruitt 1978) there are short, warm summers and long, cold winters. The winters are characterized by little wind, a marked reduction in incoming solar energy, and few incursions of maritime or tropical air masses. The result is a snow cover that comes early in the fall and lasts all winter, virtually unaffected by thaws or wind. Taiga snow cover occurs in two phases: *api,* the snow on the ground (Pruitt 1957); and *qali,* the snow on the trees (Pruitt 1958).

Api has a relatively uniform regional thickness (except for qamaniq—see table 1), but the thickness varies greatly from year to year (Pruitt 1978: fig. 5-2). Density is usually less than 0.2 and hardness less than 200 g cm^{-2}, except on large lakes where wind has reworked the cover. Over vast areas the major influences on api are the heat and moisture that rise from the earth below, pass through the api, and escape to the cold, dry supranivean air. The heat and moisture gradients result in the lower part of the api being transformed into a series of fragile, interconnected columns composed of many hollow, pyramidal crystals (fig. 1). The columnar basal layer of api is properly termed "pukak" and may be 10 or 20 cm thick; individual pukak pyramids may be as large as 10 mm across.

Fig. 1. Pukak, the fragile, lattice-like basal layer of snow cover, caused by sublimination and recrystallization. Manitoba, Agassiz Provincial Forest, 4 March 1978. Photo by Wolf Heck.

Pukak is of ecological significance not only to plants, small mammals, and some invertebrates, but to some large mammals as well. For example, caribou/reindeer (*Rangifer tarandus*) are markedly affected by pukak variations. Swedish Lapps have long recognized the variations, because they influence reindeer feeding. The Lapps have special names for the different types of pukak (table 1). Eriksson (1976) demonstrated that the named varieties of pukak correlate nicely with differences in hardness, density, and grain type.

The metamorphosis of a snow cover is a continuing, albeit irregular, phenomenon. The process of "pukakization" continues and expands upward from the soil surface, governed by (1) the amount of heat and moisture flowing from the earth, and (2) the lack of heat and moisture in the supranivean air. Therefore the colder and drier the supranivean air, the more intense and quicker the pukak expansion. One result of the phenomenon is a decrease in the hardness of the pukak layer. Hard layers or even layers of vesicular ice in the lower half of the snow cover are progressively eroded and softened. This phenomenon may affect overwintering caribou/reindeer (Pruitt 1979).

Api, in all its morphological phases, profoundly affects the amount and quality of light available to subnivean mammals. Evernden and Fuller (1972) confirmed Geiger (1961) that the pukak space under 30 to 50 cm of api is essentially dark. They also showed that when light does penetrate to the pukak level it is mostly in the red end of the visible spectrum. Because red light, even with an increased photoperiod, does not stimulate female redback voles (*Clethrionomys gapperi*) to ovulate, the api over these voles ensures that they do not breed until the api is broken in spring. Richardson and Salisbury (1977) analysed subnivean light in relation to plants. Fuller et al. (1969) also analysed subnivean temperatures in relation to overwintering small mammals.

Coulianos and Johnels (1962) discussed the importance of the pukak space to winter distribution of small rodents. By carrying out simple environmental manipulation experiments they showed that areas with a pukak space were more attractive to small mammals than were areas of the same vegetation type that lacked such a space. Pruitt (1959*b*) described the method of "trap chimneys" for studying small mammals active in the pukak space. Smirin (1970) and Merritt and Merritt (1978) described variations in construction of "trap chimneys."

The major ecological effect of api on plants is to insulate the forest floor from the vagaries of the supranivean environment. Thus the herb layer is protected from heat and moisture loss; the shrub layer is less protected,

Table 1 Glossary of some specialized snow terminology

Definitions of specialized snow terminology have been published by Pruitt (1957, 1960b, 1966, 1970, 1978: figs. 1-3, 1-4, 3-1, 5-1, 5-2) and Eriksson (1976) as well as earlier by Formozov (1946) and Nasimovich (1955).

Term	Source	English equivalent
aŋmaŋa	Kovakmuit (Inuit)	Space formed between drift and obstruction causing it.
api	Kovakmuit (Inuit)	Snow on the ground, forest.
bodni vihki	Lappish (Eriksson, 1976)	A frozen layer of snow cover next to the soil; granular structure exists but crystals are intensely bonded together.
caevvi	Lappish = tsaevvi (Eriksson, 1976)	Dense basal layer of snowcover caused by the sequence of (1) unfrozen soil; (2) warmer layer of snow resulting from high temperatures; followed by (3) thick snow layer which compresses the basal layer and results in a hard, dense mass with small crystals. Caevvi is severe for reindeer since it is suitable for growth of snow mould which prevents reindeer from smelling lichens; it probably also facilitates accumulation of CO_2 under the snow cover.
čiegar	Lappish	"Feeding trench" or a linear extension through undisturbed api of a sequential series of suov'dnji. Actual feeding on ground vegetation occurs only at the terminal end of a čiegar. Excavated snow is kicked back, partially filling the trench with snow that sinters and becomes very hard.
čuok'ki	Lappish	Layer of solid ice next to the soil.
fies'ki	Lappish	"Yard crater" or roughly circular site of thin, hard and dense snow cover caused by deer digging and extending a perimeter expanding into undisturbed api from an original suov'dnji or group of suov'dnji.

Table 1 *Continued*

Term	Source	English equivalent
kaioglaq	Kovakmuit (Inuit)	Large hard sculpturings resulting from erosion of kalutoganiq. More precise than the loose terms "zastrugi" or "skavler".
kalutoganiq	Kovakmuit (Inuit)	Arrowhead-shaped drift on top of upsik; moves downwind.
kanik	Kovakmuit (Inuit)	Ice crystals that form on cold objects when warm, moist air passes over them.
kieppi	Finnish	Hole in snow cover where an animal has plunged for shelter (pronounced "KEY-eppi").
kimoaqruk	Kovakmuit (Inuit)	"Finger drift" downwind of small obstruction.
mapsuk	Kovakmuit (Inuit)	Overhanging drift or "anvil drift."
nast'	Russian	Thickened crust on the surface of a mature snow cover.
pukak	Kovakmuit (Inuit)	Fragile, columnar basal layer of api.
qali	Kovakmuit (Inuit)	Snow on trees.
qamaniq	Kovakmuit (Inuit)	Bowl-shaped depression in api under coniferous tree.
sändjas	Lappish	Fragile, columnar basal layer of api (=pukak).
siqoq	Kovakmuit (Inuit)	Drifting or blowing snow.
suov'dnji	Lappish	"Feeding crater" or individual feeding site excavated in the api, separated from other sites by undisturbed api.
tumarinyiq	Kovakmuit (Inuit)	Small "ripple marks" which are (1) last remains of kaioglaq or (2) differential erosion of hard and soft layers.
upsik	Kovakmuit (Inuit)	Wind-hardened tundra snow cover.
vyduv	Russian	Spot blown bare of snow.
zaboi	Russian	Area of thick snow cover that persists perhaps all summer.

and only the roots of the tree layer are protected. The bowl-shaped snow shadow or qamaniq (Pruitt 1978: fig. 5-2) that forms around the base of a coniferous tree allows heat and moisture to escape, leaving the qamaniq with essentially supranivean conditions (Pruitt 1957). Viereck (1965) showed that such heat loss results in soil ice, elevation of the permafrost table, and a cycle of soil heaving and collapse that may change the vegetation type. Pruitt (1959b) observed that qamaniq were avoided by voles, which instead used most intensively those parts of their home range where the api was thickest.

Api affects animals in different ways, depending on their body mass, feet characteristics, and food habits. In general, animals in snowy regions are either subnivean or supranivean in activity. Almost all birds are supranivean, although a few such as grouse-like birds (*Tetraonidae*) become partly subnivean in cold weather (fig. 2). Subnivean roosting is rarely reported in other families. See McNicholl (1979) for a summary of records.

Supranivean mammals may be classified as either "floaters" or "stilters." Floaters such as snowshoe hares (*Lepus americana*) and lynx (*Lynx lynx*) have large feet in relation to their body mass. They are usually supported on the surface of the api if its surface hardness is greater than 40 or 50 g cm^{-2} and its density greater than 0.18. Sometimes, however, the api may be too soft to support even such specialized mammals. When this occurs the animals undergo a dramatic behavioural change and follow their own and each other's trails, which soon become hard highways. At these times carnivores such as fox (*Vulpes*), lynx, and fisher (*Martes pennanti*) also follow the hare highways. Bider (1961) pointed out that the api accumulation furnishes hares with a three-dimensional habitat, and that changes in api thickness govern the quantity of food available in both short-term and long-term considerations.

Some mammals such as caribou are too heavy to be true "floaters," but still they have greatly enlarged feet. Such mammals are notably affected by the physical characteristics of the api (Pruitt 1959c, 1979, 1981, Makridin 1963, Kelsall 1969, Stardom 1975). For example, Kelsall and Prescott (1971) pointed out that the boundary between the moose (*Alces*) and whitetail deer (*Odocoileus virginianus*) habitat in New Brunswick correlated with the chest height of the animals and the thickness of the api. "Stilters" are exemplified by moose, which possess long legs. Moose overwintering areas are those with thin snow cover (Nasimovich 1955, Edwards 1956, Edwards and Ritcey 1956). Pulliainen (1965) showed that timber wolves (*Canis lupus*) in the taiga of Finland preferred areas of thin and hard snow cover.

Fig. 2. Kieppi, made by capercaillie (*Tetrao urogallus*). Finland, Oulanka National Park, 17 March 1969.

Almost all subnivean mammals weigh less than 250 g. Because they have a large surface area in relation to their mass they are susceptible to excessive heat and moisture loss. The pukak space, with its constant, relatively high temperature and saturated air, allows such mammals to survive in regions from which they otherwise would be excluded (Pruitt 1957, Koskimies 1958). The blanket of api also protects subnivean mammals from some predation. On the other hand, a thick api exposes some small mammals to increased predation. The paradox can be explained as follows. Large "pouncing" predators (fox, lynx, raptorial birds) can catch subnivean animals only with difficulty. But because subnivean animals require bulky insulated nests they construct these in the friable pukak on top of the frozen soil. Nests in such a location are more vulnerable to predation by "tunnelling" predators (e.g., weasels, mink) than are nests surrounded by frozen soil (Formozov 1946).

In autumn air temperatures decline, approaching those of the soil and litter and eventually falling below those of the soil. The result is the autumnal thermal overturn. At some time in this progression the snow cover arrives. A thin snow cover is relatively ineffective in insulating the soil and litter from diel fluctuations in air temperature. During this period tracks, trails, and "ditches" of small mammals occur on the snow cover. At some time the snow cover reaches a thickness sufficient to insulate soil and litter from diurnal fluctuations in air temperature. This thickness is the hiemal threshold because it is the beginning of true winter for the organisms of the forest floor. Formozov (1946) quoted Nekrasov as reporting that the hiemal threshold was 20 to 25 cm, while Pruitt (1957) reported 15 to 20 cm as the hiemal threshold in the taiga of interior Alaska. The formation of the hiemal threshold is a function not only of the air temperatures of the time, but also of the density of the api (and therefore its insulating quality). Although the thermal overturn is a fairly predictable event, the arrival of a snow cover sufficient to form the hiemal threshold is highly variable.

The time between the autumnal thermal overturn and the formation of the hiemal threshold constitutes the fall critical period (Pruitt 1978: fig. 4-1). As the api disappears in the spring there is a corresponding spring critical period (Fuller 1967). These two critical periods have profound effects on the numbers and local distribution of small mammals; during them the animals are exposed to the rigours of the supranivean environment.

The mild subnivean environment allows considerable activity not only by mammals, but by ectothermal invertebrates as well (Näsmark 1964, Aitchison 1978, 1979). Such activity, as well as bacterial metabolism and some physical processes, releases carbon dioxide in the pukak space. If the api has dense layers or ice layers that retard the upward flow of air, the CO_2 may accumulate in concentration sufficient to affect small mammals (Bashenina 1956, Penny 1978). The resulting behavioural changes cause them to construct "ventilator shafts" to the api surface (fig. 3). These shafts benefit predators as well; Formozov and Birulya (1937) showed that certain species of small owls are able to overwinter in parts of the Eurasian taiga only because they catch voles at the voles' ventilator shafts.

Penny (1978) found that CO_2 accumulated each winter in certain habitats but not in others, and in concentrations up to five times ambient. She found that the accumulation was affected primarily by the density and hardness of the api. The CO_2 accumulation was sometimes accompanied by changes in small mammal distribution; the animals shifted their home range away from the affected site and returned later when the subnivean CO_2 fell to ambient concentrations.

There are, of course, some mammals that exhibit a combination of supranivean and subnivean activity. One such mammal is the North American red squirrel (*Tamiasciurus hudsonicus*). Pruitt and Lucier (1958)

Fig. 3. "Ventilator" shaft made by *Microtus oeconomus*, possibly as a reaction to subnivean CO_2. Alaska, Fairbanks region, 27 January 1963.

showed that its activity below the api surface correlated with the presence of a critical ambient air temperature of about −30 C. Pulliainen (1973) noted that in northeastern Lapland, European red squirrels (*Sciurus vulgaris*), which weigh about 320 g, sheltered in supranivean nests, while the less heavy North American red squirrel, which weigh about 230 g, sheltered below the api. Pauls (1978) confirmed air temperature as the most important environmental parameter affecting North American red squirrel activity in southern Manitoba.

Qali is the snow that collects on trees. (Pruitt 1978: fig. 5-2). In temperate regions it is of only transitory aesthetic interest. In the taiga, however, qali is a significant ecological factor (Pruitt 1958). It has been a factor in the evolution of the shape of spruce trees; it is a powerful influence over vegetation type, because it governs some aspects of forest succession (Gill et al. 1973); and it affects man-made structures such as powerlines, by breaking them. Qali can be measured objectively by means of qalimeters, but the study of qali is in its infancy at present and the standardization of observations now would be premature (Pruitt 1973).

Qali (fig. 4) forms best under conditions of continuous, light snowfall, reduced incoming solar radiation, low temperatures, and no wind. A

Fig. 4. Qali, the snow that collects on trees. Alaska, Fairbanks region, 12 February 1962.

superficially confusing phenomenon called kanik (fig. 5) forms when warm, moisture-laden air strikes cold objects. It forms best under conditions that are near freezing, with light winds. Qali forms in varying amounts on horizontal surfaces, while kanik forms a layer of more-or-less uniform thickness on vertical surfaces.

Qali affects animals such as birds and arboreal mammals by interfering with their feeding and travel. Thus during periods of heavy qali accumulation birds such as pine grosbeak *(Pinicola enucleator)*, chickadee *(Parus atricapillus)*, and crossbill *(Loxia curvirostra)* forage on windy hilltops, where qali is blown off the trees. Pine marten *(Martes martes)* and American red squirrel find their arboreal activities affected by heavy qali accumulation. On the other hand, some small birds such as chickadee protect themselves from excessive radiant heat loss by huddling under clumps of qali (Steen 1958). Snowshoe hares use snow caves formed under qali-bent shrubs so that their radiant heat will not be lost to the infinite heat sink of the night sky (fig. 6). On the other hand, qali bends shrubs over, so that their tender growing tips are brought within reach of the hares. This presents the hares with a supplementary source of food.

Fig. 5. Kanik, or ice crystals that form on cold objects exposed to warmer moist air. Finland, Värriötunturi region, 4 February 1977.

Cultural Effects

Human attitudes toward the formation of api vary from a welcoming acceptance to near-hysteria. In taiga aboriginal cultures the arrival of api signalled a complete change in behaviour and activity patterns. Specialized clothing and equipment allowed the people to travel and to use regions unavailable during the snowless times of the year. The most important material adaptations they evolved were the woodland toboggan and the pulka; snowshoes; and skis, or "snowboots." The woodland toboggan, with its high, curled prow and snow-shedding carriole, is ideally suited for traversing api. The pulka can be hauled by man, reindeer, or dogs, and because of its shape it tends to follow the trail without being guided. Also, its cover is designed to shed snow. There is a close correlation between types of snowshoes—their size, shape, fineness and density of netting, and number of supporting cross-pieces—and the thickness, hardness, and density of the api of the region in which they evolved. The snowshoes with the best workmanship, finest netting, most cross-pieces for support, and other flotation features are found in the Mackenzie Valley, a region with thick, soft, and light api. Similarly, skis vary according to api type, from long and wide Lapp forest skis to the short, broad Siberian "snow boots."

Fig. 6. Qali cave, formed under qali-bent shrubs. Alaska, Fairbanks region, 10 April 1962.

An important discovery of aboriginal man in the taiga was the process of sintering. When soft, light api is stirred and mixed, it recrystallizes and rebonds into a hard mass. The process of stirring destroys the temperature gradient, and "warmer" crystals are thrust into contact with "colder" ones, to which they form a molecular bond. Precontact Athapaskans regularly made snug, warm hunting shelters, called "quin-zhee," by heaping api into a mound, waiting an hour or two for natural sintering to occur, and then hollowing out the interior for a living space. My students and I regularly construct quin-zhees and have noted that the hardness changes naturally from, say, between 5 and 20 g cm^{-2} to 600 g cm^{-2} or more in two hours at ambient temperatures of -30 C. When all the snow is brushed from the floor, so that the earth heat can flow upward unimpeded, the temperature of the enclosed living space rises from ambient to just below freezing. Application of this knowledge made the Athapaskans masters of the taiga winter environment; like the subnivean creatures they could be free of thermal stress while sleeping (Pruitt 1973).

Tundra Snow Cover

The lands beyond the tree line have a complex of vegetation types collectively called tundra (Alexandrova 1970). The only real unifying factor among the various types of tundra is the constant lack of trees. The tundra-taiga ecotone may be narrow or it may be a wide belt of forest-tundra of several types (Pruitt 1978: figs. 1-5, 1-6, 1-7). Most tundra landscapes are rolling or even rugged. In only a few restricted areas are they flat.

The climate of almost all types of tundra is characterized by long, cold winters and short, cool summers (Hare 1970). Perhaps the most important climatic factor, particularly in relation to snow ecology, is wind (Hare 1971). It is the important factor in two great groupings of snow characteristics. An understanding of these two dichotomies is necessary for understanding many aspects of tundra ecology (Pruitt 1966, 1970).

The first dichotomy is that of the upsik-siqoq (Pruitt 1978: figs. 1-3, 1-4). In the physical sense, there are two phases of tundra snow. The wind-reworked snow cover becomes consolidated into a hard mass called upsik. Above the snow cover in the air is another phase, the moving snow

Fig. 7. Kalutoganiq, or arrowhead-shaped, moving accumulations of siqoq. Alaska, Cape Thompson region, 17 April 1960.

Fig. 8. Kaioglaq, heavily sculptured remains of kalutoganiq. Alaska, Cape Thompson region, 19 April 1960.

or siqoq. This phase either moves along, governed by wind force and direction, or is consolidated into a succession of drift forms.

On a flat, relatively unobstructed surface these particles move in drifts that have characteristic arrowhead shapes pointing upwind, with the greatest thickness at the tang. These drifts are known popularly as barkhans, but more accurately, in the Inuit language, as *kalutoganiq* (fig. 7). *Kalutoganiq* travel downwind, but whenever the wind slackens they become consolidated through the processes of sublimation and recrystallization.

Later winds, if of sufficient force, will erode the *kalutoganiq*, producing sculptured forms that are beautiful but exceedingly difficult to traverse. The sculpturings are widely known by the terms *zastrugi* (Russian) or *skavler* (Norwegian), but more accurately as *kaioglaq* (Inuit) (fig. 8). *Zastrugi* or *skavler* refer to surface sculpturings in general. *Kaioglaq* refers to large, hard sculpturings, while the word *tumarinyiq* (Inuit) (fig. 9) refers to small *zastrugi* or "ripple marks," which may be the last remains of *kaioglaq*.

Kaioglaq may eventually be eroded away completely, and the particles regrouped downwind again into *kalutoganiq*. A late stage of *kaioglaq* is the formation of overhanging drifts or *mapsuk* (fig. 10). The windward

Snow and Living Things 65

Fig. 9. Tumarinyiq, or "ripple marks" caused by differential erosion of harder or softer layers; the last remains of kalutoganiq. Alaska, Cape Thompson region, 8 February 1960.

Fig. 10. Mapsuk, or "anvil drift." Manitoba, Winnipeg region, 6 February 1973.

point of a ridge of *kaioglaq* is eroded faster at base level than above it, thus forming the characteristic anvil horn, which points upwind (Pruitt 1970).

The second dichotomy is that of zaboi-vyduv (Pruitt 1978: fig. 1-3). A concave topographic surface tends to be filled by siqoq; the resulting mass of upsik is called a zaboi. A convex surface is blown clear and is called a vyduv. Valleys of small streams may be occupied completely by zaboi that may not melt until late in the following summer. From the periphery of a zaboi to its centre the growing season is progressively shorter; consequently the number of species and degree of ground cover is likewise reduced. In extreme cases zaboi may prevent all plants from growing on the sites where they form. These bare spots are then subject to intense cryopedological processes, and may erode faster than surrounding non-zaboi surfaces.

It can be seen, then, that tundra snow cover is dense, hard, irregular in thickness, and has a moveable component. In contrast to taiga api, the thickness of tundra upsik at any one site is quite predictable from year to year, since to a large extent its local distribution is topographically controlled. Wind direction also influences upsik morphology. Tundra winds usually blow from varying directions, so that topographic concavities of different orientations are partially scoured-out or filled in irregular sequence. The result is normal hard and dense upsik. In a few atypical and restricted regions (usually because of topographic funnelling) the winds consistently blow from one or a few related directions. Under these conditions topographical concavities become filled early in winter and are not scoured-out periodically. Thus the moisture and heat gradients from the soil have time to act on the upsik, and it may be transformed into a hard, thin, "wind-slab" top layer with soft pukak underneath (Pruitt 1966: 556-557). Such a restricted region may become suitable for overwintering caribou or muskox.

Tundra plant communities are closely correlated with the distribution, thickness, and duration of the upsik. Hardness of upsik may reach several thousands of grams per square centimetre; its density may be as much as 0.7 or 0.8, although it is usually around 0.5. Consequently heat transmission is relatively great, and thick upsik is required to protect tundra plants from the supranivean environment. On the other hand, in spring light transmission is sufficient to enable plants to begin to grow under the upsik (Kil'dyushevskii 1956, Yashina 1961, Richardson and Salisbury 1977).

Upsik, particularly in alpine regions, plays perhaps an even greater role by protecting tundra plants from other aspects of the supranivean environment—the desiccation by the cold, dry wind and the abrasion by siqoq

(Pruitt 1978: fig. 1-5). Yoshino (1966) attributed his types 2 and 3 of wind-shaped trees to abrasion by siqoq and to desiccation. Indeed Křivský (1958) demonstrated how such wind-shaped trees near the altitudinal treeline could be used as indicators of main wind direction. Treeline itself (not only alpine, but arctic as well) may be controlled by siqoq abrasion and desiccation.

Zaboi affect tundra vegetation in different ways, depending on the orientation of the prevailing winds and on the spring-summer energy source (Billings and Bliss 1959, Gjaervoll 1950, 1956). The ideal overwintering situation for plants is a thick zaboi on a relatively steep, south-facing slope; this affords maximum protection during the winter, yet melts early enough to allow a full growing season. The worst situation would be a steep north or northeasterly facing slope.

When a zaboi melts it releases its moisture slowly and causes the formation of mesic habitats downhill from the site. A zaboi collects siqoq from a wide area and immobilizes it; therefore, contaminants in the siqoq are concentrated. Osburn (1963) has shown that in alpine tundra zaboi, radioactive contaminant concentrations may be remarkably dense. As the zaboi melts it releases the contaminants, which are trapped in the mesic sedge mat below the zaboi. The outlines of the problem have been worked out, but clearly much more research is needed. The problem assumes significance when one realizes that the water supplies for many cities come from alpine zaboi that are far away.

Tundra animals are also either supranivean or subnivean in habits (Pruitt 1978: fig. 3-1). Almost all birds migrate from the tundra before upsik forms. Of the few that remain, only ptarmigan (*Lagopus*) take advantage of the snow cover and find soft patches into which they burrow. Supranivean mammals—caribou, muskox (*Ovibos moschatus*), wolf, white fox (*Alopex lagopus*), and Arctic hare—are markedly influenced by nival conditions. I have, on several occasions, trailed white fox and noted how they consistently deviated to avoid soft areas in the otherwise hard upsik.

Most caribou or reindeer migrate from tundra to taiga in the fall. Some remain on the tundra even in the snow season, in restricted areas of topographically protected, relatively soft snow cover, or of so much wind that the vegetation remains exposed. Such areas are well-known as good hunting grounds (Pruitt 1959c, Henshaw 1968).

An obstruction, such as a large rock that interferes with the wind, causes the upsik to be scooped out where the wind speeds up around the obstruction. Such a scooped-out hollow is known to the Inuit as an

Fig. 11. Aŋmaŋa, or cavity scooped out in upsik by wind. Alaska, Cape Thompson region, 31 March 1961.

"aŋmaŋa" (fig. 11); it may be of importance to tundra birds and mammals. Aŋmaŋa may be the only places within many kilometres where the substrate is exposed. Here ptarmigan obtain gravel for their gizzards, and early migrants, such as snow bunting (*Plectrophenax nivalis*) and redpoll (*Acanthis*) congregate in spring. I have noted that white fox regularly visit aŋmaŋa in their hunting range, and I have also noted signs of their success there. Sulkava (1964) showed that aŋmaŋa around field hay barns (Pruitt 1978: fig. 5-1) were important in the winter ecology of partridge (*Perdix*) and brown hare (*Lepus europaeus*) on the Ostrobothnian plain in west-central Finland, with its upsik-like, wind-blown snow cover. I have noted that Arctic hare (*Lepus arcticus*) is associated almost invariably with aŋmaŋa or vyduv.

Tundra subnivean animals are influenced by the fall and spring critical periods, just as are taiga subnivean animals (Kuksov 1969). Because of the greater heat transmission of dense upsik compared with tenuous api, the hiemal threshold requires a thicker layer of upsik than of api. Perhaps the best analyses of tundra fall and spring periods have been by Fuller (1967) and Fuller et al. (1975). Fay (1960) described a method of live-trapping small mammals under tundra upsik.

Zaboi have as strong an effect on animals as they do on plants. In an area of Low Arctic tundra I found that singing vole (*Microtus gregalis*) occurred only in zaboi sites, and that it was the only small mammal found in these sites (Pruitt 1966). The reasons for such restrictions are not known, but are possibly related to subnivean CO_2.

An exception to this observation is arctic ground squirrel (*Citellus parryi*) which choose their hibernacula where thick zaboi form.

Perhaps because of an increased rate of heat loss (primarily because of wind chill), tundra mammals must be larger than those in other landscape zones to withstand the supranivean enironment (Hagmeier and Stults 1964; Baker et al. 1978). Even mammals as large as red fox and wolverine (*Gulo gulo*) become subnivean at times. The Inuit carefully search inside the curl of snow cornices along cutbanks, since this is a favourite site for such animals.

Pruitt (1957, 1970, 1978: fig. 4-1) and Payne (1974), among others, have shown that the popular and traditional concept of "winter" is insufficient for biological purposes. Pruitt (1957) defined "true winter" as occurring from the hiemal threshold in the autumn until the hiemal termination in the spring. In the taiga true winter comes to most of a region at about the same time. The major exception is in qamaniq, as we have seen. In contrast, in steppes, alpine tundra, or arctic tundra (or any site affected by wind) true winter arrives and persists at different times, depending on the degree of convexity or concavity of the substrate surface, or the presence and orientation of any obstruction to the wind. In other words, in a tundra region the wind sweeps the snow from convex surfaces and deposits it in concavities where the hiemal threshold may be achieved rather quickly, in contrast to nearby sites where the hiemal threshold never occurs. We realize that the region we traditionally define as having the most severe winter (the tundra) actually has this season only in isolated patches (if we maintain our original definition). What occurs between the patches of true winter conditions is environmentally rigorous indeed, but it is actually a greatly extended or continuous "critical period." Payne (1974) pointed out that in the Newfoundland maritime taiga with its fluctuating snow cover, the entire "calendar winter" was a critical period for small mammals. Many authors have discussed the winter restriction of tundra and steppe small mammals to topographic concavities or to zaboi sites where true winter comes early and stays late. Pruitt (1953, 1957, 1959*a*) compared microclimates and local distributions of small mammals in two regions that had different winter regimes.

It is obvious that the relationships between the size of populations of small mammals, the fall and spring critical periods, and conditions during true winter are far more complex than this simple exposition implies. For example, Fuller (1977) denied that winter mortality of northern redbacked voles (*Clethrionomys*) in the taiga was governed by the severity of conditions in the subnivean zone, while MacLean et al. (1974) concluded that subnivean temperatures were a major factor in determining the population size of brown lemming (*Lemmus sibiricus*) in tundra. In temperate zone subalpine habitats Vaughan (1969) and Stinson (1977) reported that the fall critical period resulted in great mortality in some mammals, while Merritt and Merritt (1978) observed that the fall critical period "did not prove to be a time of extreme hardship although some mortality was evident." Vaughan (1969), Fuller (1967, 1969), Stinson (1977), and Merritt and Merritt (1978) all agreed that the spring critical period coincided with a marked increase in the mortality of small mammals. These widely differing conclusions point up the great need for a co-ordinated series of long-term studies that encompass the entire spectrum of snow cover and climate regimes.

Cultural Effects

Among human cultures, that of the Inuit showed the most extreme adaptation to tundra snow cover. Perhaps that culture's main feature was the psychological welcoming of the snow cover in the fall. This attitude, along with a knowledge of how to make caribou-skin clothing and how to construct the *iglu* from blocks of upsik (Elsner and Pruitt 1959), gave the Inuit a supreme self-confidence in their ability to use their land.

The Inuit komatik and the Alaska basket sled have yet to be improved on as vehicles for traversing upsik. Because upsik is hard and dense, the Inuit have not evolved such high standards of workmanship in their snowshoes as have the taiga peoples. The Lapps, on the other hand, evolved long and narrow tundra skis.

Snow Ecology Models

During a study of the ecology of terrestrial mammals inhabiting an area of Low Arctic tundra (Pruitt 1966) I worked out a mathematical relationship between some basic parameters of the snow cover and the distribution and numbers of small mammals. I called the resulting number a "snow index," thus:

$$SI = C \ (\Sigma TD)$$

Where C = snow cover of a plot expressed as a percentage of the area of the plot covered by it, T = thickness (in cm), and D = density of each discrete layer of the cover as measured in a vertical profile. The snow index enabled me to differentiate several classes of mammal habitats and to correlate population levels with the snow parameters of the preceding winter.

We need more research along these lines to enable us to include time as a term in the snow index, since we have seen how important are the fall and spring critical periods.

In an attempt to derive a snow index that would model the winter activity of caribou, I combined the various aspects of their wintering ecology and behaviour in successively different ways. I then compared the results to the actual distribution and feeding activities of the feral reindeer in the vicinity of Värriö Subarctic Research Station in northeastern Finland (Pruitt 1979). The combination that best models their winter activity I call the Värriö Snow Index (VSI):

$$VSI = \frac{(H^{1/2} (H_b T_b) + (VT_s) + (H_h T_h)) T_{ta}}{1000}$$

where
$H^{1/2}$ = hardness of hardest layer more than halfway between the substrate and the top of the snow cover.
$H_b T_b$ = hardness times thickness of basal layer.
VT_s = vertical hardness of surface times thickness of surface layer.
$H_h T_h$ = hardness times thickness of hardest layer (if not $H_b T_b$). If basal layer is the hardest, then term $H_h T_h$ drops out.
T_{ta} = total thickness of the api.

This study showed that api through which reindeer fed had lower VSI than did api through which they did not feed. The VSI also modeled the api-relations of overwintering barren-ground caribou in northern Saskatchewan (Pruitt 1981). I noted that the Finnish feral reindeer have a significantly greater threshold of sensitivity to the factors encompassed by the VSI than do Saskatchewan caribou.

According to my comparison of VSIs, the reported difference between New and Old World caribou-api relations appears to be real. Indeed, the difference can be viewed as a measure of the amount of domestication or departure from the wild type. We can conclude that the VSI is a potentially useful tool for determining nivally suitable areas for caribou overwintering. VSI can probably also be used to delimit migration routes to and from the nivally suitable areas.

Our knowledge of caribou/reindeer-api relations has progressed to the point where we can classify stages or "seasons" of the snowy period of the year in relation to reindeer activity (Pruitt 1979: table 1). The VSI models these relations during all periods but the api-maturation period.

Zones of Snow Influence and Evolution

Our knowledge of snow ecology in general has progressed to the point where we can divide North America into zones of snow influence. At present we can recognize two general zones: one where snow cover is so infrequent or of such short duration as to be unimportant in the evolution of mammals, and one where snow cover has influenced mammalian evolution.

Hall (1951) published a map showing the distribution of regions where brown and white winter pelages are found in longtail weasel (*Mustela frenata*). South of the northern, winter-white region, the southern boundary of which extends as a fluctuating band across North America from New England–northern Virginia to central California and thence north to southern British Columbia, weasels always remain in brown pelage. This is a genetic difference; snow-country weasels transplanted to the south continue to moult brown to white to brown. Because obliterative colouration is the most logical explanation for white winter pelage (see Hammel 1956), we can conclude that snow cover has influenced the evolution of this particular species.

Another species that has evolved a morphological specialization to tundra upsik is collared lemming (*Dicrostonyx groenlandicus*), which has bifid, enlarged claws on the third and fourth digits of its front feet in winter. Sutton and Hamilton (1932) observed that, while brown lemming excavated tunnels in the pukak at the base of upsik, only collared lemming dug tunnels up through the hard and dense mass. Just as the shape of caribou hooves is governed by differential wear from snow or bare ground (Pruitt 1959c), so the bifid claws of collared lemming wear away in summer, when the animal digs soil instead of snow (Hansen 1957).

The great Russian naturalist A. N. Formozov (1946) classified animals on the basis of their ecological relations to snow: *Chionophobes* — species that do not inhabit snowy regions, and avoid snow; *Chioneuphores* — species that can withstand winters with considerable snow; and *Chionophiles*

—species whose ranges lie completely or almost completely in regions of hard and continuous winters. These latter species have characteristic adaptations (e.g., winter-white colouration, winter peculiarities of foot-coverings), which undoubtedly were perfected by snow cover taking part in selection. Subsequent research has served to confirm this classification.

Conclusion

In boreal regions the snow cover is an integral part of the environment of living things. With study, many ecological phenomena previously explained by way of tortuous, involved physiological or historical processes will likely be found to be the result of simple reactions to snow cover. Many human activities and schemes are influenced by snow cover. A knowledge of snow ecology, especially the principles of quin-zhees, would be valuable to any outdoorsman, or even any winter-time motorist.

References

Aitchison, C. W. 1978. Spiders active under snow in southern Canada. Symp. Zool. Soc. London 42: 139–148.

———. 1979. Winter active subnivean invertebrates in southern Canada. I Collembola: 113–120; II Coleoptera: 121–128; III Acari: 153–160; IV Diptera and Hymenoptera: 153–160. Pedobiologia 19.

Alexandrova, V. D. 1970. The vegetation of the tundra zones in the U.S.S.R. and data about its productivity. *In* W. A. Fuller and P. G. Kevan, eds. Proc. Conf. Productivity and Conservation in Northern Circumpolar Lands. IUCN Pub. New Ser. 16: 93–114. 344 pp.

Baker, A. J., R. L. Peterson, J. L. Eger, and T. H. Manning. 1978. Statistical analysis of geographical variation in the skull of the Arctic Hare (*Lepus arcticus*). Can. J. Zool. 56(10): 2067–2082.

Bashenina, N. V. 1956. Influence of the quality of subnivean air on the distribution of winter nests of voles. Zool. Zh. 35(6): 940–942.

Bider, J. R. 1961. An ecological study of the hare *Lepus americanus*. Can. J. Zool. 39: 81–103.

Billings, W. D., and L. C. Bliss. 1959. An alpine snowbank environment and its effect on vegetation, plant development and productivity. Ecology 40(3): 388–397.

Coulianos, C. C., and A. G. Johnels. 1962. Note on the subnivean environment of small mammals. Ark. Zool. 15(24): 363–369.

Edwards, R. Y. 1956. Snow depths and ungulate abundance in the mountains of western Canada. J. Wildl. Manage. 20(2): 159–168.

———, and R. W. Ritcey. 1956. The migrations of a moose herd. J. Mammal. 37(4): 486–494.

Elsner, R. W., and W. O. Pruitt, Jr. 1959. Some structural and thermal characteristics of snow shelters. Arctic 12(1): 20–27.

Eriksson, O. 1976. Snöförhållandenas inverken på renbetningen. Meddelanden från Växtbiologiska institutionen. No. 2. Uppsala. 19 pp. + append.

Evernden, L. N., and W. A. Fuller. 1972. Light alteration caused by snow and its importance to subnivean rodents. Can. J. Zool. 50(7): 1023–1032.

Fay, F. H. 1960. Technique for trapping small tundra mammals in winter. J. Mammal. 41: 141–142.

Formozov, A. N. 1946. Snow cover as an environmental factor and its importance in the life of mammals and birds. Moscow Society of Naturalists, Materials for Fauna and Flora U.S.S.R. Zoology Section, new series 5: 1–152. [Translation 1963 by W. Prychodko and W. O. Pruitt, Jr. Occ. Pap. No. 1, Boreal Institute, Univ. Alberta.]

———. 1962. A study of mammalian ecology under the conditions of snowy and cold winters in northern Eurasia. Symp. Theriol. (Brno) 1960: 102–111.

———, and N. B. Birulya. 1937. Supplementary data on the question of interrelationships of raptorial birds and rodents. "Student Notes" Moscow State Univ. 13: 17–84.

Fuller, W. A. 1967. Ecologie hivernale des lemmings et fluctuations de leurs populations. La Terre et la Vie 1967(2): 97–115.

———. 1969. Changes in numbers of three species of small rodents near Great Slave Lake, N.W.T., Canada, 1964–1967 and their significance for general population theory. Ann. Zool. Fennici 6: 113–144.

———. 1977. Demography of a subarctic population of *Clethrionomys gapperi*: Numbers and survival. Can. J. Zool. 55(1): 42–51.

———, A. M. Martell, R. F. C. Smith, and S. W. Speller. 1975. High Arctic lemmings (*Dicrostonyx groenlandicus*). I. Natural History. Can. Field-Natur. 89: 223–233.

———, L. I. Stebbins, and G. R. Dyke. 1969. Overwintering of small mammals near Great Slave Lake, Northern Canada. Arctic 22(1): 34–55.

Geiger, R. 1961. The climate near the ground. Harvard University Press, Cambridge, Mass. xiv + 611 pp. [Translation 1965 from 4th German ed.]

Gill, D., J. Root, and L. D. Cordes. 1973. Destruction of boreal forest stands by snow loading: Its implication to plant succession and the creating of wildlife habitat. Kootenay Collection Res. Stud. Geogr., British Columbia Geogr. Ser. No. 18, Occ. Pap. in Geogr.: 55–70.

Gjaervoll, O. 1950. The snow-bed vegetation in the vicinity of Lake Torneträsk, Swedish Lapland. Svensk Bot. Tidskrift 44(2): 387–440.

———. 1956. Plant communities of the Scandinavian alpine snow-beds. Trondheim, Norske Videnskabers Selskab, Skrifter No. 1. 405 pp.

Hagmeier, E. M., and C. D. Stults, 1964. A numerical analysis of the distributional patterns of North American mammals. Syst. Zool. 13(3): 125–155.

Hall, E. R. 1951. American weasels. Univ. Kansas Pub., Mus. Natur. Hist. 4: 1–466.
Hammel, H. T. 1956. Infrared emissivity of some Arctic fauna. J. Mammal. 37 (3): 375–378.
Hansen, R. M. 1957. Remarks on the bifid claws of the varying lemming. J. Mammal. 38(1): 127–128.
Hare, F. K. 1970. The tundra climate. Trans. Royal Soc. Can. (Series 4) 8: 393–400.
———. 1971. Snow-cover problems near the Arctic tree-line of North America. Rep. Kevo Subarctic Res. Sta. 8: 31–40.
Henshaw, J. 1968. The activities of the wintering caribou in northwestern Alaska in relation to weather and snow conditions. Int. J. Biometeorol. 12(1): 21–27.
Kelsall, J. P. 1969. Structural adaptation of moose and deer for snow. J. Mammal. 50(2): 302–310.
———, and W. Prescott. 1971. Moose and deer behaviour in snow in Fundy National Park, New Brunswick. Can. Wildl. Rep. Ser. 15: 1–25.
Kil'dyushevskii, I. D. 1956. Subnivean growth of some species in the flora of Malyy Yamal. Bot. Zh. 41(11): 1641–1646.
Koskimies, J. 1958. Lumipeitteen merkityksesta elainten lamposuojana. Suomen Riista 1958: 137–140.
Křivský, L. 1958. Bestimmung der vorherrschenden windrichtung als windfahnenbaumen. Meteorol. Rundschau 11(3): 86–90.
Kuksov, V. A. 1969. Influence of some climatic factors on the numbers of mouselike rodents in Western Taimyr. Min. Agr. RSFSR, Sci. Res. Inst. of Agr. in the Far North. Trudy 17: 176–179.
MacLean, S. F., B. M. Fitzgerald, and F. A. Pitelka. 1974. Population cycles in arctic lemmings: Winter reproduction and predation by weasels. Arctic and Alpine Res. 6: 1–12.
Makridin, L. N. 1963. Distribution and migration of wild reindeer on Taimyr Peninsula. All-Union Agr. Inst., Extension Education. Trudy 15: 150–159.
McNicholl, M. 1979. Communal roosting of song sparrows under snowbank. Can. Field-Natur. 93: 325–326.
Merritt, J. F., and J. M. Merritt. 1978. Population ecology and energy relationships of *Clethrionomys gapperi* in a Colorado subalpine forest. J. Mammal. 59(3): 576–598.
Nasimovich, A. A. 1955. The role of snow cover conditions in the life of ungulates in the U.S.S.R. Inst. Geogr., Academy of Science Press, Moscow. 402 pp.
Näsmark, O. 1964. Vinteraktivitet under snön hos landlevande evertebrater. Zool. Revy 26: 5–15.
Osburn, W. S. 1963. The dynamics of fallout distribution in a Colorado alpine tundra snow accumulation ecosystem. pp. 51–71. *In* V. Schultz and A. W. Klement, eds. Radioecology. Reinhold Pub. Corp. and Amer. Inst. Biol. Sci. (Inst. Arctic and Alpine Res., Univ. Colorado, Contribution No. 8).
Pauls, R.. W. 1978. Behavioural strategies relevant to the energy economy of the red squirrel (*Tamiasciurus hudsonicus*). Can. J. Zool. 56: 1519–1525.

Payne, L. E. 1974. Comparative ecology of introduced *Clethrionomys gapperi proteus* (Bangs) and native *Microtus pennsylvanicus terranovae* (Bangs) on Camel Island, Notre Dame Bay, Newfoundland. M.Sc. thesis. Univ. Manitoba. 136 pp.

Penny, C. E. 1978. Subnivean accumulation of CO_2 and its effects on winter distribution of small mammals. M.Sc. thesis. Univ. Manitoba. 106 pp.

Pruitt, W. O., Jr. 1953. An analysis of some physical factors affecting the local distribution of the short-tail shrew (*Blarina brevicauda*) in the northern part of the lower peninsula of Michigan. Misc. Pub., Mus. Zool., Univ. Michigan. No. 79. 39 pp.

———. 1957. Observations on the bioclimate of some taiga mammals. Arctic 10(3): 130–138.

———. 1958. Qali, a taiga snow formation of ecological importance. Ecology 39(1): 169–172.

———. 1959a. Microclimates and local distribution of small mammals on the George Reserve, Michigan. Misc. Pub., Mus. Zool., Univ. Michigan. No. 109. 27 pp.

———. 1959b. A method of live-trapping small taiga mammals in winter. J. Mammal. 40(1): 139–143.

———. 1959c. Snow as a factor in the winter ecology of barren ground caribou (*Rangifer arcticus*). Arctic 12(3): 158–179.

———. 1960. Animals in the snow. Sci. Amer. 202(1): 60–68.

———. 1966. Ecology of terrestrial mammals. pp. 519–564 (chapter 20). *In* Environment of the Cape Thompson region, northwestern Alaska. U.S. Gov. Printing Office, Washington, D.C.

———. 1970. Some ecological aspects of snow. Proc. 1966 Helsinki Symp. on Ecology of the Subarctic Regions, UNESCO ser. Ecology and Conservation 1: 83–99. Paris. 364 pp.

———. 1973. Techniques in boreal ecology. Part A—Environmental analysis. Part B—Animal populations and activity. 16 mm colour teaching films produced by Instructional Media Centre, Univ. Manitoba.

———. 1978. Boreal ecology. Studies in biology No. 91. Edward Arnold (Pub.) Ltd., London. iv + 73 pp.

———. 1979. A numerical "snow index" for reindeer (*Rangifer tarandus*) winter ecology. (Mammalia: Cervidae.) Ann. Zool. Fennici 16(4): 271–280.

———. 1981. Application of the Värriö snow index to overwintering North American barren-ground caribou (*Rangifer tarandus arcticus*). Can. Field-Natur. 95(3): 363–365.

———, and C. V. Lucier. 1958. Winter activity of red squirrels in interior Alaska. J. Mammal. 39(3): 443–444.

Pulliainen, E. 1965. Studies on the wolf (*Canis lupus*) in Finland. Ann. Zool. Fennici 2(4): 214–259.

———. 1973. Winter ecology of the red squirrel (*Sciurus vulgaris* L.) in northeastern Lapland. Ann. Zool. Fennici 10: 487–494.

Richardson, S. G., and F. B. Salisbury. 1977. Plant responses to the light penetrating snow. Ecology 58: 1152–1158.

Shvarts, S. S. 1963. Adaptations of terrestrial vertebrates to the environmental conditions of the Subarctic. Academy of Science, U.S.S.R. Ural Affiliate, Inst. Biol. Trudy 33. 19 pp.
Smirin, Y. M. 1970. On the biology of small forest rodents in the winter period. Fauna and Ecol. of Rodents 9: 134–150.
Stardom, R. R. P. 1975. Woodland caribou and snow conditions in southeast Manitoba. Proc. 1st Int. Reindeer-Caribou Symp.: 324–334.
Steen, J. 1958. Climatic adaptations in some small northern birds. Ecology 39(4): 625–629.
Stinson, N. S. 1977. Species diversity, resource partitioning and demography of small mammals in a subalpine deciduous forest. Ph.D. thesis. Univ. Colorado. 238 pp.
Sulkava, S. 1964. On the living conditions of the partridge (*Perdix perdix* L.), and the brown hare (*Lepus europaeus* L.) in Ostrobothnia. Aquilo, Ser. Zool. 2: 17–24.
Sutton, G. M., and W. J. Hamilton, Jr. 1932. The mammals of Southampton Island. Memoirs Carnegie Mus. XII, Part II, Sect. 1: 1–109.
Vaughan, T. A. 1969. Reproduction and population densities of a montane small mammal fauna. pp. 51–74. *In* J. K. Jones, ed. Contributions in mammalogy. Univ. Kansas, Mus. Nat. Hist., Misc. Pub. No. 51. 428 pp.
Viereck, L. A. 1965. Relationship of white spruce to lenses of perennially frozen ground, Mount McKinley National Park, Alaska. Arctic 18: 262–267.
Yashina, A. V. 1961. Subnivean growth of plants. pp. 137–165. *In* M. I. Iveronova, ed. Role of snow cover in natural processes. Academy of Science Press, Moscow. 272 pp.
Yoshino, M. M. 1966. Wind-shaped trees as indicators of micro and local climatic wind situation. Proc. 3rd Int. Biometeorol. Congress (Sept. 1963): 997–1005. Pergamon Press, Oxford.

Animal Communities

Population Dynamics of the Pine Marten (*Martes americana*) in the Yukon Territory

W. R. Archibald
R. H. Jessup

Abstract

Population dynamics of pine marten (*Martes americana*) were studied on a 14 km² study area in south-central Yukon from 1978 to 1981. Fall and spring densities for resident marten were .6 and .4/km², respectively. Male home ranges averaged 6.2 km²; female home ranges averaged 4.7 km². Home ranges were exclusive within sexes. There appeared to be two periods of dispersal, one for the young-of-the-year marten and one for overwintering marten. The conception rate for yearling marten (3.3) was significantly different from that of older marten (3.8). Population growth rates in a harvested population appear to be a function of trapping frequency and trapping intensity. Management implications of the results are discussed.

Introduction

The pine marten (*Martes americana*) is a species associated with climax coniferous forests (Koehler et al. 1975, Lensink 1953, Soutiere 1979). The present distribution of marten in North America is considerably smaller than its known historical distribution (Hagmeier 1956, Koehler et al. 1975, Mech and Rogers 1977). The decline in marten distribution has been attributed to the destruction of mature coniferous forests by fire and logging (Bergerud 1969, Soutiere 1979) and to trapping (de Vos 1952, Dodds and Martell 1971).

Marten have always figured significantly in the Yukon fur harvests. During the period 1920–1950, when trapping was a significant contributor to the economy of the Yukon (McCandless 1977), marten harvests ranged from 4,300 to less than 100 animals (Klassen 1975). In recent years a revitalization of trapping in the Yukon has resulted in increasingly higher marten harvests, with each year's harvest representing a new record high. The ease with which this animal is trapped, coupled with its recent high market value, has meant that it has provided one of the highest financial returns per unit effort of any Yukon furbearer.

The upward trend in marten harvests and the growing potential for habitat destruction through logging, road building, and mining and hydro development clearly demonstrated the need for a population and habitat management strategy for this species. The purpose of this paper is to report information on the popultion dynamics of marten in the Yukon Territory, based on an analysis of data obtained between November 1977 and November 1981.

Study Area

The 14-km^2 intensive study area is located in the south-central Yukon (approximately 64°97'N, 133°15'W), in the Evelyn Creek area of the Nisutlin River drainage (fig. 1). The area has a northerly aspect and varies in altitude from 1,220 m in the southwest corner to 875 m in the northeast corner. The maximum mean temperature from May through October is 15 C; from November through April it is – 6 C. The mean annual precipitation is 364 mm, with maximum snow accumulations averaging 156 cm during the 3 years of investigation. Fires burned portions of the study area in 1810, 1870, and 1920. The fires and the rugged topography have resulted in considerable vegetative diversity. Lodgepole pine (*Pinus contorta*) is the dominant early successional species. White spruce (*Picea glauca*) and subalpine fir (*Abies lasiocarpa*) dominate the mid and late successional stands. Climax stands are predominantly subalpine fir. Prevalent understory species include *Empetrum nigrum, Vaccinium vitis-idaea, Rosa acicularis,* and *Potentilla fruticosa*. Common in the herb layer are *Cornus canadensis, Lupinus arcticus, Mertensia paniculata,* and *Carex* spp.

The study area is within the boundary of a registered trapline. Commercial trapping in this area ceased after the winter of 1977–78, although reg-

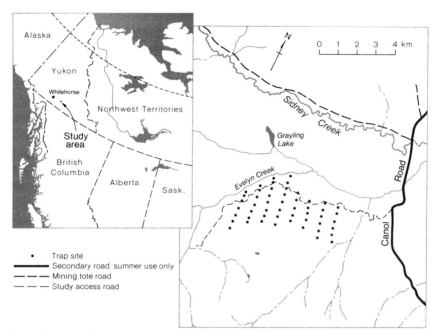

Fig. 1. Study area location and permanent trap sites used in this study.

istered trappers continued to trap around the periphery of the intensive area to the north and east.

In addition to carrying out intensive field work, the researchers collected marten carcasses from four registered traplines scattered throughout the Yukon. General trapline locations were the southeast Yukon, south-central Yukon (the trapline containing the study area), and two in the west mid-Yukon. Trapline selection was based primarily on historical marten harvest records and trapper co-operation.

Methods and Materials

Study Area Design, Capturing, and Marking

Eight parallel transects were cut across the study area. Transects were positioned 0.8 km apart using aerial photographs, and averaged 2.0 km in length (range 1.5 to 2.7). Transects were run with a compass and brushed to an average width of 2 m.

Permanent trap sites were established at 0.4-km intervals along each transect, for a grid of 52 traps (fig. 1). Traps (Model 205, Tomahawk Live

Trap Co.) were set on the ground and covered with boughs to protect captured marten from inclement weather. They were prebaited with a combination of strawberry jam and oatmeal 24 hours before the commencement of a trapping session. Trapping sessions were 3 days in duration. It was assumed that all marten in the study area during the 3-day session would be captured. This assumption was tested with radio-collared marten. Only once in 3 years was a collared marten known to be within the trapping grid and not captured during the 3-day session.

Live-trapping commenced in October 1978. Trapping recommenced in June 1979 and was conducted at approximately 3-week intervals until November 1979. Trapping sessions for 1980 and 1981 roughly paralleled those in 1979.

Traps were checked daily during each session. Captured marten were run into a holding cone (Hawley unpublished) and ear-tagged with numbered metal tags (Style 4-1005, Size 1, National Band and Tag Co.). Sex was determined by palpation for the baculum.

Captured marten were categorized as transients (captured only once), temporary residents (captured over a time span of less than 3 months), or residents (captured over a time span exceeding 3 months) (Weckwerth and Hawley 1962). Resident marten were radio-collared (AVM Instrument Co.). Martens to be fitted with radio collars were immobilized with a mixture of 20 mg of Ketamine hydrochloride and 4 mg of Rompum. Induction time was rapid, averaging 1 minute, 53 seconds ($n = 40$), and full recovery was generally accomplished within 90 minutes.

A premolar tooth was extracted from each immobilized marten for age determination.

Relocations

Radio-collared marten were relocated by triangulation of two or more fixes of the sharpest signal readings from hand-held antennas.

Relocations were attempted daily during the period of intensive investigation (June to September in both 1980 and 1981). Unreliable telemetry equipment and difficult travel conditions resulted in relatively few relocations during the remainder of the study.

All locations were recorded on a 1:25,000 scale map of the study area. Home range sizes were estimated by using Mohr's (1947) minimum-perimeter-polygon method.

Carcass Analysis

Based on preliminary carcass collections in the 1977–78 trapping season, four heavily trapped traplines were selected for further intensive investigation. Complete carcass collections were gathered from these traplines in 1978–79, 1979–80, and 1980–81. Trappers tagged each marten, indicating capture location and date.

Female reproductive tracts were examined fresh. Ovaries were hand-sectioned with a razor blade, and the presence of corpora lutea and follicular activity recorded. Uterine horns were examined for implantation sites, and the condition of the uterus recorded (parous versus non-parous). The tracts were then fixed and preserved in A.F.A.

Premolars were decalcified in a buffered 25% formic acid solution. Decalcified teeth were washed for 12 to 15 hours in running tap water. Teeth were imbedded in Lab-tech compound, then cut on a cryostat at −15 C into longitudinal sections 10 μ thick. Cut sections were affixed to a

Table 1 Marten live-trapping success during intensive sessions in 1979, 1980, and 1981

Session no.	Dates	No. trapnights	No. captures	Success*
		1979		
1	Jun 6-8	156	4	39.0
2	Jun 26-28	156	6	26.0
3	Jul 16-18	156	16	9.8
4	Aug 10-12	156	18	8.7
5	Aug 29-31	156	15	10.4
6	Sep 19-21	156	12	13.0
7	Nov 1-3	156	17	9.2
		1980		
1	Jun 4-6	156	11	14.2
2	Jun 24-26	156	16	9.8
3	Jul 15-16	104	9	11.6
4	Aug 5-7	156	10	15.6
5	Aug 26-28	156	14	11.1
6	Sep 24-26	156	10	15.6
		1981		
1	Jun 3-5	156	13	12.0
2	Jun 23-25	156	14	11.1
3	Jul 15-16	104	11	9.5
4	Aug 3-5	156	15	10.4
5	Aug 24-25	104	7	14.9
6	Oct 28-30	139	6	23.2

* No. trapnights/no. captures

Table 2 Occupancy status of martens live-trapped during intensive sessions in 1979, 1980, and 1981

Year	Status*	Trapping session						
		1	2	3	4	5	6	7
1979								
	Transient	0	1	3	3	3	0	1
	Temporary resident	0	0	1	2	4	3	0
	Resident	3	3	4	4	4	5	8
1980								
	Transient	2	0	1	0	3	0	
	Temporary resident	0	2	2	2	2	0	
	Resident	6	6	6	6	6	8	
1981								
	Transient	1	1	2	3	1	1	
	Temporary resident	0	1	2	3	3	0	
	Resident	6	5	5	5	5	8	

* After Weckwerth and Hawley (1962).

glass slide with egg albumen. Sections were stained in a filtered 0.032% aqueous solution of Toluylene Blue.

Ages for marten were determined by counting annuli in the cementum.

Results and Discussion

Density

Live-trapping success (number of trapnights per capture) on the study area is presented in table 1. Trapping success was significantly different in the 3 years of investigation ($X^2 = 21.75$, $df = 8$). A similar test in 1980–81 revealed no significant difference ($X^2 = 1.98$, $df = 4$). Close inspection of table 1 suggests that in 1979 the significant difference was due to the low capture success in the first two sessions. This likely resulted from the presence of only three resident marten on the study area during the first two sessions in 1979 (table 2) (five marten actually overwintered, but ♂ 12–13 was a trap fatality and ♀ 22–23 was not recaptured until session 3). By comparison, six resident marten were on the study area during the same sessions in 1980 and 1981 (table 2).

The fall resident marten populations of eight in 1979, 1980, and 1981 (table 2) was believed to represent saturation density for the study area. Inspection of table 2 reveals that once the number of residents plus the

Table 3 Sex ratio and age structure of martens resident in the intensive study area during spring and fall seasons for 1979, 1980, and 1981

Tag no.	Sex	1979 Spring	1979 Fall	1980 Spring	1980 Fall	1981 Spring	1981 Fall
12–13	♂	2*					
10–11	♂	1	1⁺	2	2⁺		
22–23	♀	2	2⁺	3	3⁺		
16–17	♀	1	1⁺	2	2⁺	3	3⁺
78–79	♂	2	2⁺	3	3⁺	4	4⁺
103–105	♀		1⁺				
108–109	♂		0⁺	1	1⁺	2	
116–117	♂		0⁺				
113–114	♀		0⁺	1	1⁺	2	2⁺
176–177	♂				1⁺	2	2⁺
128–129	♀				1⁺	2	2⁺
374–375	♂						1⁺
376–377	♂						2⁺
326–327	♂						3⁺
Sex ratio		3♂:2♀	4♂:4♀	3♂:3♀	4♂:4♀	3♂:3♀	5♂:3♀

* Age based on cementum annuli.

number of temporary residents equalled eight, no additional transient martens were able to attain even temporary resident status. The resultant fall density of .6 resident martens per km² is identical to that reported by Francis and Stephenson (1972) in south-central Ontario, and less than 1.2 adult residents per km² reported by Soutiere (1979) in Maine.

Carrying capacity was defined as the number of resident marten that successfully overwintered on the study area. Five marten from an unknown fall 1978 resident population overwintered on the study area during the winter of 1978–79 (table 3). Six of eight resident marten successfully overwintered during the winters of 1979–80 and 1980–81 (table 3). The remaining marten moved off the study area. Male 116–117 and female 103–105 were killed 8.4 and 10 km respectively from their known home ranges. Male 10–11 and female 22–23 established new home ranges beyond the area of intensive study. It appears, then, that the carrying capacity of the study area was six marten, or .4 marten per km².

The sex ratio of the resident population was usually equal. Exceptions occurred in the spring of 1979 (3♂:2♀) and fall of 1981 (5♂:3♀). Equal sex ratios for resident populations have also been reported by Newby and Hawley (1954) and Francis and Stephenson (1972). Weckwerth and Hawley (1962) reported equal numbers of resident males and females except during the initial phase of a population decline, when the ratio favoured males. Population fluctuations similar to those reported by Weckwerth and Hawley (1962) did not occur on our study area.

Home Range

Home range was defined as the total area through which a marten moved. The average home range size for resident male and female marten as determined by radio telemetry and trapping locations was 6.2 km² and 4.7 km² respectively (table 4). These home ranges were greater than those reported by Hawley and Newby (1957), 2.4 km² for males and 0.7 km² for females; Francis and Stephenson (1972), 3.6 km² for males and 1.1 km² for females; and by Soutiere (1979), 0.1 to 7.6 km² for males and 0.1 to 2.3 km² for females. The differences are likely attributable to the fact that home range determinations for these earlier studies were based solely on

Table 4 Home range sizes of resident marten as determined by telemetry and live-capture locations from June through November, 1980 and 1981

Tag no.	Sex	1980			1981		
		Age (yrs)	No. of Locations	Home Range Size (km²)	Age (yrs)	No. of Locations	Home Range Size (km²)
16–17	♀	2⁺	36	3.7	3⁺	49	5.2
113–114	♀	1⁺	42	7.7	2⁺	14	2.0
78–79	♂	3⁺	40	8.7	4⁺	21	2.8
176–177	♂	1⁺	42	6.5	2⁺	34	7.2
22–23	♀	3⁺	30	8.0			
108–109	♂	1⁺	58	5.7			
128–129	♀				1⁺	31	2.1
374–375	♂				1⁺	24	5.9

trapping locations. Major (1979) and Steventon (1979) employed field procedures similar to those used in the present study and reported home ranges of 7.6 km² and 9.2 km² for males and 1.0 km² and 2.1 km² for females respectively. The home range sizes for males in these two studies are comparable to those in the Yukon.

The ratio of male home range size to female home range size was 1.3:1. Other investigators have reported this ratio to lie between 3:1 and 7:1 (Hawley and Newby 1957, Francis and Stephenson 1972, Mech and Rogers 1977, Major 1979, Soutiere 1979, Steventon 1979).

At first examination these results appear to conflict with the results of the present study. The period of intensive telemetry investigation during the present study bracketed estrus (June to September in both 1980 and 1981). All of the resident marten were radio-collared and tracked during this period. In these respects this study was unique. For example, Mech and Rogers (1977) and Steventon (1979) conducted their research during winter months. Major (1979) conducted summer telemetry research on four resident marten (three males, one female), but limited his home range determinations to areas of intensive use only.

We suggest that adult female marten expand their home ranges during estrus to the point where they approach the size of male home ranges. We observed male and female marten travelling together during estrus, as determined by the condition of the vulva (Enders and Leekley 1941). For example, female 113–114 travelled with male 78–79 on the morning of 18 July 1980. These marten were separated by evening, but were recorded travelling together again on 22 July 1980. No evidence was found of a female travelling with more than one male. It appears, then, that the results of the present study may not conflict with previous studies.

In the section dealing with density it was reported that the sex ratio of the resident marten population was usually equal. Intuitive interpretation of the published ratio of male home range size to female home range size (mentioned earlier in this section) would be that a wild marten population would have a preponderance of females. We suggest that the mechanism for maintaining an even sex ratio is the expanded female home range during estrus.

Within each sex marten appear to have exclusive home ranges. The three resident males occupying the study area concurrently in 1980 had little (<3%) or no overlap of home ranges (fig. 2). In 1981 there was a small region of range overlap between males 176–177 and 374–375 (fig. 3). The home range of male 78–79 overlapped by 25% the range of male 374–375 (fig. 3).

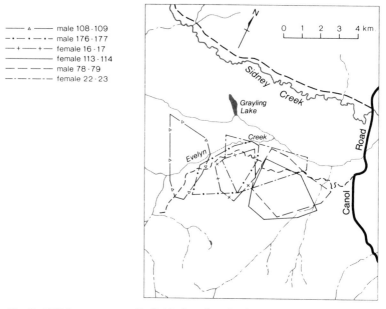

Fig. 2. 1980 home ranges of individual study animals.

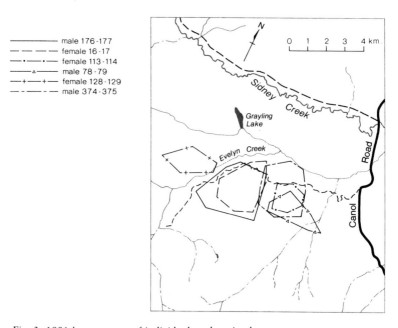

Fig. 3. 1981 home ranges of individual study animals.

Population Dynamics of the Pine Marten 91

There was little overlap of female home ranges in 1980 (fig. 2). There was no overlap of ranges between neighbouring female marten 16–17 and 113–114. Female 22–23 overlapped both of the above ranges during part of the summer, but eventually abandoned this range and established a new home range beyond the boundaries of the ranges of females 16–17 and 113–114. There was no overlap of female home ranges in 1981 (fig. 2).

Male and female home ranges demonstrated considerable overlap (figs. 2 and 3). These results concur with those of Hawley and Newby (1957) and Francis and Stephenson (1972).

The mechanism for establishing and maintaining home ranges is unclear at present. We handled 44 different marten a total of 263 times and found no overt evidence of intraspecific antagonistic behaviour. We also observed temporal separation within sexes along common boundaries. This suggests that some mechanism other than active border defence exists for the development and maintenance of exclusive home ranges. Pulliainen (1980) believes that scent marking plays a vital role in the maintenance of home range boundaries for European pine marten (*Martes martes*). Lockie (1966) contends that weasels (*Mustela nivalis*) use droppings to mark the boundaries of their home ranges. We suggest that a similar mechanism exists for pine marten. High numbers of marten scats were found in certain areas, such as on promontories. These areas are similar to the traditional bathrooms reported by Lockie (1966) for weasels. Also, many scats were found along the transects. Inspection of figs. 1 and 2 reveals that transect lines occasionally constituted home range boundaries. For example, in 1981 one side of the home range polygon for martens 176–177, 374–375, and 16–17 lay roughly along one of the mid-transects (figs. 1 and 3).

Dispersal

There appeared to be two periods of dispersal. The first period occurred mostly between mid-July and mid-September (trapping sessions 3 to 5) (table 2). Nineteen of 26 transients were captured during this period (table 2). Four of these transients attained residency status. Three of these four were young-of-the-year marten. The onset of dispersal of young-of-the-year marten coincided with the onset of estrus.

Dispersal of overwintering marten also occurs. Five of 26 transients were captured in the first two sessions (table 2). All of these marten attained residency status. In two instances, resident marten were apparently displaced from their known home range by dispersing overwintering marten. Resident male 10–11, a 2-year-old marten, moved off its home range in the winter of 1979–80 and established a new home range on the northern periphery of the study area. His former range was occupied sometime prior to the first trapping season in early June 1980 by male 176–177. Resident male 78–79, a 3-year-old marten, had a much-reduced home range in 1981 as compared with 1980 (table 4). A large portion of this marten's former range was occuped by male 374–375 (figs. 2 and 3).

The timing of the dispersal of overwintered marten is not clear. Telemetry equipment failure precluded the documentation of the initiation of this movement. However, this movement likely occurs early in the spring, since familiarity with home range during the rigours of winter is probably a prerequisite to overwintering success. Table 2 suggests that this movement may be completed prior to the onset of estrus.

Age and Sex Structure of a Harvested Sample

A total of 839 carcasses were collected from four registered traplines in the Yukon for 1978–79, 1979–80, and 1980–81 (table 5). The proportion of each age class in the harvest did not differ between years ($X^2 = .0617$, $df = 4$). The resultant average proportions of .67 for young-of-the-year, .23 for yearlings, and .10 for older martens appear to be representative of heavily harvested traplines.

The age structure of a harvested marten sample will be a function of trapping frequency. When resident marten are removed from an area, the opportunity exists for dispersing marten to establish themselves in the vacant areas. Since the majority of dispersing marten are young-of-the year, the resultant population will have a reduced mean age. Alternatively, if a population is untrapped for a period of time, the mean age will in-

Table 5 Age and sex structure of marten harvested from four Yukon traplines for 1978–79, 1979–80, and 1980–81

Age (yrs)	1978–79 ♂	♀	N	Proportion	1979–80 ♂	♀	N	Proportion	1980–81 ♂	♀	N	Proportion	Average proportion
0⁺	137	88	225	.64	103	78	181	.64	103	55	158	.76	.67
1⁺	49	44	93	.27	36	37	73	.26	14	16	30	.14	.23
≥ 2⁺	22	10	32	.09	17	10	27	.10	11	9	20	.10	.10
Total	208	142	350		156	125	281		128	80	208		

crease. For example, the age structure of the fall 1981 resident population in the study area after 3 years of no trapping was .13 yearlings and .87 older (table 3).

A higher percentage of male marten was harvested in each of the 3 years of carcass submissions (table 5). The ratio of males to females was 59:41 for 1978–79, 59:41 for 1979–80, and 61:39 for 1980–81. These results are comparable to those reported by Yeager (1950) and de Vos (1952).

Yeager (1950) collected data from 22 marten ranches and reported that the marten sex ratio at birth is even. The present study and the work of Francis and Stephenson (1972) suggest that the sex ratio of a resident marten population is also even. Weckwerth and Hawley (1962) reported that the adult resident population had an even sex ratio, except during the initial phase of a population decline, when there was a disproportionately greater number of males. In the absence of a population decline, one may therefore assume that the sex ratio of a wild population is even.

The unbalanced sex ratio of a harvested sample is probably not representative of the population. Other studies have reported males having much larger home ranges than females during winter (Mech and Rogers 1977, Steventon 1979). If a trapper enters a marten area and randomly sets one trap, the probability of his catching one sex or the other will be proportional to home range size of male versus female marten. Therefore it appears that the sex ratio of the harvested sample is a function of trapping intensity. An exhaustive trapping effort should produce a harvest with an even sex ratio. Quick (1956) reported an even sex ratio from a trapline that was intensively trapped. A late-winter reconnaisance along this trapline revealed a complete absence of marten tracks, although marten tracks were observed on excursions about 1.6 km away from the line.

Reproduction

Table 6 presents data on the age-specific pregnancy rates and conception rates for harvested marten. Chi-square analysis revealed that there was no difference in the pregnancy rate between yearling and older marten ($X^2 = 7.5068$, $df = 6$). Significant results may have been obtained with a larger sample of older animals. The yearling conception rate of 3.3 was significantly different from the 3.8 reported for older marten ($X^2 = 2.12$, $df = 87$).

Lensink (1953) reported that female marten conceived at 27 months. Quick (1956) made the same observation, although he did report that there was some evidence of breeding at 15 months. The results of the pres-

Table 6   Age-specific pregnancy rates and average conception rates for pregnant female marten harvested from four Yukon traplines during 1978–79, 1979–80, and 1980–81

	1978–79			1979–80			1980–81				
Age	No.	Preg-nancy rate	Con-ception rate	No.	Preg-nancy rate	Con-ception rate	No.	Preg-nancy rate	Con-ception rate		Average pregnancy rate	Average conception rate
0^+	88	.01	3.0	78	0.0	0.0	55	0.0	0.0			
1^+	44	.66	3.2	37	.73	3.5	16	.69	3.3			3.3
											.74	
$\geq 2^+$	10	.80	3.9	9	.67	3.3	9	.89	4.1			3.8

ent study support Quick's (1956) statement. Lensink (1953) reported an average conception rate of 2.8 over all age classes. This is considerably lower than the results of the present study.

Normally, harvest data can be used to structure a theoretical wild population and determine population growth rates. However, since the sex and age structure of a wild marten population is largely a function of trapping intensity and frequency, marten population growth rates are highly variable. The following example illustrates the point.

A large block of marten habitat has a resident population of 100 marten prior to trapping season. The harvest during the winter is exhaustive. Prior to estrus some overwintering marten move into the void created by the heavy trapping pressure. Some of these animals breed. There is no production in this area in summer I. Dispersing young-of-the-year marten raise the resident population to 100 marten by late fall. In summer II the total production of the area is from the overwintered marten that emigrated prior to summer I. As the proportion of older animals in the resident population increases, so will the population growth rate.

Imagine the same area with no trapping for several years. A population can be structured using the age structure of our study population of fall 1981 (table 3). The resultant population would consist of 13 yearlings and 87 older marten. Applying the age-specific pregnancy rates and conception rates to this population and assuming an equal sex ratio would result in a theoretical population increase of 138 martens in summer I.

The above example presented the two extremes: no production and maximum production. This example is valuable in that it reveals rather dramatically that population production will vary by several degrees of magnitude, depending on the sex and age structure of the resident population.

Management Application

The results of the present study reveal that population growth rate is highest among unharvested populations. Young-of-the-year marten will disperse in search of suitable unoccupied habitat. These results suggest a management strategy of setting aside blocks of land as reservoirs from which dispersing marten would be harvested. Reservoir size and configuration will likely be dictated by local physiographic features. Trapping trails through or around the perimeter of a reservoir must be no less than three marten home ranges apart to preserve a resident population. The results of the present study suggest that in the Yukon, neighbouring traplines need to be 10 km apart.

The clear advantage of this type of management strategy is that there is no need for quotas. The size of the harvest will be regulated by the productivity of the reservoirs. Additional points relating to this strategy include:

1. The harvest should be exhaustive along trapping trails, as it is highly unlikely that marten will move into unoccupied habitat during the winter. Also, by creating an area totally devoid of marten, habitat is being provided for dispersing marten the following summer.
2. The integrity of the reservoir must be maintained. Population declines similar to those reported by Weckwerth and Hawley (1962) may occur in the Yukon, independent of trapping pressure. During such declines, the number of marten dispersing from the reservoirs will be reduced. In these situations the trapping of reservoirs to compensate for low harvests may totally eliminate marten from the area. Since dispersal distances are probably not great and the growth rate of a newly established population is very low, it would probably take many years to re-establish a productive resident population in reservoirs.

In the strategy outlined above, the trapper has the ultimate responsibility for managing the resource. This type of system is best suited to areas where registered traplines are in place. In areas where a number of trappers are competing for a common resource, a different strategy will likely have to be employed. The amount of marten habitat for each area could be determined for aerial photographs or satellite imagery. A density estimate would be applied and a harvest rate developed. For example, if 100 km^2 of marten habitat were to be communally trapped in the Yukon, the estimated resident population would be 40 marten (.4 martens/km^2 × 100 km^2). The age structure of the resident population would be taken as that representative of a heavily trapped population (.67, .23, .10, table 5). The theoretical production from this trapline would be 34 marten. A conservative harvest rate of 50% should be set, since the theoretical maximum production figure is probably never attained. Therefore, the quota for this area should be 17 marten. In areas where this strategy is to be employed the sex ratio of the harvest may serve as an index of population status. When the ratio is skewed in favour of males, the harvest is probably proceeding in a biologically sound fashion. However, if the sex ratio approaches even, the trapping effort is probably locally exhaustive and the harvest rate should be adjusted downwards.

Acknowledgments

We are grateful to Gordon Hartman, former Director of the Yukon Wildlife Branch, for his support and encouragement throughout this study. Many individuals provided invaluable assistance in the field. We are grateful to Steve Beare, Lisa Hartman, Gavin Johnson, Rhonda Markel, Kathy McKeweon, Wes Olson, and Brian Slough. We would also like to thank Don Russell and Christine Boyd, for assistance with the habitat classification; Grant Lortie and Phil Merchant, for their assistance in the laboratory; and Anne Jessup, for preparing the figures. The Davignon family at Johnson's Crossing gave us a great deal of assistance and hospitality during our many trips to the study area. We are particularly grateful for the advice, assistance, and co-operation of the following Yukon trappers: Pius Bahm, Jack Fraser, Leo Heisz, Steve Kormendy, Mable Robson, and Jim Rose. Don Eastman and Rick Ellis offered many helpful editorial comments. Finally, this manuscript was very capably typed by Joyce Kirk and Louise Reid.

References

Bergerud, A. T. 1969. The status of the pine marten in Newfoundland. Can. Field-Natur. 83(2): 128-131.
de Vos, A. 1952. The ecology and management of fisher and marten in Ontario. Ont. Dep. Lands and Forests Tech. Bull. 90 pp.
Dodds, D. G., and A. M. Martell. 1971. The recent status of the marten, *Martes americana americana* (Turton), in Nova Scotia. Can. Field-Natur. 85(1): 61-62.

Enders, R. K., and J. R. Leekley. 1941. Cyclic changes in the vulva of the marten (*Martes americana*). Anat. Rec. 79(1): 1–5.
Francis, G. R., and A. B. Stephenson. 1972. Marten ranges and food habits in Algonquin Provincial Park, Ontario. Min. Nat. Resour. Ont. Res. Rep. 91. 53 pp.
Hagmeier, E. M. 1956. Distribution of marten and fisher in North America. Can. Field-Natur. 70(4): 149–168.
Hawley, U. D. Unpublished. A holding cone for marten, mink and other mammals. 5 pp.
_____, and F. E. Newby. 1957. Marten home ranges and population fluctuations. J. Mammal. 38(2):174–184.
Klassen, W. J. 1975. Fur-bearers of the Yukon Territory and some socio-economic effects of trapping. Yukon Game Branch Rep. 23 pp.
Koehler, G. M., W. R. Moore, and A. R. Taylor. 1975. Preserving the pine marten: Management guidelines for western forests. *In* Western wildlands summer 1975. Univ. Montana.
Lensink, C. J. 1953. An investigation of the marten in interior Alaska. M.Sc. thesis. Univ. Alaska, Fairbanks. 89 pp.
Lockie, J. D. 1966. Territory in small carnivores. Symp. Zool. Soc. London 18: 143–165.
Major, J. T. 1979. Marten use of habitat in a commercially clear-cut forest during summer. M.Sc. thesis. Univ. Maine, Orono. 32 pp.
McCandless, R. G. 1977. Trophies or meat: Yukon game management 1896 to 1976. Yukon Game Branch Rep. 122 pp.
Mech, L. D., and L. L. Rogers. 1977. Status, distribution and movements of martens in northeastern Minnesota. USDA Forest Serv. Res. Pap. NC-14-3. 7 pp.
Mohr, C. O. 1947. Tables of equivalent populations of an endangered species. Amer. Midland Natur. 37: 223–249.
Newby, F. E., and U. D. Hawley. 1954. Progress on a marten live-trapping study. Trans. North Amer. Wild. Conf. 19: 452–460.
Pulliainen, E. 1980. Winter habitat selection, home range, and movements of the pine marten in Finnish forest Lapland. Varrio Subarctic Res. Sta., Univ. Helsinki Rep. No. 110. 22 pp.
Quick, H. F. 1956. Effects of exploitation on a marten population. J. Wildl. Manage. 20(3): 267–274.
Soutiere, E. C. 1979. Effects of timber harvesting on marten in Maine. J. Wildl. Manage. 43(4): 850–860.
Steventon, J. D. 1979. Influence of timber harvesting upon winter habitat use by marten. M.Sc. thesis. Univ. Maine, Orono. 24 pp.
Weckwerth, R. P., and U. D. Hawley. 1962. Marten food habits and population fluctuations in Montana. J. Wildl. Manage. 26(1): 55–74.
Yeager, L. E. 1950. Implications of some harvest and habitat factors on pine marten management. Trans. North Amer. Wildl. Conf. 15: 319–334.

Population Dynamics and Horn Growth Characteristics of Dall Sheep (*Ovis dalli*) and Their Relevance to Management

Manfred Hoefs

Abstract

A brief review of the population dynamics of Dall sheep is presented, with particular reference to 12 years of observation of the Sheep Mountain herd in Kluane National Park. A graphic model of the structure of this population is developed. Horn growth characteristics for Dall rams are given in terms of annual growth increments in length and circumference, as well as in angular growth, expressed as "degree curl," since it is this latter criterion that is most commonly used in sheep management regulations. Demographic data and horn growth parameters are combined to work out sustainable yield figures under various trophy management goals. The opposing trends of declining abundance with age but increasing trophy quality are elaborated on, with suggestions for compromises under differing harvest regulations. The additive or compensatory nature of hunting is also discussed. Finally, conservation concerns are expressed in relation to factors other than hunting.

Introduction

The Dall sheep (*Ovis dalli dalli*) is a truly northern animal. Its range extends latitudinally from 59° 30' to 69° 01'N, covering suitable mountain ranges in Alaska, the Yukon, the Northwest Territories (Mackenzie

Fig. 1. Mature Dall rams in late winter. Note the light colour of the horns and the distinct annual rings. Age determination, an important aspect of the study of population dynamics, is relatively easy in the males of this species.

Range), and adjacent areas of British Columbia. It is the most abundant North American sheep at present; estimates range from 51,000 to 76,000 (Nichols 1978b). However, it is assumed that the bighorn sheep (*Ovis canadensis*) was more numerous when the white man first colonized the West (Buechner 1960). With few exceptions the Dall sheep still occupy most of their historic range, and the population size is near historic levels. In spite of its abundance and importance, only six long-term, detailed population studies have been carried out on this species. Heimer (1979, 1980, 1981) and Nichols (1978a, 1978b) investigated Dall sheep in the Dry Creek area and the Kenai Range of Alaska; Simmons et al. (1984) carried out a 7-year project on a population in the Mackenzie Mountains. Most is known about the Sheep Mountain population in Kluane National Park, which the writer has had under observation for the past 12 years. Most other projects have been short-term in nature, being mostly thesis investigations or biological impact assessments of proposed industrial developments.

In the following brief summary of population dynamics parameters, I will make reference primarily to the Sheep Mountain investigation, since this study covered a longer time period than any other (Hoefs and Bayer 1983).

Fig. 2. A band of ewes and lambs in early September. At this time the lambs have reached 60% of the weight of their dams, and their sex can be determined by horn growth.

Population Dynamics

Lambs are born throughout May and June, after a gestation period of about 171 days. They weigh about 3.2 to 4.1 kg at birth and grow rapidly during their first summer, reaching an average weight of 30.4 kg by fall (Nichols 1978b). On Sheep Mountain 70 to 80% of the lambs are born during the second half of May (Hoefs 1975).

In most populations females have their first lamb when 3 years old, but in some Alaskan ranges this occurs around their second birthday. Single births are the rule; twinning is rare. Females are able to give birth as long as they live. Rams may reach sexual maturity at 1.5 years of age, but in wild populations with a natural age structure, they hardly have the opportunity to breed before the age of 5 to 6 years. Great variations have been observed in the productivity of populations and between different years in the same population, as well as in the survival rates of lambs to yearling age. In the 12-year investigation of the Sheep Mountain herd, the fecundity rate, expressed as lambs per 100 ewes in reproductive age, has averaged 41:100, the range being 64:100 (1971) to 10:100 (1975). Heimer (1981) reported for the Dry Creek area in Alaska lamb-to-ewe ratios of 67:100 (1980) to 15:100 (1972), and Simmons et al. (1984) observed for the Mackenzie Mountain population a ratio of 62:100.

In spite of great variation in the number of lambs observed, the initial pregnancy rate appears to be generally high. Simmons et al. (1984) collected 107 ♀ in March and found a 75% pregnancy rate. Nichols (1978a) collected 25 ♀, with a 100% pregnancy rate for those 2 years or older, and a 75% pregnancy rate for yearling ewes.

The sex ratio at birth averages 100 ♂ :100 ♀. In Alaska the lambing success could be correlated with winter weather conditions (Murphy 1974, Nichols 1976). For Sheep Mountain, a correlation to forage production on the winter range is indicated (Hoefs 1982b). Mortality during the first year of life averages around 50% (r = 65%–12%). During the second year it is low, most studies assuming a rate of 5 to 10%. Mortality in the middle-age classes is low, with rates of 3 to 6% for the Sheep Mountain population. After the seventh year of life it begins to rise, and reaches over 50% per annum in animals over 10 years old. Sheep may reach a maximum life expectancy of 14 to 16 years. For the Sheep Mountain population the 12 to 13-year class was the maximum life expectancy. Growth is largely complete after 6 years, when rams may reach weights of 100 to 120 kg and ewes 50 to 55 kg in September. Life tables for Dall sheep have been constructed by Simmons et al. (1984), Hoefs and Bayer (1983), Bradley and Baker (1967), and Deevey (1947). These tables reveal the highest life expectancy at birth for the Sheep Mountain rams (5.46 years), even though the maximum life expectancy is the lowest reported. No generalization is possible for the adult sex ratio, because populations are exposed to differing degrees of hunting pressure. For the unhunted Sheep Mountain population this ratio averaged 86 ♂:100 ♀ over 12 years. Detailed information on the population makeup of this herd is presented in the next section.

Model of the Population Structure

Based on 12 years' demographic data for the Sheep Mountain population, a graphic model of the average population structure has been prepared (fig. 3). In the preparation of this model a number of assumptions had to be made, as well as a few minor changes in mortality rates, since an attempt was made to integrate the actual average number of sheep obsserved over a 12-year period with mortality rates derived from a life table, which was based on a much larger sample size (Hoefs and Bayer 1983). The life table was based on recorded mortalities of 241 rams; the average number of rams present in any one year was only 111 (fig. 3). This

Population Dynamics and Horn Growth Characteristics of Dall Sheep

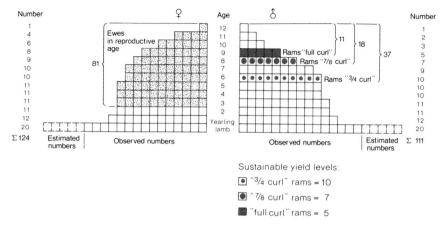

Fig. 3. Structure of Dall sheep population.

compromise resulted in a few distortions of mortality rates, since we only wanted to show "whole" animals in the model of the population structure. For accurate, age-specific, mortality rates the reader is referred to the actual life table (Hoefs and Bayer 1983).

The model shows the number of sheep in each age cohort. In this population the 12 to 13-year age class is the maximum life expectancy. The model assumes that early mortality of lambs during the first month of life was on the average the same, as was observed during only 3 years (1971, 1972, and 1976). This early mortality is shown separately from the remaining first-year mortality, which was documented for all 12 years. Very few skulls of winter-killed ewes were located, and therefore little information exists on age-specific mortality rates of the female cohort of this population. We know, however, the number of lambs and yearlings and the total number of adult females, as well as the fact that the maximum life expectancy is the same as that for rams, the 12 to 13-year age class. The exact mortality pattern of adult females, here modified from Bradley and Baker (1967), is not relevant in this context.

The model represents the population structure in early June, after all lambs have been born. As far as mature rams are concerned, it is relevant also to the following hunting season (August and September), since hardly any mortality occurs during the summer months (Hoefs 1975). Combined with information about horn growth characteristics, which will be dealt with shortly, the model can be used to evaluate various management schemes. Even though an earlier version of this model (Hoefs 1975) has been used in sheep management in the Yukon for the past 5 years, its

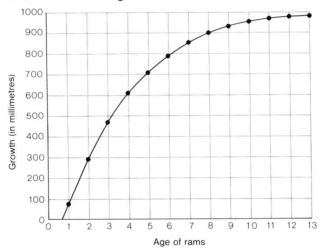
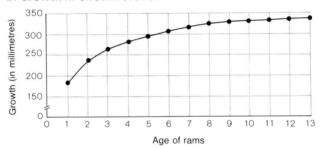
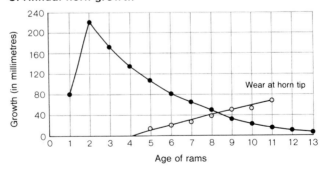

Fig. 4. Horn growth of Dall rams.

limitations must always be kept in mind. Most important in this context is the fact that it is based on average demographic data, which show considerable variations for individual years (Hoefs and Bayer 1983). Longevity and age of sexual maturity varies among different populations, as does sex ratio and horn growth characteristics. The model, therefore, cannot replace annual assessments of individual populations. However, few management agencies have the resources to carry out an intensive, population-specific, management, most depend on the use of generally applicable guidelines. It is our experience that the model predicts best when management strategies are planned over a larger area with several sheep populations or for the same population over a period of several years.

Horn Growth Characteristics

Horn growth characteristics of rams for the Sheep Mountain population are shown in figs. 4 and 5. The information is based on data of 96 rams, most of which died from natural causes (Hoefs 1975). Recent investigations by the Yukon Wildlife Branch indicate that these data are representative of other sheep populations in the southwestern Yukon (Miners Range, Coast Mountains, Champagne area, and parts of the Ruby Range). Figs. 4A and 4C show horn growth in length, and fig. 4B growth in circumference. Fig. 5 expresses horn growth as "circular" growth in "degree curl," since most North American sheep hunting regulations are based on this criterion.

Horn growth follows the pattern described in many other investigations (Taylor 1962, Simmons 1968, Geist 1971, Heimer and Smith 1975, Bunnell 1978). Very little growth occurs in the first year (the lamb growth), and most of this is worn off as the animal ages. This "wearing off" of the lamb tip is partially a gradual process, since rams use their horn tips to rub their hides or support their heads when bedding-down, and partially the result of damage from fighting. The latter type of horn length reduction results in splintered ends, referred to as "broomed tips" (Shackleton and Hutton 1971). Both types of tip wear increase with age (Bunnell 1978, Hoefs and Nette 1982). They have also been documented for bighorn sheep (Geist 1971, Shackleton and Hutton 1971), where this problem is much more severe; for desert bighorns (Monson and Summer 1980); and for the European mouflon (Uhlenhaut und Stubbe 1978, Hromas 1979, Hoefs 1982a). Most horn growth occurs during the second year in this population, and there is a continuous, more-or-less proportional reduction in annual growth increments with each succeeding year. Of the potential total growth, 80% is accomplished after the sixth growing season (fig. 4A). After the ninth growing season, when over 90% of the potential growth in horn length is accomplished, the lengths of the annual increments become insignificant.

Growth in horn circumference follows a similar pattern. Most growth is accomplished after the eighth growing season; further increases are insignificant (fig. 4B). In fig. 4C we have indicated the wear at the horn tips. Only an estimate is possible here, since very few skulls of young rams with complete first-year horn growth increments were available for measurement. Our estimate must therefore be considered as a minimum. It appears that on the average, between the eighth and ninth growing season, the wear at the horn tips exceeds the new growth put on at the bases of horns. A similar trend has been described for the European mouflon (Hromas 1979, Hoefs 1982a). In intensely hunted sheep populations where all legal rams are removed annually, tip wear has management implications, since little if any improvement in "average horn length" can be expected if rams are allowed to live longer than 9 years. In less-intensely hunted sheep populations, as in the Yukon and the Northwest Territories, where perhaps only 50 to 70% of the available legal rams are annually removed, trophy quality continues to improve with age, since hunters select rams with the best trophies—those with the least-damaged tips (Hoefs et al. 1984).

Even though the horn growth characteristics shown here are representative in general for Dall rams in the southwestern Yukon, considerable variation has been observed for Dall rams in their entire range. We know, for instance, that northern sheep have slower horn growth rates, with the third horn-increment very often being the largest. They often reach a certain trophy quality, expressed as total length or "degree curl," 1 or sometimes 2 years later than rams in southern populations. On the other hand, there are indications that they make up for it with a longer life expectancy. The most thorough work on horn growth characteristics of Dall rams has been carried out by Heimer and Smith (1975) for Alaskan populations. The authors combined length and circumference data for horn growth into an assessment of volume, and compared various Alaskan populations. They found that, generally speaking, horn growth quality in rams is inversely related to population density, and that the physiography, glacial history, and present glaciation of sheep habitat also appear to have some influence. The aforementioned retarding influence of latitude was substan-

Population Dynamics and Horn Growth Characteristics of Dall Sheep 107

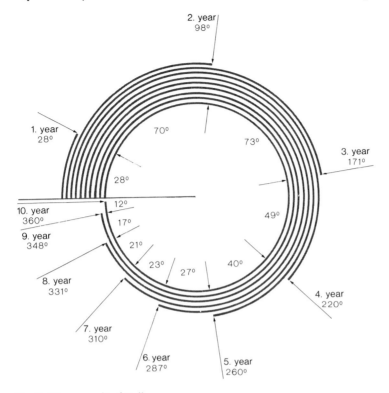

Fig. 5. "Degree curl" of Dall rams.

tiated by their work. Bunnell (1978) hypothesizes that there are years in which horn growth rates depart significantly from the expected mean value, and that these departures are related to rainfall and thereby indirectly to range productivity. Bunnell (1978) assumes that the natality rate of the sheep population follows the same pattern.

North American sheep hunting regulations use "degree curl" almost exclusively as the criterion to define "legal rams." Fig. 5 gives average values for rams from the southwestern Yukon. A graphic, two-dimensional presentation of angular horn growth is an oversimplification and is particularly misleading for older Dall rams, because of the flaring, spiralling nature of their horn growth. However, we are confronted with these diagrams in every hunting regulation brochure, and we cannot ignore them in this discussion. For sheep species whose horns grow more or less in a single plane (*Ovis musimon, O. cycloceros, O. ammon hodgsoni*), this type of assessment of "degree curl" is more appropriate. Most rams reach horn growth of "270° curl" (3/4 curl) during their sixth growing season, when they are about 5 years and 3 months old. Rams with "3/4 curl" horns are in many areas defined as legal sheep for hunting. More restrictive regulations require that rams must have horns describing an angle of 315° (7/8 curl) before being legal game. Even though the Yukon Hunting Regulations use the term "full curl" for these rams, the graphic presentation in the hunting brochure is that of a "7/8 curl" ram (Yukon Wildlife Branch 1981). This horn growth is reached by most rams in the southwestern Yukon during their eighth growing season, when they are about 7 years and 3 months old (fig. 5). Similar observations have been made for southern Alaska Dall rams (Heimer and Smith 1975). Rams whose horns describe an angle of 360° or more are rare.

The flaring nature of the horns and damage to the horn tips become important characteristics in the old-age classes. In the sample size investigated for fig. 5, rams reached "full curl" (360°) only in their tenth growing season. There are, however, many other populations that have a lower density, less flaring of horns, and less damage at the tips, where "full curl" is reached during the ninth growing season. This may indeed be more often the case. It is a paradox in North American sheep management that hunting regulations in general define "legal rams" in terms of "degree curl," while the subsequent trophy assessments (the scoring according to Boone and Crocket standards) ignore "degree curl" and consider only the length and circumference measurements of the horns.

Management Implications

Referring back to the model of the population structure (fig. 3), we can estimate the number of rams with specific horn characteristics in a population as well as predict what kind of sustainable yield a population can be exposed to under various hunting regulations.

There are in this population 37 rams in their sixth growing season or older. These have horns that describe an angle of 270° (3/4 curl) or larger. They represent approximately 50% of the rams in the population that are in ram bands separated from nursery groups, assuming that about 30 to 40% of the 2 to 3-year-old rams are still in nursery bands (Hoefs 1975). These 37 rams amount to 19% of the total population (lambs excluded). If this population were severely hunted under "3/4 curl" regulations, it can be assumed that all rams older than those in their sixth growing season would be missing. In this case "sustainable harvest" becomes equal to the recruitment of 5 to 6-year-old rams. Of these rams, 10 have been in this population on the average. It is therefore reasonable to estimate that, under "3/4 curl" regulations, in a population with 81 ewes in reproductive age about 10 rams would become available per year. From the practical point of view a more meaningful ratio would be that of "nursery sheep" to legal rams per year. Assuming that 5 of the 2 to 3-year-old rams are still in nursery bands, which in addition include all ewes and yearlings, one can estimate that 121 "nursery sheep" would produce 10 legal rams (3/4 curl) per year, or 12 "nursery sheep" 1 legal ram.

Examples with other management regulations can be calculated from figs. 5 and 3. For instance, if a population is managed under "7/8 curl" regulations, and has been hunted severely for a number of years so that no older rams are left, the following ratios would apply. The number of rams available decreases because of natural mortality between the 6th and 8th year. In the 8 to 9-year age class, 7 rams can be expected to be available on the average. We can therefore say that a population with 81 ewes in reproductive age produces annually 7 rams with "7/8 curl" horns, or that 121 "nursery sheep" accomplished this. This means that 17 "nursery sheep" are needed to make available 1 legal ram per year.

We have already made reference to the fact that populations differ with respect to productivity, age of sexual maturity, adult sex ratio, and longevity, as well as horn growth characteristics. These factors will influence the ratios given in the above examples. A few citations from the literature are given here to demonstrate these differences.

The natural mortality of rams between their sixth growing season, when their horns reach 270° curl, and rams in their ninth season, when their horns may reach full curl, appears to be higher in Alaska than in the Kluane area. Heimer and Smith (1975) calculated from a hunter-killed sample a reduction in available rams from 1,900 (3/4 curl) to 1,200 (full curl), which translates into a reduction of 37%. Calculations with life table data for the Sheep Mountain population will reveal that this reduction amounts to only 27% in this herd. Even greater mortality rates are described for a sheep population in the Brooks Range (Alaska), where Summerfield (1974) observed reduction in 3/4 curl to full curl rams components of the population between 61% and 69%. Simmons et al. (1982) also attempt to calculate trophy yield figures for the Mackenzie Mountains sheep population. They write: "The incidence of full curl rams in the male population is 10.5%. Production of full curl rams is 7.3% of male births." The incidence of full curl rams (by Simmons et al.'s definition, "Rams over 8 years old") amounts to 14% for the Sheep Mountain population, if 65% of the 2 to 3-year-old males are part of the male population. The lower rate in the Mackenzie Mountains population is expected, since it is subject to hunting. The production of full curl rams (8+ years old) amounts to 34.8% of male births for the Sheep Mountain population. This difference from the Mackenzie rams is brought about primarily by a much lower mortality rate during the first year of life (65.3% vs. 42.0%), as well as lower mortality rates ($\bar{x}$ = 10%) up to the 8th year.

The relationship between increasing age of rams, the corresponding reduction in numbers through natural mortality, and the improving trophy quality can be demonstrated in a different and more appropriate manner, by using horn length as the criterion instead of degree curl. In fig. 6 we have demonstrated this relationship by using as a definition of "trophy" rams those whose horns equal or exceed a length of 38 inches (96.5 cm). We have started with a hypothetical population of 1,000 rams in their sixth growing season, since this is the age at which they usually become legal under "3/4 curl" regulations. The curve showing the number of rams alive in the older age classes is derived from life tables (Hoefs 1975). The curve showing the percentage of trophies (Horns ≥ 38 inches) is derived from inspections of 884 hunter-killed rams from the southwestern Yukon. By this definition of "trophy," there are none in the 5 to 6 and 6 to 7-year age classes and only about 8% in the 7 to 8-year cohort. The proportion of trophies then rises with age, reaching 50% and more in the old-age classes.

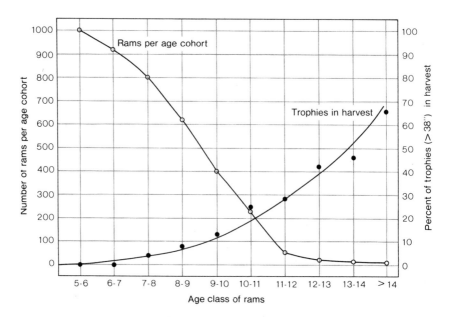

Fig. 6. Relation between age, abundance, and trophy quality in Dall rams.

However, there are very few rams left in that category, and it would be unrealistic to have their harvest as the management goal. In fig. 6 these two opposing trends of abundance and quality intersect at the 10 to 11-year age class. This class, as well as the 9 to 10-year cohort, constitute the best compromise, if the management goal is a high percentage of trophies by our definition. In the 9 to 10-year age class 418 rams can be expected in this hypothetical population; of these, 13.3% or 56 are "trophies." In the 10 to 11-year age class only 214 rams are left, but 23.5%, or 50, are trophies. Even though the number of trophies in this latter cohort is smaller, they contain a higher number of skulls whose horns exceed 40 inches (101 cm) in length: 4 (9 to 10-year cohort), 12 (10 to 11-year cohort) (Hoefs and Cowan 1979).

This correlation between age and trophy quality does not contradict earlier statements about the increase of horn tip wear with age. The wear information presented average values, whereas the percentage of "trophies" per cohort are those exceptional rams with above-average horn growth rates and below-average horn tip damage.

In fig. 6 we used a trophy size of ≥ 38 inches as the example. In areas where sheep hunting pressure is high, a less restrictive definition of "trophy" will have to be used. During the past 10 years the Yukon sheep harvest has been around 300, from an estimated total huntable population size of 15,000 to 16,000. At a harvest rate of only 2%, translating into a removal rate of "legal" rams of only 50 to 70% annually, fairly stringent management goals can be set.

On the Compensatory or Additive Nature of Sheep Hunting

The assessment of this question is complicated by the fact that sheep hunting, in contrast to the hunting of most other big game species, has been limited almost exclusively to the taking of older rams. Hunting pressure was therefore relatively light. Only in Alaska has an ewe harvest been carried out on an experimental basis.

Investigations to establish at what level of harvest hunting mortality would become additive to natural mortality have not been carried out. On the other hand, there are a number of indirect means through which the impact of hunting can be assessed. These are evidence of a population decline, a reduction in the frequency of older rams in the population, a change in the sex ratio, and—under extreme hunting pressure—abandonment of part of the historical range of the population, with consequently greater natural mortality.

The Yukon Wildlife Branch monitors the population dynamics of a number of Dall sheep herds routinely every year. It was observed that fluctuations in population size from one year to the next are primarily brought about by variations in the number of lambs and yearlings. The adult cohorts varied little, and no negative impact of hunting could be detected, as long as hunting was the only human influence. It has to be considered, though, that hunting pressure in the Yukon is low, amounting to only 2 to 3% of the total population and being restricted largely to rams of "315° curl," or 9 years of age.

In a detailed assessment of hunted and unhunted Dall sheep populations in the Kluane area, southwest Yukon, Hoefs et al. (1984) found that a harvest rate of 2.5%, which translates into a removal rate of annually 60 to 70% of the "legal" rams, did not result in a population reduction or a change in the sex ratio. On the contrary, the hunted population exhibited a higher fecundity rate and appeared to be increasing. Both these populations were exposed to "normal" predator pressure, no controls being carried out. It is therefore reasonable to assume that a harvest rate for rams of up to 3% is compensatory and not additive. It should be understood that "compensatory" need not necessarily mean that hunters select those rams that would have died for other reasons, even though there are indications that this is to some degree the case (Hoefs et al. 1984); it could also mean that removal by hunting reflects itself in a higher fecundity rate, a higher survival rate of the remaining population members, and a higher recruitment rate of rams into the legal age classes. Nichols (1978a) also observed that on the Kenai Peninsula in Alaska two heavily hunted populations had a higher rate of increase than an unhunted one.

We are aware of only one published case in which over-harvesting is assumed to have contributed to a population decline. The Dall sheep in the northern Richardson Mountains, particularly those that winter on Mount Goodenough at the Mackenzie Delta, are hunted by residents of Aklavik. Native hunting is for meat and not for trophies, and generally speaking is not selective. Data published by Simmons (1973) reveal that more ewes than rams were taken, and that in 1973 the harvest was 11% in excess of the assumed annual increment rate. Repeated estimates of the population size and the concerns that the observed trend precipitated are elaboated on in Hoefs (1978).

If ewes are taken as well as rams, an assessment of the impact of hunting becomes more complicated, since the reproductive potential of a population is tampered with. Based on his long-term study of the Dry Creek population in Alaska and on his experience with an experimental ewe season,

Heimer (1981) writes: "Even when populations are known to be lower than historically observed and all conditions for population growth are favourable, ewe harvest should never exceed 3% of the ewe segment, unless a population reduction is desired." If we add to this the normal complement of hunter-killed rams, we are dealing with a total harvest rate of 5 to 6%.

Similar rates are advocated for the management of mountain goats, a species with a very similar reproductive potential. Hebert and Turnbull (1977) suggest that for population maintenance not more than 4% of the total population or less than 5% of the adult population can be taken. Goat regulations allow for either-sex hunting.

Bergerud (1979), in his review of population data of over 30 caribou herds, another species with a reproductive potential similar to that of sheep, found that a harvest rate of below 5% did not result in population declines, even under normal predator pressure.

It is therefore not unreasonable to assume that Dall sheep populations could be exposed to a hunting pressure of up to 5% (under either-sex regulations) without adverse effects on population size.

Conservation Concerns

In our discussion we have concentrated on the harvest of trophy rams, since that is the primary sheep-management objective at the present time. Hunting, however, can be regulated by the wildlife management agencies to reduce or eliminate negative impact on a population. There are a number of other factors that have or can have adverse influence. Most of these are not under the jurisdiction of wildlife management agencies to regulate. A few should be mentioned here.

First, there are the many undesirable by-products of "northern development." The building of roads, trails, railroads, pipelines, power dams, and powerlines may destroy "critical" areas such as winter ranges, lambing sites, mineral licks, or important migration trails, or interfere with their normal use. Improved access will cause greater disturbance, particularly since it is accompanied by a build-up of recreational activities. The search for minerals in the back country by helicopter-supported exploration crews has a great potential for negative impact on wildlife through harassment and the disruption of normal activity patterns and range use strategies.

Second, there is the potential for unregulated meat hunting by native people. Tradition and lack of access have so far directed this activity toward species such as caribou and moose, but the possibility for damage exists under present law, and it may become a real threat once more-desirable species become rarer.

To ensure the future conservation of the Dall sheep, a magnificent wilderness animal, these latter questions will have to be addressed more thoroughly. Government, industry, and public interest groups must deal with them more adequately than has so far been the case.

References

Bergerud, T. 1979. Population dynamics of North American caribou. Paper submitted to the 2nd Internat. Caribou-Reindeer Symp., Oslo, Norway.
Bradley, W. E., and D. P. Baker. 1967. Life tables for Nelson bighorn sheep on the Desert Game Range. Trans. Desert Bighorn Council 11: 142–169.
Bunnell, F. L. 1978. Horn growth and population quality in Dall sheep. J. Wildl. Manage. 42(4): 764–775.
Buechner, H. K. 1960. The bighorn sheep in the United States: Its past, present and future. Wildl. Monogr. 4. 174 pp.
Deevey, E. S. 1947. Life tables for natural populations of animals. Quart. Rev. Biol. 22: 283–314.
Geist, V. 1971. Mountain sheep: A study in behaviour and evolution. Univ. Chicago Press, Chicago. 363 pp.
Hebert, D. M., and W. G. Turnbull. 1977. A description of southern interior and coastal mountain goat ecotypes in British Columbia. Proc. 1st Int. Mountain Goat Symp. 1: 126–146. British Columbia Fish and Wildlife Service, Victoria.
Heimer, W. E. 1979. Interior sheep studies. Alaska Dep. Fish and Game, Proj. W-17-11.
———. 1980. Interior sheep studies. Alaska Dep. Fish and Game, Proj. W-17-12.
———. 1981. Interior sheep studies. Alaska Dep. Fish and Game, Rep. W-21-1/6.92.
———, and A. S. Smith III. 1975. Ram horn growth and population quality: Their significance to Dall sheep management in Alaska. Alaska Dep. Fish and Game. Wildl. Tech. Bull. 5.
Hoefs, M. 1975. Ecological investigation of Dall sheep and their habitat. Ph.D. thesis. Univ. British Columbia. 499 pp.
———. 1978. Dall sheep in the Richardson Mountains, Yukon Territory: Distribution, abundance, and management concerns. Yukon Game Branch, Rep. No. 78-2.
———. 1982a. Beitrag zur morphometrie und wachstumsdynamik der schnecken des muffelwidders. Zeitschrift für Jagdwissenschaft. 28(3): 145–162.

———. 1982b. The importance of Sheep Mountain, Kluane National Park, as winter range for Dall sheep, and recommendations for its conservation. Report on file with the National and Historic Parks Branch of Canada.

———, and M. Bayer. 1983. Demographic characteristics of an unhunted Dall Sheep (Ovis dalli dalli) population in southwest Yukon, Canada. Can. J. Zool. 61(6): 1346–1357.

———, and T. Nette. 1982. Horn growth and horn wear in Dall rams and their relevance to management. Bienn. Sym. North Wild Sheep and Goat Counc. 3: 143–156.

———, N. Barichello and T. Nette. 1984. Comparison of a hunted and an unhunted Dall Sheep population. Submitted to the 4. Bienn. Sym. North Wild Sheep and Goat Counc.

———, and I. McT. Cowan. 1979. Ecological investigation of a population of Dall sheep. Syesis 12 (Supp.1): 1–81.

Hromas, J. 1979. On the wear of mouflon horns (in CSSR). Folia venatoria 9: 53–66.

König, R., and M. Hoefs. 1982. Längenmesswerte und bewertungspunkte als index des grösse oder stärke von schafgehörnen. Allg. Forst. Zeitsehr. 51/52: 1557–1559.

Monson, G., and L. Sumner, eds. 1980. The desert bighorn. Univ. Arizona Press. 370 pp.

Murphy, E. C. 1974. An age structure and reevaluation of the population dynamics of Dall sheep. M.Sc. thesis. Univ. Alaska. 113 pp.

Nichols, L. 1976. An experiment in Dall sheep management, progress report. Trans. North Amer. Wild Sheep Conf. 2.

———. 1978a. Dall sheep reproduction. J. Wildl. Manage. 42: 570–580.

———. 1978b. Dall's sheep. In Schmidt, J. L. and D. L. Gilbert., eds. Big game of North America. Stackpole Books, Harrisburg, Pa. 494 pp.

Shackleton, D., and D. A. Hutton. 1971. An analysis of the mechanism of brooming of mountain sheep horns. Zt. Säugetierk. 36: 342–350.

Simmons, N. M. 1968. Non-resident big game hunting in Game Management Zone 12, Mackenzie Mountains, N.W.T. Unpub. rep., Can. Wildl. Service.

———. 1973. Dall's sheep harvest in the Richardson Mountains, N.W.T. Unpubl. rep., Can. Wildl. Service.

———, M. B. Bayer, and L. O. Sinlay. 1984. Dall's sheep demography in the Mackenzie Mountains. Submitted for publication to J. Wildl. Manage.

Summerfield, B. L. 1974. Population dynamics and seasonal movement patterns of Dall sheep in the Atigun Canyon area, Brooks Range, Alaska. M.Sc. thesis. Univ. Alaska, Fairbanks. 109 pp.

Taylor, R. A. 1962. Characteristics of horn growth in bighorn rams. M.Sc. thesis. Univ. Montana.

Uhlenhaut, K., and M. Stubbe. 1978. Kampfbeschädigungen bei muffelwiddern. Beitr. zur Jagd- und Wildforsch. 11: 151–169.

Yukon Wildlife Branch. 1981. Brochure of synopsis of hunting regulations 1981/82.

Winter Range Ecology of Caribou (*Rangifer tarandus*)

D. E. Russell
A. M. Martell

Abstract

Caribou (*Rangifer tarandus*) are examined as organisms that are highly adapted to life in arctic and subarctic winters. Throughout their circumpolar range, caribou have in common many behavioural and physiological adaptations to ensure their survival in winter. In addition, each geographic population has further specialized to cope with its unique environmental situation. Adaptations to snow and forage are discussed in terms of activity, diet, nutritional physiology, and energetics. Despite caribou's large, supporting hooves, their movements and activities are affected by the changing conditions of the snowpack. Caribou cratering strategy reflects both the distribution of lichens and the quality and quantity of the snowpack. As winter progresses, caribou generally spend more time lying down and less time moving about, and more of their active period is spent feeding. Energy-rich but nutrient-poor lichens form the major component of the winter diet. Lichens are supplemented with more nutrient-rich foods, such as winter-green vegetation, in some regions, but in other regions caribou have adapted to utilize less energy-rich forage, such as moss. Caribou are highly specialized in their ability to compensate, at least partially, for dietary deficiencies by recycling nitrogen, by resynthesizing and recycling glucose, and by storing and cycling many of the macroelements. Caribou have many physical, physiological, and behavioural adaptations to reduce heat loss, and also reduce energy expenditure behaviourally by altering their activity patterns and cratering strategies. They draw on fat reserves to aid in balancing their energy budget, but reserves are small; therefore they must minimize weight loss and energy expended in obtaining forage and travelling through snow.

Introduction

At the beginning of the Pliocene Epoch (12 to 13 million years ago), tundra began to develop for the first time as an open environment at the northern edge of the taiga. Caribou, or reindeer (*Rangifer tarandus*),* likely evolved at that time from taiga ancestors to exploit the newly developing tundra habitat (Kowalski 1980). Caribou evolved many specialized adaptations to tundra conditions and to the use of lichens as a primary forage (Klein 1970). They are the large herbivores most often associated with arctic and subarctic habitation and are often cited in terms of their ability to survive in spite of harsh northern winters. A close examination of the ecology of the species soon reveals that, rather than simply coping, caribou thrive and are productive due to a vast array of behavioural and physiological adaptations that characterize every aspect of their existence.

Nasimovich (1955) felt that scientists would gain in their understanding of the ecology of caribou by ignoring traditional taxonomic considerations. He treated *Rangifer* as four ecological races, or ecotypes, based on differences in their seasonal mode of life: marine, tundra (migratory), forest, and mountain (alpine) races. Those ecotypes, derived from Russian experience, are analogous to some subspecies and races in North America.

The marine ecotype, similar to the North American Peary caribou, inhabits the Arctic Islands and remains relatively sedentary. The tundra ecotype normally spends summer on the forage-rich and mosquito-poor treeless tundra. After a long migration it winters in the taiga, or boreal forest, regions far to the south. Those migrations are made primarily in response to a need for an energy-rich diet, which is less accessible on the tundra in winter. Most North American barren-ground caribou fall into this category. A few herds of the tundra ecotype, however, spend winter north of tree line, as do the central arctic herd and portions of the western arctic herd in Alaska. The forest ecotype remains in the taiga all year, normally migrating short to long distances in response to snow, food, and perhaps

*Throughout this paper we will use the North American common name "caribou" for all circumpolar members of the species. In Eurasia, "reindeer" is the term applied to both domestic and wild forms.

other factors such as herd size. Most woodland caribou in North America follow that strategy. The mountain ecotype is analogous to some woodland, or mountain, caribou in western North America, and to some Grant's caribou in Alaska and the Yukon. In North America the mountain ecotype spends most of the year in the mountains and may make annual migrations in response to snow depth and hardness (Edwards and Ritcey 1959, Freddy and Erickson 1975).

Various authors divide winter into between three and seven periods, based on snow characteristics and on movement rates of caribou. However, criteria vary among authors (Bergerud 1974b, Miller 1974, Pruitt 1979, Roby 1978, Skogland 1978). For purposes of comparison, we have divided winter into three calendar periods: early winter (November and December), mid-winter (January and February), and late winter (March and April). Those periods usually have distinctly different mean snow conditions. They approximate the periods of other authors; however, we have experienced a winter in the central Yukon when "early winter" snow conditions were still present in January and February.

Snow

Caribou are true chionophiles (Formozov 1946, Pruitt 1959), which means that they are highly adapted to snow. Caribou feet are unlike those of any other member of the deer family. They have blunt toes, crescent-shaped hooves with a sharp edge for grip on hard snow and ice, and functional lateral digits, or dew claws. A heavy growth of bristle-like hairs surrounds the hoof, and the joints of the middle toes can be bent sharply to assume an almost horizontal position. In fact, the caribou hoof can be considered in a state of transition toward a plantigrade foot (Nasimovich 1955). Those adaptations greatly increase the surface area of the hoof and help the caribou to "float" on soft snow. Caribou foot loading is only about 125 to 180 g/cm^2 when standing (Nasimovich 1955, Thing 1977), 500 g/cm^2 when walking, and 1000 g/cm^2 when trotting, compared with 390 to 659 g/cm^2 for standing moose (Telfer and Kelsall 1979).

Despite these adaptations, the movements and activities of caribou are heavily influenced by snow conditions. Caribou select areas with favourable snow conditions for feeding (Bergerud 1974b) and avoid areas of unfavourable snow (Pruitt 1959). Favourable areas for digging feeding craters (cratering areas) for tundra and mountain caribou generally have snow depths of less than 50 to 60 cm and densities of less than 0.35 g/cm^3

(Baskin 1970, Formozov 1946, Henshaw 1968, LaPerriere and Lent 1977, Lent and Knutson 1971, Nasimovich 1955, Pruitt 1959, Skogland 1978, Thing 1977), but threshold depths and densities are greater (65 to 74 cm, 0.40 g/cm³) for sedentary forest caribou (Bergerud 1974b, Helle 1981, Stardom 1975). Pruitt (1979, 1981) has presented a snow index based on the hardness and thickness of several layers of the snow cover. The index is a potentially useful tool for determining suitable areas for caribou to winter.

Nasimovich (1955) notes that the critical depth of 50 to 60 cm applies to solitary animals and that the critical depth for a herd is 80 to 90 cm, because mobility is facilitated by the movements of the group. Forest caribou will crater in 70 to 90 cm of snow, provided that the sinking depth is <70 cm (Helle 1981). Caribou may sometimes crater in snow over 100 cm deep (Skoog 1968, personal observations in central Yukon), but deep, soft snow can temporarily immobilize caribou and restrict them to small pockets of range (Bergerud 1974b, Edwards and Ritcey 1959). When snow depths exceed 80 cm and density and hardness are sufficient for support, caribou may shift their feeding from terricolous lichens to arboreal lichens if they are available in the area (Nasimovich 1955). That shift usually occurs in late winter (Bergerud 1974b, Edwards and Ritcey 1960, Helle 1981, Miller 1974). Caribou in mountainous areas migrate from taiga to alpine tundra when snow depth reaches 40 to 60 cm (Nasimovich 1955). On alpine tundra, snow depths are usually less than they are in the taiga, but density and hardness are greater. Caribou wintering on tundra occasionally suffer from a lack of food when unusually dense snow cover or ground icing occur, and large numbers may die (Nasimovich 1955, Parker et al. 1975, Vibe 1967).

LaPerriere and Lent (1977) indicate that the selection for feeding sites takes place at three levels. Caribou first choose regional areas (a watershed), then feeding areas (a specific sidehill), and finally a cratering site (the side of a sedge tussock). Data from the Porcupine herd in the central Yukon (unpublished observations of the authors) indicate a change in the relative use of three major terrain types from early winter to late winter in a normal snow year, but in a low snow year the three types were used in the same proportions throughout the winter. Therefore, general habitats chosen in early winter remain favoured throughout winter unless snow restricts feeding and movement.

The third level of selection, microsite selection of cratering sites, is undoubtedly related to the ability to detect lichens beneath the snow cover. For example, Kelsall (1968) noted that caribou have an uncanny ability to locate food in craters. Caribou can detect lichens through 15 to 18 cm of undisturbed snow cover (Bergerud and Nolan 1970), but use air vents adjacent to emergent stems of tall shrubs to detect lichens at greater snow depths (Bergerud 1974a). Helle (1981) found that caribou can detect lichens at a depth of at least 72 cm. He suggested that the ability to detect food under snow by scent is related not only to the presence of air vents, but also to the abundance of food resources; the scent transmitted through the air vents on the snow surface correlates with the biomass of food at that point. Sablina (1960) noted that while a band of reindeer cratered only 18% of a feeding area it obtained 37 to 61% of the available lichens.

Caribou paw craters in the snow with their broad forehooves. As the density and hardness of the snow cover increase, the pawing rate decreases but more energy is used per blow. Skogland (1978) observed the pawing rate to decrease from 16.4 strokes/min in early winter to 12.1 strokes/min in mid-winter and 9.3 strokes/min in late winter. Likewise, we observed a change from 12.0 to 8.0 to 9.8 strokes/min in the taiga and from 15.5 to 10.3 to 11.7 strokes/min on the tundra during the same phases of winter (unpublished data, central Yukon).

Crater size also changes with snow conditions. As conditions become more difficult for cratering, caribou spend more time in each crater and dig larger craters. Helle (1981) found that crater size for forest caribou increased from 0.13 m³ in early winter to 2.20 m³ in mid-winter to 2.92 m³ in late winter, with increasing snow depth. We found that crater size increased from 0.09 m³ in early winter to 1.03 m³ in mid-winter to 1.14 m³ in late winter in the taiga, and from 0.13 m³ to 0.20 m³ to 0.82 m³ on the tundra, with increasing snow depth and density in both habitats (unpublished data, central Yukon). In contrast, Thing (1977) found that crater volume decreased from 0.07 m³ in early winter to 0.03 m³ in late winter, but in his case snow depth also decreased because the caribou shifted habitats. We also found that the time caribou spent in each crater increased from 75 sec in early winter to 112 sec in mid-winter to 218 sec in late winter in the taiga and from 26 to 156 to 340 sec on the tundra. The proportion of the active period spent in craters also increased from 50 to 69 to 88% throughout the winter, but the time spent feeding as a proportion of the total time spent in the crater stayed relatively constant at about 84%. Caribou usually crater an area only twice before the snow is too hard to recrater, but at very low temperatures (below −45 C) the disturbed snow in craters will set harder and force caribou to move more frequently than they would in warmer weather (Pruitt 1959, 1960).

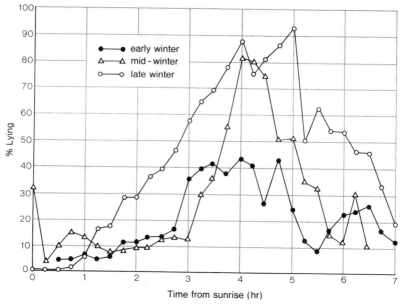

Fig. 1. The proportion of time caribou in central Yukon spend lying in winter, in relation to the time of sunrise.

Activity

In winter, caribou are active for five or six relatively evenly spaced periods during each 24 hr period, with alternating periods of rest (Baskin 1970, Segal 1962). In mid-winter, caribou feed at dawn and dusk, with a lying period in midday (Roby 1978, Segal 1962, Thomson 1973, 1977), but by late winter they move away from that pattern and have two or three rest periods during the daylight hours (Thomson 1973, 1977). Roby (1978) found that caribou in northern Alaska keyed their daily activity to sunrise, and we found the same pattern in the central Yukon (fig. 1). Caribou cueing of activity to sunrise produces a conspicuous midday peak in mid-winter lying, which has been observed by many authors. The length of lying periods is relatively constant throughout the winter, averaging 110 min (Boertje 1981, Roby 1978, Segal 1962), but the length of active periods declines from 210 min in early winter to 150 min in mid-winter to 145 min in late winter (Boertje 1981). Segal (1962) found that in mid-winter the average lying period was shorter in the light (112 min) than in the dark (164 min) while the average active period was longer in the light (128 min) than in the dark (97 min), but combined periods were similar in length

(240 min in the day, 243 min at night). However, because feeding time is related to food availability and lying time is related to food quality, the lengths of both periods may vary depending on the quality of the winter range.

As can be seen from table 1, there is a general similarity in the winter activity patterns of diverse caribou populations. Although the percentage of time spent feeding varies among populations, there is a common pattern of increase in feeding intensity as winter advances. The proportion of time spent lying generally increases throughout the winter. We do not think that the latter is only a result of restricting the sampling period to daylight hours, because even in the first active-rest cycle of the day we found an increase in the proportion of time spent lying as the winter advanced (fig. 1). The proportion of time spent standing varies among caribou populations, but is generally stable or declines slightly throughout the winter. The proportion of time caribou spend moving (walking, trotting, or running) also declines throughout the winter. Caribou are relatively sedentary in the winter, moving only about 5 km/day (Boertje 1981, Roby 1978, 1980, Thomson 1973).

Caribou in West Greenland and on Svalbard are on poor winter range (Reimers 1980, Roby 1980), without predators, and are particularly sedentary in winter (table 1). Caribou in both areas show a very high feeding intensity, and West Greenland animals also show a high proportion of time spent lying. Roby (1980) suggested that feeding intensity, the proportion of time spent lying, and the mobility of animals reflects range quality, particularly in mid- and late winter. By Roby's criteria, caribou in central Alaska and in Hardangervidda would be considered to be on poor quality range in late winter (table 1). That conclusion is supported by data on the diet of caribou in central Alaska (Boertje 1981). By the same criteria, caribou in the central Yukon would be on the best winter range of the populations examined (table 1) because feeding intensity and the proportion of time spent lying were the lowest, and the proportion of time spent moving was the greatest.

Caribou are relatively insensitive to all but the most severe environmental conditions. Temperatures as low as -50 C have little effect on caribou activity (Baskin 1970, Henshaw 1968, Roby 1978, Skoog 1968, Thomson 1973), although movement may be somewhat reduced below -35 C (Roby 1978). However, caribou become quite sedentary at temperatures below -50 C (Baskin 1970, Skoog 1968). Moderate wind speeds (less than 15 km/hr) have little effect on caribou activity, but caribou movement increases as the animals begin to lose heat due to increasingly greater

Table 1 Activity patterns of caribou in early, mid, and late winter.

			Percent of time						Feeding
Population[1]	Habitat[2]	Season	Feed	Lie	Stand	Walk	Trot/run	Other	intensity[3]
Central Yukon, Canada[a]	M	early	48	20	6	22	1	3	60
		mid	55	30	6	7	1	1	79
		late	45	39	5	11	<1	<1	74
Central Alaska, U.S.A.[b]	M	early	48	32	8	11	<1	<1	70
		late	53	42	2	4	<1	<1	91
Hardangervida, Norway[c]	M	early	41	38	11	9	1	0	70
		mid	45	34	10	11	1	0	82
		late	48	46	2	9	<1	0	92
Northern Alaska, U.S.A.[d,e]	T	early	52	31	3	12	1	<1	75
		mid	61	28	3	8	1	<1	84
		late	46	44	3	6	<1	<1	82
West Greenland[e]	T	early	59	25	4	12	<1	<1	78
		mid	51	44	1	4	0	0	92
		late	47	48	2	2	0	<1	91
Svalbard, Norway[f]	T	all	61	34	4	2	0	0	92

Notes

1. Source: a. Unpublished data, Porcupine Herd; b. Boertje, 1981; c. Gaare et al., 1975; d. Roby, 1978; e. Roby, 1980; f. Remiers, 1980.
2. Wintering habitat: F, taiga; M, mountain (includes taiga and tundra); T, tundra.
3. Feeding intensity = % feed ÷ (100-% lie)

wind speeds (Thomson 1977). In high winds (greater than 30 to 40 km/hr) activity is disrupted; caribou aggregate and eventually lie down in order to conserve heat (Henshaw 1968, Thomson 1977). This also occurs under blizzard conditions (Baskin 1970).

Diet

"The single nutritional characteristic of reindeer that distinguishes them from all other large herbivores and that has enabled them to use holarctic rangelands is their preference for and the ability to survive on lichens during the long (6- to 8-months) winter grazing period" (Luick 1977). Lichens are prominent in the diet of caribou wintering in taiga and in mountainous regions (table 2). In early and mid-winter, when snow depths are not limiting, the preferred lichens are species of *Cladina* (*C. alpestris, C. mitis, C. rangiferina*). These are the reindeer lichens, or "reindeer moss"; they are eaten throughout the circumpolar range of caribou. In late winter, as snow becomes deeper and harder, caribou wintering in taiga often shift to arboreal lichens (Bergerud 1972, 1974a, Edwards and Ritcey 1960, Helle 1981, Miller 1974, Nasimovich 1955, Segal 1962). Also in late winter, caribou wintering in mountainous areas may shift from taiga to alpine tundra, where the lichens eaten change from *Cladina* to species of *Cetraria* (*C. cuculata, C. islandica, C. nivalis*) (Gaare and Skogland 1975, Parker 1981, Roby 1978, unpublished data, central Yukon). However, the winter diet in taiga and in mountainous regions is not exclusively lichens (table 2). Green foods, including the bases of sedges (*Carex*), evergreen horsetails (*Equisetum*), and evergreen shrubs (*Vaccinium vitis-idaea*) are actively sought, especially in early and late winter (Aleksandrova and Andreyev 1964, Baskin 1970, Edwards and Ritcey 1960, Kelsall 1968, Miller 1974, 1976, Skoog 1968). In some areas of mountain tundra, grasses, sedges, and dried forbs make up the majority of the diet for at least part of the winter (Michurin and Makhaeva 1962, Parker 1981). Moss is normally only a minor component of the diet of caribou wintering in taiga and in mountainous regions, but is prominent in the diet of caribou wintering in mountain tundra in central Alaska (Boertje 1981) and in Taimyr (Michurin and Vakhtina 1968).

Lichens are less abundant on most tundra than in taiga; consequently they are often only a minor component of the diet of caribou wintering on the tundra (table 2, Holt 1980, Reimers 1980, Scotter 1967, Shank et al. 1978). However, lichens are taken when available and make up 41 to

Table 2 Diets of caribou in early, mid, and late winter based on rumen analysis (R) and fecal pellet analysis (F)

				Percent composition						
Population[1]	Habitat[2]	Sample Type	Season	Moss	Lichen	Mush-room	Grasses and Horse-tails	Forbs	Decid-uous shrubs	Ever-green shrubs
Kuhmo, Finland[a]	F	F	early	15	27	0	34	4	15	5
			mid	11	21	tr.	35	1	17	14
			late	12	15	0	28	3	35	8
Newfoundland, Canada[b]	F	R	early	5	40	12	10	12	9	12
			late	6	56	0	7	0	3	23
Manitoba, Saskatchewan, N.W.T., Canada[c]	F	R	all	3.1	57.5	0.4	2.9	<0.1	0.8	19.1
Central Yukon, Canada[d]	M	F	early	6.3	67.8	0	15.7	0	1.1	9.1
			mid	3.6	85.1	1.0	0.6	0	0.3	9.2
			late	4.7	70.5	0	10.6	0.2	0.4	13.5
Central Alaska, U.S.A.[e]	M	F	early	6.4	80.1	0.2	5.8	0.7	0	6.9
			mid	22.6	65.6	0.3	2.0	0.5	0.2	8.9
			late	21.7	58.6	2.1	3.5	1.0	1.6	11.5
Hardangervidda, Norway[f]	M	R	early	2.6	52.0	0	32.2	0	8.0	
			mid	2.5	56.0	0	27.9	0	8.9	
			late	12.0	34.0	0	21.8	0	22.0	

Location	M	R									
Snöhetta, Norway[g]	M	R	early	4	38	0	27	3	28		
			mid	9	31	0	28	4	28		
			late	10	34	0	28	2	26		
Northern Keewatin, Canada[h]	T	F	all	48.8	41.1	0	5.0	1.2	2.3	1.6	
Taimyr, U.S.S.R.[i]	T	R	all	30.5	12.2	0	29.4	0	15.8		
Arctic Islands, Canada[j,k]	T	F	all	56.4	20.1	0	2.9	11.6	5.4	3.6	
		R	late	24.0	6.0	0	39.0	12.0	6.0	9.0	

Notes

1. *Source:* a. Helle, 1981; b. Bergerud, 1972; c. Scotter, 1967; d. Unpublished data, Porcupine Herd; e. Boertje, 1981; f. Gaare and Skogland, 1975; g. Gaare, 1968; h. Fischer et al., 1977; i. Michurin and Vakhtina, 1968; j. Fischer and Duncan, 1976; k. Parker et al., 1975, Thomas et al., 1976, Thomas et al., 1977, Thomas and Broughton, 1978.
2. Wintering habitats: F, taiga; M, mountain (includes taiga and tundra); T, tundra.

Table 3 Nutrient content of winter forage of caribou[1]

		Moss	Lichen	Mush-rooms	Grasses and Horsetails	Forbs	Deciduous shrubs	Evergreen shrubs
Digestibility	(%)	5–19	54–83	84–92	25–74	–	26–49	15–64
Crude protein[2]	(%)	2–16	<1– 8	18–46	2– 8	10–11	4– 8	6–10
Crude fat[3]	(%)	2– 3	<1– 7	–	<1– 1	5	4– 9	3–10
Crude carbohydrate[4]	(%)	50–57	25–95	–	40–58	62	53–67	53–71
Crude fiber	(%)	28–35	4–69	–	36–55	12	16–36	13–36
Ash	(%)	2–37	1– 3	–	1– 8	10	2– 5	2– 3
Cell contents	(%)	16–32	13–57	57–73	24–55	50	65–74	64
Hemicellulose	(%)	14–34	28–83	2–28	1–37	23	2	3
Cellulose	(%)	20–30	<1–12	5–11	22–34	20	12–21	20
Lignin	(%)	16–27	<1– 7	1–10	4–12	5	11–12	12
Nitrogen	(%)	0.3–2.6	0.2–1.3	2.8–7.3	0.4–1.3	1.6	0.6–1.0	1.0–1.6
Phosphorous	(%)	0.1–0.2	<0.1–0.2	0.4–0.9	<0.1–0.3	0.1–0.2	0.1	0.1–0.2
Potassium	(%)	0.4	<0.1–0.7	2.4–3.6	0.2–1.6	–	0.1–0.3	0.3–0.4
Sodium	(%)	<0.1	<0.1–0.3	<0.1–0.4	<0.1	<0.1	<0.1	<0.1
Calcium	(%)	0.2–0.5	<0.1–2.8	<0.1–0.1	0.1–1.3	0.3	0.5–1.0	0.2–0.7
Magnesium	(%)	0.2	<0.1–0.3	0.1	0.1–0.5	–	0.1	0.1

Copper	(ppm)	—	0.7–2.2	—	3.7–4.6	4.0
Molybdinum	(ppm)	—	0.2–0.3	—	0.2–0.3	0.3
Iron	(ppm)	—	71–1200	—	47–332	80
Manganese	(ppm)	—	16–100	—	78–121	118
Zinc	(ppm)	—	8–50	—	160–206	15

Notes

1. Based on: Aleksandrova and Andreyev, 1964; Bergerud, 1972, 1977; Boertje, 1981; Cameron, 1972; Hanson et al., 1975; Hyvarinen et al., 1977; Jacobsen and Skenneberg, 1975; Kelsall, 1968; Luick, 1977; Miller, 1976; Nieminen et al., 1980; Pakarinen and Vitt, 1974; Parker, 1975; Pegau, 1968; Person et al., 1980 a,b; Pulliainen, 1971; Rundel, 1978; Scotter, 1965, 1972; Scotter and Miltmore, 1973; Thomas and Kroeger, 1980, 1981; Trudell et al., 1980; White et al., 1975; Williams et al., 1978.
2. Nitrogen x 6.25
3. Ether extract
4. Nitrogen – free extract

61% of the diet in Keewatin (Fischer et al. 1977, Thompson et al. 1978) and 40% in the coastal areas of West Greenland (Holt 1980), compared with only 2 to 3% in the Canadian Arctic Islands (Thomas and Broughton 1978) and in Svalbard (Reimers 1980). Grasses, sedges, horsetails, dried forbs, shrubs, and moss are prominent in the diet of caribou wintering on tundra (table 2, Parker 1978, Reimers 1980, Shank et al. 1978).

A trade-off exists between total food biomass intake and the amount of time spent digesting the food. Because rumen volume is limited, the longer the time spent digesting food, the less the total food intake per unit time. Conversely, the shorter the digesting time (or rumen turnover time), the more food per unit time can be passed through the system. Rumen turnover time is defined as the length of time required to replace an amount of dry matter equal to rumen dry matter (Person et al. 1975). Physiological alteration of rumen turnover time ensures the optimum digestion of forage that varies both in quality and availability.

White and Trudell (1980) found that rumen turnover time changes in caribou from 11 hr on a summer diet to 23 hr on a lichen-dominated winter diet. Based on the relationships they presented, White and Trudell concluded that altering retention time for lichen from 11 hr to 23 hr would increase the digestibility by 62%, whereas if caribou retained their summer diet (forbs, for example) for 23 hr instead of 11 hr, the increase in digestibility would be only 5%. One can see that when lichen is scarce, and particularly when much energy is expended in acquiring it, there is an advantage in increasing rumen retention time, thus optimizing the energy derived from a given amount of food. Indeed, when fed a solely lichen diet of restricted quantity, caribou can have rumen turnover times as long as 3 to 5 days (Person et al. 1975).

Lichens are highly digestible by caribou (table 3) because the caribou are able both to increase their rumen retention time and to ferment lichens (White and Gau 1975). That high digestibility, combined with a high carbohydrate content (table 3), makes lichens extremely high in digestible energy when compared with other winter foods. Boertje (1981) concluded that energy was probably the compelling requirement for caribou in winter.

Winter-green vegetation, most of which has a greater nitrogen content but a lower digestibility than most lichens (table 3), may be essential to maintaining the caribou's condition during the winter. In the Canadian Arctic Islands Parker (1978) found that the amount of marrow fat (an index of condition) correlated positively with the amount of woody material in the diet and negatively with the amount of moss in the diet. Winter-green forage is especially important to pregnant cows in late winter; experimental studies have shown that a relatively small amount of supplementary protein and minerals in late pregnancy led to a nearly 50% increase in lichen intake (which leads to better maintenance of body weight) (Jacobsen and Skjenneberg 1975, Jacobsen et al. 1981). This was followed by an increase in milk yield and in the birth weight and subsequent growth of calves. In poorly nourished cows the mean birth date is delayed and the behavioural development of the calf retarded, with a negative effect on the cow-calf relationship (Espmark 1980).

Caribou on Svalbard have apparently adapted to a high moss content (26.8%) in their diet (Reimers 1980) by increasing their efficiency in digesting mosses (Trudell et al. 1980). They have a relatively larger cecum/colon complex, when the whole digestive system and body weight are taken into account, than caribou in Norway (Staaland et al. 1979). Also, that system relative to the whole digestive system is larger in winter than in summer in Svalbard but not in Norway. Staaland et al. noted that the relatively large cecum/colon complex could allow for the enhanced absorption of water, ions, and products of cecal fermentation when caribou consume foods of extremely low digestibility, such as mosses. A large cecum, therefore, could be particularly advantageous to Svalbard caribou in winter, maximizing the digestibility of forage and assisting in nitrogen recycling.

Although optimizing the supply of metabolizable energy is the key to the caribou's winter survival, mechanisms that enhance the supply of other components in the diet require some mention. The ability of caribou to optimize the limited nitrogen component of their winter diet (table 3), to maintain constant levels of blood glucose, to finely tune the process of water metabolism, to optimize the deposition and utilization of fat reserves, and to ensure an adequate supply of macroelements set caribou apart from other ruminants.

Nutritional Physiology

The absolute percentage of protein in the diet is not the only factor determining the nitrogen balance of wintering caribou. Protein is made up of amino acids, and only certain combinations of amino acids are useful to caribou. The biological value of protein relates to the percentage of digested protein that is retained for bodily functions. Protein is required in the summer diet for growth and for tissue replacement. However, growth in caribou ceases in winter, and as a result the protein requirement is greatly reduced. Although the digestible nitrogen requirement in winter is only 0.46 g/kg$^{0.75}$ · d (McEwan and Whitehead 1970), daily intake on a lichen-base diet is frequently less than that. The primary mechanism for nitrogen conservation is urea recycling. Most ruminants can recycle urea when the ingested protein decreases, but caribou are even more efficient than other ruminants because they can minimize urinary losses of urea (nitrogen), increase the availability of nitrogen for synthesis processes, and, unlike other ruminants, reduce nitrogen loss in feces (Nieminen 1980, Wales et al. 1975). White (1980) stressed the need for further study of the importance of raising the biological value of digested protein via microbial synthesis from recycled urea.

Glucose is important in the growth and development of the fetus. Therefore, during late winter and early spring the glucose demands of pregnant females are higher than at any other time except lactation, and, unlike lactation, coincide with a time of limited food resources. Because ruminants absorb very little dietary glucose, glucose must be either synthesized from dietary precursors or resynthesized from protein and lipid metabolites (Luick and White 1975, McEwan et al. 1976). The rate of glucose synthesis is, therefore, highly correlated with food intake (McEwan et al. 1976). Except in severely starved animals, blood glucose levels in caribou remain relatively constant throughout the winter (Nieminen 1980), and higher than in other ruminants (Luick and White 1975). During winter, when dietary precursors may not be available, caribou have the ability to recycle up to 70% of all glucose carbon, a higher level of recycling than that recorded for other ruminants (Luick and White 1975). Luick et al. (1973) speculate that the higher blood glucose levels, which are largely dependent upon recycling glucose carbon in late winter, may be a mechanism that enables caribou to survive periods of malnutrition in adverse winters. However, it is not known how significant that reserve would be in the event of malnutrition.

Whether it is in the form of free water or snow, water intake by wintering caribou is directly correlated to crude protein intake. It is normally reduced in winter (Cameron and Luick 1972). If large amounts of water are ingested, caribou can apparently dispose of the excess without affecting the nitrogen balance by excreting diluted urine (Syrjala et al. 1980). On a lichen diet caribou do not normally need to ingest any water, as the moisture content of the lichens is sufficient to meet body demands (Valtonen 1980). Cameron et al. (1975) found that the total body-water volume of lichen-fed animals increased from 70% to between 84 and 89% between December and May, while body weight remained relatively constant. That was primarily due to an expansion of rumen volume, which was related to the change in diet. Cameron et al. noted that a larger body-water volume favours survival in a cold environment. The high conductivity of water enhances the rapid and uniform distribution of body heat, and provides thermal buffering against fluctuations in body temperature associated with the consumption of snow and frozen forage. The energy required to raise the temperature of frozen ingesta to deep body temperature has been estimated at approximately 25% of the resting metabolic rate (Cameron 1972, Luick 1977), a considerable cost to wintering caribou.

The requirements of caribou for the macroelements (phosphorus, potassium, sodium, calcium, and magnesium) are largely undetermined. Lichens are poor in most of the macroelements (table 3); consequently caribou on a lichen-base diet likely have a negative phosphorus balance (Aleksandrova and Andreyev 1964, Boertje 1981) and may have a negative calcium balance (Aleksandrova and Andreyev 1964, Jacobsen and Skjenneberg 1975, Nieminen 1980). However, severe calcium deficiencies may not occur when food intake is not limited (Bjarghov et al. 1976, Boertje 1981). Although Boertje (1981) concluded that the macroelements other than phosphorus were not limited in winter, Nieminen (1980) found that concentrations of most of the serum macroelements declined through the winter. Caribou probably compensate for the poor mineral content of the winter forage by using their skeletons and muscles as a mineral bank (Hyvärinen et al. 1977, Nieminen 1980), which is then replenished by a short period of high mineral intake through selective grazing in summer (Staaland et al. 1980).

Less is known of the importance of microelements and vitamins in the winter diet of caribou (table 3). Caribou on a lichen-base diet may be deficient in copper and zinc (Scotter and Miltmore 1973), but can likely obtain an adequate amount of selenium (0.06 to 0.20 ppm in forage lichens) (Westermarck and Kurkela 1980). Lichens have a low vitamin C content (1.2 to 19.6 mg%) (Aleksandrova and Andreyev 1964) and a very low vitamin A content (<0.01 to 0.29 mg%) (Aleksandrova and Andreyev 1964, Scotter 1972). However, the vitamin content of other winter foods is probably higher (Scotter 1972), and rumen bacteria may be able to synthesize vitamins and therefore compensate.

Energetics

Caribou have a wide repertoire of physiological and behavioural mechanisms that contribute to lower energy requirements in the winter months. The basal metabolic rate for caribou has been reported to be 20 to 30% lower in winter than in summer (McEwan 1970, Segal 1962), which means that caribou require 20 to 30% less energy to carry on basic non-productive, non-activity body functions.

Caribou are able to further reduce energy requirements in winter by reducing heat loss, altering activity budgets, and minimizing energy expended in obtaining forage and travelling through snow. During much of the winter caribou must maintain an 80 C gradient between the ambient environment and their body core. Under normal activity and nutrition, Hart et al. (1961) determined that the critical temperature for a 9-month-old caribou was lower than −55 C in still air. That means that, at −55 C or warmer, caribou do not need to specifically produce heat (shiver) to maintain their body temperature. Segal (1980) noted that oxygen consumption and energy expenditure are not related to external temperatures down to −45 C in winter.

The tremendous insulating qualities of the caribou winter coat have been known for centuries. The insulation of the caribou winter coat is about 15.9 kJ/m^2 · hr · C (Hammel 1955, Moote 1955), therefore, the well-insulated coat can greatly reduce heat loss due to conduction and convection. Conductive heat loss while sniffing for lichens under the snow is further reduced by the caribou's densely haired muzzle. Furthermore, caribou have the ability to control evaporative heat loss by cooling exhaled air from 38.4 C (winter body temperature) to 14 C or less before expelling the air (Hammel et al. 1962, Langman and Taylor 1980). Similarly, inhaled air is heated in the nasal passages before entering the lungs. Subcutaneous fat reserves, however, do not appear to play a role in reducing heat loss (Ringberg et al. 1980).

Although low temperatures do not cause discomfort under normal conditions, wind in combination with low temperatures can result in sufficient heat loss to stress caribou. Earlier we outlined a number of behavioural modifications exhibited by caribou when wind speed increased at low ambient temperatures. As wind chill increases, lying decreases and feeding increases, but at high wind chills that pattern is reversed. Roby (1978) hypothesized that, at very high wind chills, lying is more effective than continued feeding in reducing heat loss.

As previously discussed, caribou alter their activity budgets in response to deep snow in late winter. This results in reduced energy expenditure. Data from the central Yukon showed significantly reduced daily movement rates of radio-collared caribou in a deep-snow winter, as compared with normal winters (unpublished observations of the authors). We noted that the percentage of time spent feeding, walking, and trotting decreases from early to late winter, while the percentage of time spent lying increases. Boertje (1981) noted that one would expect that the high energy costs of pregnancy in late winter would result in increased energy expenditures from mid- to late winter. However, based on estimates of the energy costs of activities (table 4), Boertje (1981) calculated that the energy requirements of an adult female caribou in central Alaska were 637 kJ/ $kg^{0.75} \cdot d$ both in early and late winter. Those estimates are low, compared with the high energy requirements during calving and post-calving (926), summer (1163), and rut (939).

We have already discussed the ability of caribou to detect lichens under

Table 4 Estimated energy costs of activities of adult female caribou in winter, adapted from Boertje 1981

Activity	Energetic cost	Reference
Standing	505kJ/Kg$^{0.75} \cdot$d	Makarova and Segal, 1958 McEwan, 1970
Lying	459kJ/Kg$^{0.75} \cdot$d	Brody, 1945 White and Yousef, 1978
Lying to standing and return	0.10kJ/Kg	Blaxter, 1962 — cattle
Rumination	0.13kJ/Kg·h	Osuji, 1974 — sheep
Foraging/eating	1.88kJ/Kg·h	Osuji, 1974 — sheep
Cratering	0.94kJ/Kg·h	Boertje, 1981[1]
Walking in snow	2.64kJ/Kg·km	Boertje, 1981[1]
Climbing +9% grade -9% grade	 31.0kJ/Kg·km (vertical) -5.7kJ/Kg·km (vertical)	 White and Yousef, 1978 White and Yousef, 1978
Trotting/galloping	35kJ/Kg·h	Boertje, 1981[1]
Pregnancy	369 MJ/Calf	Boertje, 1981[1]
Fattening	-365 MJ/winter	Boertje, 1981[1], energy gain from catabolism

Note
1. Theoretical values; no empirical data.

the snow, thereby economizing on the energy required during the cratering process. Segal (1962) indicated that the energy involved in cratering for lichen under the snow is equal to the foraging energy expenditure in the summer, because greater energy is required for the high-summer movement rates. Caribou are known to further economize while cratering in deep snow by trenching and communal digging. After a crater is dug in deep snow, caribou will either trench or enlarge the feeding crater, rather than dig a new crater. It is not uncommon to observe a band of caribou in a communal crater feeding side-by-side on a wide front. Both trenching and communal cratering would reduce energy expenditure per unit food intake under deep snow conditions.

Like other northern cervids, caribou essentially cease to grow during the winter (McEwan 1975). McEwan and Whitehead (1970) noted that caribou, even those penned and fed *ad libidum,* reduce their voluntary food intake in winter. That reduction translates into a 35 to 45% lower caloric intake in winter compared with summer.

White and Trudell (1980) noted a number of intrinsic and extrinsic factors influencing food intake in caribou. In winter, food availability (affected by forage biomass and snow parameters), food quality, harassment, and social interactions combine to extrinsically determine food intake. Caribou intrinsically adjust food intake to balance the energy budget. However, this is usually not possible by mid- or late winter, and caribou must draw on fat reserves. Young and adult female caribou lose 6 to 11% of their autumn weight during winter (Dauphine 1976), primarily by metabolizing fat reserves. Boertje (1981) estimated that adult female caribou gain 365 MJ of energy from those fat reserves, or about 9% of the daily energy requirements in mid- and late winter. Greater weight losses occur in males (Dauphine 1976) and under severe winter conditions. Winter, therefore, is a time when caribou must minimize weight loss.

Estimates of the winter lichen intake of caribou vary widely (White et al. 1981). Recent studies using the fallout radionuclide Cesium-137 estimate that free-ranging caribou consume an average of 4.9 kg dry-weight of lichens per day (Hanson et al. 1975, Holleman et al. 1979, 1980). That intake level translates into a metabolizable energy intake (the energy available to satisfy basic growth, fattening, and activity requirements) of 1.4 MJ/$kg^{0.75}$ · d, an intake that is greater than the normal requirements for early winter (Holleman et al. 1980). Therefore caribou may be able to deposit fat in early winter, although the evidence is equivocal (White et al. 1981). Fat deposit would be of particular importance to bulls, which may lose 30% of their autumn weight during the rut (Dauphine 1976).

Summary

Caribou exhibit many behavioural and physiological adaptations to ensure their survival during the arctic and subarctic winter. Each ecotype and geographic population, however, has further specialized to cope with its own unique environmental situation. Specific adaptations to cold and snow include a superb insulating coat, large, supporting hooves, and an uncanny ability to detect lichens under snow. The winter activity pattern of caribou reflects the quality and quantity of the snowpack and is a constant effort to balance energy expenditure with metabolizable energy intake. Energy is probably the compelling requirement for caribou in winter, and energy-rich lichens form the major component of their winter diet. Lichens, however, are poor in nutrients; therefore other foods, such as winter-green vegetation, may be essential to maintaining the condition of caribou during the winter. The availability of lichens varies with geographic location and snowpack, and some caribou populations have adapted to utilize less energy-rich forage, such as moss. Alteration of rumen turnover time appears to be a key mechanism in ensuring the optimum digestion of forage that varies both in quality and availability. Caribou are highly specialized in their ability to compensate, at least partially, for dietary deficiencies by recycling nitrogen, by resynthesizing and recycling glucose, and by storing and cycling many of the macroelements. The total energy balance of caribou is enhanced by the species' ability to reduce heat loss, alter activity patterns, and minimize energy expended in obtaining forage and travelling through snow.

Acknowledgments

We would like to thank the many people who have worked with us in our studies on the Porcupine Caribou Herd, and also our colleagues in Alaska, Alberta, and the Northwest Territories, who have contributed ideas and critical discussions. This paper benefited from comments on an earlier draft by D. R. Flook, F. E. Geddes, D. R. Klein, D. C. Thomas, and R. G. White.

Don Gill devoted his life to exploring the ecological relationships in the North and to introducing others to their mystery and importance. We hope that this review will help people further understand the relationships of one animal to what is perhaps the most important season in the North — winter.

References

Aleksandrova, V. D., and V. N. Andreyev. 1964. Forage plants of reindeer. pp. 10-52. In V. D. Aleksandrova, V. N. Andreyev, T. V. Vakhtina, R. A. Dydina, G. I. Karev, V. V. Petrovski, and V. F. Shamurin, eds. Forage characteristics of the plants in the far north: Vegetation of the extreme north of the U.S.S.R. and its use. Issue 5. Soviet Academy of Sciences, Moscow. [Translated from Russian, Can. Wildl. Serv., Ottawa.]

Baskin, L. M. 1970. Reindeer: Their ecology and behavior. A. N. Seventsov Institute of Evolutionary Morphology and Ecology of Animals. Nauka Pub. House, Moscow. [Translated from Russian, Can. Wildl. Serv., Ottawa.]

Bergerud, A. T. 1972. Food habits of Newfoundland caribou. J. Wildl. Manage. 36: 913-923.

———. 1974a. Relative abundance of food in winter for Newfoundland caribou. Oikos 25: 379-387.

———. 1974b. The role of the environment in the aggregation, movement and disturbance behaviour of caribou. pp. 552-584. In V. Geist and F. Walther, eds. The behaviour of ungulates and its relation to environment. Int. Union for Conservation of Nature and Natural Resources Pub., New Ser., No. 24.

———. 1977. Diets for caribou. pp. 243-294. In M. Recheigl, Jr., ed. CRC handbook series in nutrition and food. Section G: Diets, culture, media, and food supplements. Vol. 1. CRC Press, Cleveland, Ohio.

———, and M. J. Nolan. 1970. Food habits of hand-reared caribou *Rangifer tarandus* L. in Newfoundland. Oikos 21: 348-350.

Bjarghov, R. S., P. Fjellheim, K. Hove, E. Jacobsen, S. Skjenneberg, and K. Try. 1976. Nutritional effects on serum enzymes and other blood constituents in reindeer calves (*Rangifer tarandus tarandus*). Comp. Biochem. and Physiol. 55A: 187-193.

Blaxter, K. L. 1962. The energy metabolism of ruminants. C. C. Thomas, Springfield, Illinois.

Boertje, R. D. 1981. Nutritional ecology of the Denali caribou herd. M.Sc. thesis. Univ. Alaska, Fairbanks.

Brody, S. 1945. Bioenergetics and growth. Reinhold Pub. Co., New York.

Cameron, R. D. 1972. Water metabolism by reindeer (*Rangifer tarandus*). Ph.D. thesis. Univ. Alaska, Fairbanks.

———, and J. R. Luick. 1972. Seasonal changes in total body water, extracellular fluid and blood volume in grazing reindeer. Can. J. Zool. 50: 107-116.

———, R. G. White, and J. R. Luick. 1975. The accumulation of water in reindeer during winter. pp. 374-378. In Luick et al. 1975.

Dauphine, T. C., Jr. 1976. Biology of the Kaminuriak population of barren-ground caribou. Part 4: Growth, reproduction and energy reserves. Can. Wildl. Serv., Rep. Ser. No. 38.

Edwards, R. Y., and R. W. Ritcey. 1959. Migration of caribou in a mountainous area in Wells Gray Park, British Columbia. Can. Field-Natur. 73: 21-25.

———, and R. W. Ritcey. 1960. Foods of caribou in Wells Gray Park, British Columbia. Can. Field-Natur. 74: 3-7.

Espmark, Y. 1980. Effects of maternal pre-partum under-nutrition on early mother-calf relationships. pp. 485–496. *In* Reimers et al. 1980.

Fischer, C. A., and E. A. Duncan. 1976. Ecological studies of caribou and muskoxen in the Arctic Archipelago and northern Keewatin. Polar Gas Project, Toronto.

———, D. C. Thompson, R. L. Wooley, and P. S. Thompson. 1977. Ecological studies of caribou on the Boothia Peninsula and in the District of Keewatin, N.W.T., 1976. Polar Gas Project, Toronto.

Formozov, A. N. 1946. Snow cover as an integral factor of the environment and its importance in the ecology of mammals and birds. Materials for Fauna and Flora of the U.S.S.R., New Ser., Zool. 5. [Translated from Russian, Boreal Institute for Northern Studies, Univ. Alberta, Occ. Pub. No. 1.]

Freddy, D. J., and A. W. Erickson. 1975. Status of the Selkirk Mountain caribou. pp. 221–227. *In* Luick et al. 1975.

Gaare, E. 1968. A preliminary report on winter nutrition of wild reindeer in the Southern Scandes, Norway. Symp. Zool. Soc. London 21: 109–115.

———, and T. Skogland. 1975. Wild reindeer food habits and range use at Hardangervidda. pp. 195–205. *In* F. E. Wielgolaski, ed. Fennoscandian tundra ecosystems. Springer-Verlag, New York.

———, B. R. Thomson, and O. Kjos-Hanssen. 1975. Reindeer activity on Hardangervidda. pp. 206–215. *In* F. E. Wielgolaski, ed. Fennoscandian tundra ecosystems. Springer-Verlag, New York.

Hammel, H. T. 1955. Thermal properties of fur. Amer. J. Physiol. 182: 369–376.

———, T. R. Houpt, K. L. Anderson, and S. Skjenneberg. 1962. Thermal and metabolic measurements on a reindeer at rest and exercise. Arctic Aeromedical Laboratory, Tech. Doc. Rep. AAL-TDR-61-54.

Hanson, W. C., F. W. Whicker, and J. F. Lipscomb. 1975. Lichen forage ingestion rates of free-roaming caribou estimated with fallout cesium-137. pp. 71–79. *In* Luick et al. 1975.

Hart, J. S., O. Heroux, W. H. Cottle, and C. A. Mills. 1961. The influence of climate on metabolic and thermal responses of infant caribou. Can. J. Zool. 39: 845–856.

Helle, T. 1981. Habitat and food selection of the wild forest reindeer (*Rangifer tarandus fennicus* Lönn.) in Kuhmo, eastern Finland, with special reference to snow characteristics. Res. Inst. Northern Finland, Univ. Oulu A2: 1–33.

Henshaw, J. 1968. The activities of the wintering caribou in north-western Alaska in relation to weather and snow conditions. Int. J. Biometeorol. 12: 21–27.

Holleman, D. F., J. R. Luick, and R. G. White. 1979. Lichen intake estimates for reindeer and caribou during winter. J. Wildl. Manage. 43: 192–201.

———, R. G. White, J. R. Luick, and R. O. Stephenson. 1980. Energy flow through the lichen-caribou-wolf chain during winter in northern Alaska. pp. 202–206. *In* Reimers et al. 1980.

Holt, S. 1980. Vegetation patterns and effects of grazing on caribou ranges in the Sondre Stromfjord area, West Greenland. pp. 57–63. *In* Reimers et al. 1980.

Hyvärinen, H., T. Helle, M. Nieminen, P. Väyrynen, and R. Väyrynen. 1977. The influence of nutrition and seasonal conditions on mineral status in the reindeer. Can. J. Zool. 55: 648–655.

Jacobsen, E., K. Hove, R. S. Bjarghov, and S. Skjenneberg. 1981. Supplementary feeding of female reindeer on a lichen diet during the last part of pregnancy. Acta Agr. Scandinavica 31: 81–86.

———, and S. Skjenneberg. 1975. Some results from feeding experiments with reindeer. pp. 95–107. In Luick et al. 1975.

Kelsall, J. P. 1968. The migratory barren-ground caribou of Canada. Can. Dep. Indian Affairs and Northern Development, Can. Wildl. Serv., Monogr. No. 3.

Klein, D. R. 1970. Tundra ranges north of the boreal forest. J. Range Manage. 23: 8–14.

Kowalski, K. 1980. Origin of mammals of the arctic tundra. Folia Quaternaria 51: 3–16.

Langman, V. A., and C. R. Taylor. 1980. Nasal heat exchange in a northern ungulate, the reindeer (Rangifer tarandus). [Abstract.] p. 377. In Reimers et al. 1980.

LaPerriere, A. J., and P. C. Lent. 1977. Caribou feeding sites in relation to snow characteristics in northeastern Alaska. Arctic 30: 101–108.

Lent, P. C., and D. Knutson. 1971. Muskox and snow cover on Nunivak Island, Alaska. pp. 50–62. In A. O. Haugen, ed. Proc. Snow and Ice in Relation to Wildlife and Recreation Symp. Iowa State Univ., Ames, Iowa.

Luick, J. R. 1977. Diets for freely grazing reindeer. pp. 267–278. In M. Recheigl, Jr., ed. CRC handbook series in nutrition and food. Section G: Diets, culture, media, and food supplements. Vol. 1. CRC Press, Cleveland, Ohio.

———, P. C. Lent, D. R. Klein, and R. G. White, eds. 1975. Proceedings of the first international reindeer and caribou symposium, 9–11 August 1972, University of Alaska, Fairbanks, Alaska. Biol. Pap. Univ. Alaska, Special Rep. No. 1.

———, S. J. Person, R. D. Cameron, and R. G. White. 1973. Seasonal variations in glucose metabolism of reindeer (Rangifer tarandus) estimated with carbon-14 glucose and tritiated glucose. Brit. J. Nutrition 29: 245–259.

———, and R. G. White. 1975. Glucose metabolism in female reindeer. pp. 379–386. In Luick et al. 1975.

Makarova, A. R., and A. N. Segal. 1958. Physiological characteristics of the winter grazing of reindeer. Experimental Research on the Regulation of Physiological Function 4: 29–35. [Translated from Russian, Can. Wildl. Serv., Ottawa.]

McEwan, E. H. 1970. Energy metabolism of barren-ground caribou. Can. J. Zool. 48: 391–392.

———. 1975. [The adaptive significance of the growth patterns in cervid compared with other ungulate species.] Zool. Zh. 54: 1221–1231. [In Russian.]

———, and P. E. Whitehead. 1970. Seasonal changes in the energy and nitrogen intake in reindeer and caribou. Can. J. Zool. 48: 905–913.

———, P. Whitehead, R. G. White, and J. O. Anvik. 1976. Effects of digestible energy intake on glucose synthesis in reindeer and caribou. Can. J. Zool. 54: 737–751.

Michurin, L. N., and L. V. Makhaeva. 1962. Feeding of wild reindeer on Taimyr. Zool. Zh. 41: 1883–1888. [Translated from Russian, Can. Wildl. Serv., Ottawa.]

———, and T. V. Vakhtina. 1968. Winter feeding of wild reindeer (Rangifer tarandus L.) in arctic tundras of Taimyr. Zool. Zh. 47: 477–479. [Translated from Russian, Can. Wildl. Serv., Ottawa.]

Miller, D. R. 1974. Seasonal changes in the feeding behaviour of barren-ground caribou on the taiga winter range. pp. 744–755. *In* V. Geist and F. Walther, eds. The behaviour of ungulates and its relation to management. Int. Union for Conservation of Nature and Natural Resources Pub., New Ser., No. 24.

———. 1976. Biology of the Kaminuriak population of barren-ground caribou. Part 3: Taiga winter range relationships and diet. Can. Wildl. Serv., Rep. Ser. No. 36.

Moote, I. 1955. Insulation of hair. Textile Res. J. 25: 832–837.

Nasimovich, A. A. 1955. The role of the regime of snow cover in the life of ungulates in the U.S.S.R. Soviet Academy of Sciences, Moscow. [Translated from Russian, Can. Wildl. Serv., Ottawa.]

Nieminen, M. 1980. The composition of reindeer blood in respect to age, season, calving and nutrition. Acta Univ. Ouluensis, Series D, Medica No. 54, Pharmacol. et Physiol. No. 11.

———, S. Kellokumpu, P. Väyrynen, and H. Hyvärinen. 1980. Rumen function of the reindeer. pp. 213–223. *In* Reimers et al. 1980.

Osuji, P. O. 1974. The physiology of eating and the energy expenditure of the ruminant at pasture. J. Range Manage. 27: 437–443.

Pakarinen, P., and D. H. Vitt. 1974. The major organic components and caloric contents of high arctic bryophytes. Can. J. Bot. 52: 1151–1161.

Parker, G. R. 1975. An investigation of caribou range on Southampton Island, Northwest Territories. Can. Wildl. Serv., Rep. Ser. No. 33.

———. 1978. The diets of muskoxen and Peary caribou on some islands in the Canadian High Arctic. Can. Wildl. Serv., Occ. Pap. No. 35.

———. 1981. Physical and reproductive characteristics of an expanding woodland caribou population (*Rangifer tarandus caribou*) in northern Labrador. Can. J. Zool. 59: 1929–1940.

———, D. C. Thomas, E. Broughton, and D. R. Gray. 1975. Crashes of muskox and Peary caribou populations in 1973–74 on the Parry Islands, Arctic Canada. Can. Wildl. Service, Progress Notes No. 56.

Pegau, R. E. 1968. Reindeer range appraisal in Alaska. M.Sc. thesis. Univ. Alaska, Fairbanks.

Person, S. J., R. E. Pegau, R. G. White, and J. R. Luick. 1980*a*. *In vitro* and nylon bag digestibilities of reindeer and caribou forages. J. Wildl. Manage. 44: 613–622.

———, R. G. White, and J. R. Luick. 1975. *In vitro* digestibility of forages utilized by *Rangifer tarandus*. pp. 251–256. *In* Luick et al. 1975.

———, R. G. White, and J. R. Luick. 1980*b*. Determination of nutritive value of reindeer-caribou range. pp. 224–239. *In* Reimers et al. 1980.

Pruitt, W. O., Jr. 1959. Snow as a factor in the winter ecology of the barren-ground caribou (*Rangifer arcticus*). Arctic 12: 158–179.

———. 1960. Behavior of the barren-ground caribou. Biol. Pap. Univ. Alaska, No. 3.

———. 1979. A numerical "snow index" for reindeer (*Rangifer tarandus*) winter ecology (Mammalia, Cervidae). Ann. Zool. Fennici 16: 271–280.

———. 1981. Application of the Värriö snow index to over-wintering North American barren-ground caribou (*Rangifer tarandus arcticus*). Can. Field-Natur. 95: 363–365.

Pulliainen, E. 1971. Nutritive values of some lichens used as food by reindeer in northeastern Lapland. Ann. Zool. Fennici 8: 385-389.
Reimers, E. 1980. Activity pattern; the major determinant for growth and fattening in *Rangifer*. pp. 466-474. *In* Reimers et al. 1980.
―――, E. Gaare, and S. Skjenneberg, eds. 1980. Proceedings of the Second International Reindeer/Caribou Symposium, 17-21 September 1979, Roros, Norway. Direktoratet for Vilt og Ferskvannsfisk, Trondheim.
Ringberg, T., K. Nilssen, and E. Strom. 1980. Do Svalbard reindeer use their subcutaneous fat as insulation? pp. 392-395. *In* Reimers et al. 1980.
Roby, D. D. 1978. Behavioral patterns of barren-ground caribou of the Central Arctic Herd adjacent to the Trans-Alaska oil pipeline. M.Sc. thesis. Univ. Alaska, Fairbanks.
―――. 1980. Winter activity of caribou on two arctic ranges. pp. 537-544. *In* Reimers et al. 1980.
Rundel, P. W. 1978. The ecological role of secondary lichen substances. Biochem. Syst. and Ecol. 6: 157-170.
Sablina, T. B. 1960. The feeding habits and ecologico-morphologic characteristics of the digestive system of the reindeer of Karelia. Trans. Seventsov Institute of the Morphology of Animals 32: 215-258. [Translated from Russian, Can. Wildl. Serv., Ottawa.]
Scotter, G. W. 1965. Chemical composition of forage lichens from northern Saskatchewan as related to use by barren-ground caribou. Can. J. Plant Sci. 45: 246-250.
―――. 1967. The winter diet of barren-ground caribou in northern Canada. Can. Field-Natur. 81: 33-39.
―――. 1972. Chemical composition of forage plants from the Reindeer Preserve, Northwest Territories. Arctic 25: 21-27.
―――, and J. E. Miltmore. 1973. Mineral content of forage plants from the Reindeer Preserve, Northwest Territories. Can. J. Plant Sci. 53: 263-268.
Segal, A. N. 1962. Reindeer in the Karelian, A.S.S.R. Soviet Academy of Sciences, Moscow. [Translated from Russian, Can. Wildl. Serv., Ottawa.]
―――. 1980. [Thermoregulation in the reindeer (*Rangifer tarandus*).] Zool. Zh. 59: 1718-1725. [In Russian.]
Shank. C. C., P. F. Wilkinson, and D. F. Penner. 1978. Diets of Peary caribou, Banks Island, N.W.T. Arctic 31: 125-132.
Skogland, T. 1978. Characteristics of the snow cover and its relationship to wild mountain reindeer (*Rangifer tarandus tarandus* L.) feeding strategies. Arctic and Alpine Res. 10: 569-580.
Skoog, R. O. 1968. Ecology of the caribou (*Rangifer tarandus granti*) in Alaska. Ph.D. thesis. Univ. California, Berkeley.
Staaland, H., E. Jacobsen, and R. G. White. 1979. Comparison of the digestive tract in Svalbard and Norwegian reindeer. Arctic and Alpine Res. 11: 457-466.
―――, R. G. White, J. R. Luick, and D. F. Holleman. 1980. Dietary influences on sodium and potassium metabolism of reindeer. Can. J. Zool. 58: 1728-1734.
Stardom, R. R. P. 1975. Woodland caribou and snow conditions in southeast Manitoba. pp. 324-334. *In* Luick et al. 1975.
Syrjala, L., J. Salonen, and M. Valtonen. 1980. Water and energy intake and nitrogen utilization in reindeer. pp. 252-261. *In* Reimers et al. 1980.

Telfer, E. S., and J. P. Kelsall. 1979. Studies of morphological parameters affecting ungulate locomotion in snow. Can. J. Zool. 57: 2153-2159.

Thing, H. 1977. Behavior, mechanics and energetics associated with winter cratering by caribou in northwestern Alaska. Biol. Pap. Univ. Alaska, No. 18.

Thomas, D. C., and E. Broughton. 1978. Status of three Canadian caribou populations north of 70° in winter 1977. Can. Wildl. Serv. Progress Notes No. 85.

———, and P. Kroeger. 1980. In vitro digestibilities of plants in rumen fluids of Peary caribou. Arctic 33: 757-767.

———, and P. Kroeger. 1981. Digestibility of plants in ruminal fluids of barren-ground caribou. Arctic 34: 321-324.

———, R. H. Russell, E. Broughton, E. J. Edmunds, and A. Gunn. 1977. Further studies of two populations of Peary caribou in the Canadian Arctic. Can. Wildl. Serv. Progress Notes No. 80.

———, R. H. Russell, E. Broughton, and P. Madore. 1976. Investigations of Peary caribou populations on Canadian Arctic Islands, March-April 1975. Can. Wildl. Serv. Progress Notes No. 64.

Thompson, D. C., G. H. Klassen, and C. A. Fischer. 1978. Ecological studies of caribou in the southern District of Keewatin, 1977. Polar Gas Project, Toronto.

Thomson, B. R. 1973. Wild reindeer activity, Hardangervidda: 1971. Rep. Grazing Project of the Norwegian IBP Committee, Trondheim.

———. 1977. The behaviour of wild reindeer in Norway. Ph.D. thesis. Univ. Edinburgh, Edinburgh.

Trudell, J., R. G. White, E. Jacobsen, H. Staaland, K. Ekern, K. Kildemo, and E. Gaare. 1980. Comparison of some factors affecting the in vitro digestibility estimate of reindeer forages. pp. 262-273. In Reimers et al. 1980.

Valtonen, M. H. 1980. Effects of dietary nitrogen and sodium chloride content on water intake and urine excretion in reindeer. pp. 274-277. In Reimers et al. 1980.

Vibe, C. 1967. Arctic animals in relation to climatic fluctuations. Medd. om Grønl., Bd. 170, No. 5: 1-227.

Wales, R. A., L. P. Milligan, and E. H. McEwan. 1975. Urea recycling in caribou, cattle, and sheep. pp. 297-307. In Luick et al. 1975.

Westermarck, H., and P. Kurkela. 1980. Selenium contents in lichen in Lapland and south Finland and its effect on the selenium values in reindeer. pp. 278-285. In Reimers et al. 1980.

White, R. G. 1980. Nutrient acquisition and utilization in Arctic herbivores. pp. 13-50. In L. S. Underwood, L. L. Tieszen, A. B. Callahan, and G. E. Folk, eds. Comparative mechanisms of cold adaptation. Academic Press, New York.

———, F. L. Bunnell, E. Gaare, T. Skogland, and B. Hubert. 1981. Ungulates on arctic ranges. pp. 397-483. In L. C. Bliss, J. B. Cragg, D. W. Heal, and J. J. Moore, eds. Tundra ecosystems: A comparative analysis. Cambridge Univ. Press, Cambridge.

———, and A. M. Gau. 1975. Volatile fatty acid (VFA) production in the rumen and cecum of reindeer. pp. 284-289. In Luick et al. 1975.

———, B. R. Thomson, T. Skogland, S. J. Person, D. E. Russell, D. F. Holleman, and J. R. Luick. 1975. Ecology of caribou at Prudhoe Bay, Alaska. pp. 151-201. In J. Brown, ed. Ecological investigations of the tundra biome in the

Prudhoe Bay region, Alaska. Biol. Pap. Univ. Alaska, Special Rep. No. 2.

———, and J. Trudell. 1980. Patterns of herbivory and nutrient intake of reindeer grazing tundra vegetation. pp. 180–195. *In* Reimers et al. 1980.

———, and M. K. Yousef. 1978. Energy expenditure in reindeer walking on roads and on tundra. Can. J. Zool. 56: 215–223.

Williams, M. E., E. D. Rudolph, E. A. Schofield, and D. C. Prasker. 1978. The role of lichens in the structure, productivity, and mineral cycling of the wet coastal Alaskan tundra. pp. 185–206. *In* L. L. Tieszen, ed. Vegetation and production ecology of an Alaskan arctic tundra. Springer-Verlag, New York.

Circumpolar Distribution
and Habitat Requirements
of Moose (*Alces alces*)

E. S. Telfer

Introduction

Moose (*Alces alces*) are the largest surviving land mammals in the circumpolar boreal forests and the largest living representatives of the deer family (*Cervidae*). The weights of adult bulls in Alberta averaged 477 kg in early winter when adult females averaged 402 kg (Canadian Wildlife Service unpublished data). The average weight of moose of all ages and sexes killed in a herd reduction at Elk Island National Park was 303 kg. The seven contemporary subspecies of *Alces alces* (fig. 1) are said to vary in size as well as in skeletal features (Peterson 1955), although few comparable data on weights within sex and age classes are available. Antler sizes unquestionably vary among subspecies (Geist 1971, Nesbitt and Parker 1977), and body weights also may well vary.

Although moose from all parts of their vast range look surprisingly similar, there are noticeable differences among races, especially among the European and west Siberian populations on the one hand, and those of eastern Siberia and America on the other. While all have dark brown to blackish winter coats shading to grey on legs and flanks, and a lighter summer coat (Van Wormer 1972), American and east Asian moose have a "saddle" of lighter-toned hair and darker legs. The color of the face is sexually dimorphic in American and east Asian moose, with cows exhibiting a light brown shade and bulls a black coloration. Both sexes of European moose have black faces (Geist 1983). The "bell" of American and east Asian moose also differs from that of the European moose by being larger.

146 E. S. TELFER

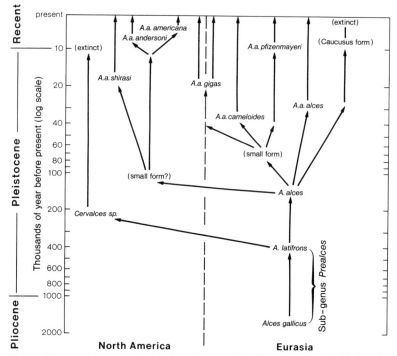

Fig. 1. Schematic diagram of radiation of genus *Alces* (based on Kurtén 1968 and Peterson 1955).

Antlers of European populations have three tines on each side. Tines often divide into two points, supported by a flattened beam but without broad palms. In contrast, American and east Asian populations have four-tined antlers. Their tines also divide into additional points. The back, or top, two customarily branch into two or four points and coalesce to form a broad supporting palm.

As large, evolutionarily advanced, and long-lived mammals, moose are capable of responding to their environment with a wide repertoire of behaviours. It is thus not surprising that they take advantage of the landscape elements in their vast circumpolar range in a variety of ways. It is the purpose of this paper to describe the habitat aspect of mooses' utilization of their environment. Habitat considerations will be integrated with other environmental parameters so that certain hypotheses can be put forth to explain the location of the boundaries to the range of moose as a species and the local distribution of animals within that range.

I use the term "habitat" to designate the subset of an animal's environment that comprises the physical structure on which, or within which, the animal lives and acts. Habitat includes those places where the animal rests, sleeps, feeds, and travels. It does not include food or feeding habits, social behaviour, predators, diseases, competitors, disturbance by humans, or climate. These are other extremely important environmental factors that determine the area a species is able to occupy and the extent to which it can occupy the ecological niche within which it is able to make its living. In the following discussion of the limitations to moose distribution some consideration will be given to all the factors that seem relevant.

Distribution

Pleistocene

According to Kurtén (1968), moose-like mammals first appear in the fossil record during the Villafranchian Age (late Pliocene and early Pleistocene geologic epochs), in the form of *Alces gallicus* Azzaroli, whose fossil remains have been found in England, France, and Germany. *A. gallicus* had wide-spreading, spoon-shaped antlers. In the Villafranchian Age the climate was cooling and the glacial advances were beginning. The succeeding ice ages saw the rise of another moose form with narrower and more palmate antlers, which had been classed a separate species, *A. latifrons* Johnson. These early moose are sometimes placed in a subgenus called *Prealces*. The modern moose, *Alces alces* Linné, first appeared in European fossil beds in the 3-Riss Climatic Phase, approximately between 100,000 and 200,000 years ago. Contemporary with *A. latifrons* and *A. alces* was a related genus in North America, *Cervalces*. This genus appeared at the end of the Yarmouth Interglacial (Flint 1971) and survived as the only moose form in North America until mid-Wisconsin times, when *A. alces* arrived. *Cervalces* became extinct at the end of the Wisconsin glaciation, as did many other North American mammal groups (Flint 1971).

In Europe, *A. alces* appeared in the late Pleistocene and did not overlap with *A. latifrons*. Kurtén (1968) believed *A. latifrons* to be ancestral to *A. alces*; however, fossils that would demonstrate the transition have not yet been found. Similarly, *A. gallicus* was probably ancestral to *A. latifrons*, although intermediate fossils have not been found. The ancestry of *Cervalces* is uncertain.

Geist (1971) has hypothesized that the distribution of ungulates into glaciated regions as the glaciers melted resulted in the development of new forms that were larger, had larger horns or antlers, and were more active socially. Moose now existing near hypothetical glacial refugia (*A. a. cameloides*, *A. a. shirasi*) are considered to be small forms (Peterson 1955), while other subspecies in glaciated regions (*A. a. pfizenmayeri*, *A. a. andersoni*, *A. a. americana*) are larger. The largest form of all, by antler measurements, is *A. a. gigas*, which is found in Alaska and eastern Siberia. It is thought to be descended from moose that survived the Wisconsin glaciation on the Bering land bridge (Beringia) and adjacent areas, and were cut off from other populations by cordilleran ice sheets (Peterson 1955). However, a substantial portion of the *A. a. gigas* population lives on formerly glaciated land. *A. a. alces* occupies glaciated regions but is reputed to be smaller, at least in antler size, than other races in glaciated habitats. Another factor affecting this subspecies is long-term, and often intensive, trophy hunting that may have amounted to selective pressure in favour of small antlers. *A. a. alces* are reported to vary in size in Scandinavia; northern specimens are very large (Markgren pers. com.). Most mammal genera periodically produced large forms during the Pleistocene epoch (Kurtén 1968). This may have been due either to their invasions of productive deglaciated habitats according to Geist's dispersal hypothesis or as an expression of Bergman's Rule, which states that forms of a species are usually larger in cold climates because the lower surface-mass ratio enables them to conserve heat. With cold-climate mammals like moose, the evolution of smaller, heat-losing forms might permit higher reproduction and survival during warmer integlacial periods and encourage them to invade warmer regions. A hypothetical scheme of the development of the genus *Alces* is presented in fig. 1.

The one modern species of moose, *Alces alces* appears to have evolved in the general region of Siberia. During the late Pleistocene, it probably dispersed from there into Europe and, by way of the Bering land bridge, into North America (Peterson 1955). Moose appear always to have been associated with cold regions. During periods of glacial maxima much of the present-day moose range was covered by ice, tundra, or cold steppe. However, changes in sea level and southerly advances of cold-temperate climatic conditions toward the equator created suitable climatic and vegetative conditions for moose in regions from which they are now absent (fig. 2). Whereas the modern distribution of moose is largely north of 50°N (with the important exception of eastern North America), during the Pleistocene the southern limit of boreal vegetation types was 40°N in Europe, 35°N in eastern North America (Flint 1971), and down to 30°N in east Asia (Frenzel 1968). The presence of moose in the southerly Pleis-

Circumpolar Distribution and Habitat Requirements of Moose   149

tocene extension of the boreal forest in North America has been substantiated by fossil evidence (Peterson 1955). It may be hypothesized that moose occupied suitable Pleistocene vegetation zones on other continents as well.

Following the retreat of the ice sheets of the last glaciation, moose dispersed into the formerly glaciated regions (Geist 1971, Kelsall and Telfer 1974). Historic distributions largely coincide with regions that were either formerly ice-covered, or occupied by tundra or cold, dry steppe (as in central Asia, the Ukraine, and central Alaska).

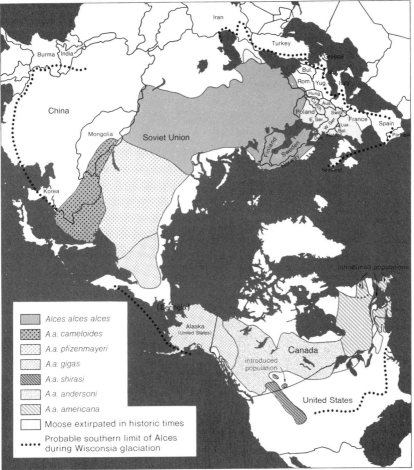

Fig. 2. Map of present-day distribution of extant subspecies of *Alces*, with historic distribution (since Roman times in Eurasia and since the 1600s in North America), and hypothetical southern limits of the genus during the last glacial maximum, based on distribution of boreal vegetation as mapped by Flint (1957) and Frenzel (1968).

Historic and Modern Distribution

At the dawn of the historic period (*ca* 2,000 years B. P.) moose had much the same distribution as they have today with the exception of a significant additional range north of the Alps and in northern France (fig. 2). Moose remains have also been reported to have been found at Roman sites in southern Scotland (Reynolds 1934, Steven and Carlisle 1959). In North America, moose have held most of their historic range, with minor exceptions such as northern Pennsylvania, Vermont, northwestern Massachusetts, northern New York, the adjacent heavily developed St. Lawrence lowlands, and northern parts of Wisconsin and Michigan. North American moose distribution has been significantly extended by man, who introduced the species into the island of Newfoundland in the early years of this century as well as into Anticosti and Labrador. Significant natural range expansion (or population increase) has also occurred in northern Ontario, northern Quebec, Isle Royale in Lake Superior, central and southern British Columbia, Montana, Wyoming, Idaho, and Utah, and along the coastal fjords of southeast Alaska.

After the seventeenth century, moose numbers were drastically reduced in the Russian territories and distribution was limited to regions of better habitat. Increasing human populations, the spread of Russian trappers and settlers to Siberia, and "cattle diseases" are mentioned as causes by Syroechkovskiy and Rogacheva (1974). However, with protection from overhunting in the twentieth century, moose have reoccupied much of their old range. They also now occur in a vast region bounded on the south by the middle Ob basin, on the east by the upper Enisei and Lena basins, and on the north by the tundra that was not reported to have been previously inhabited by moose. The historic absence of moose in that region is attributed to overhunting (Syroechkovskiy and Rogacheva 1974) but Formazov (1946) thought that the average snow accumulations there — the highest in the Soviet Union — effectively precluded colonization by moose.

On the southwestern flanks of Eurasian moose distributions, moose have also reoccupied their historic range, moving into Poland, Romania, the Ukraine, and the Caucasus in recent decades (Filonov and Zykov 1974).

The fact that moose have retained such a large proportion of their historic range into the late twentieth century is due in part to the unsuitability of much of the boreal forest regions for intensive human settlement. However, it also reflects the behavioural flexibility of moose. This flexibility has allowed the species to adjust to the vast changes in the vegetation of its habitat, changes that were produced by agriculture and forest management. It has also enabled moose to tolerate considerable disturbance by man and his dogs, livestock, and industrial operations.

Factors Limiting Distribution

Living things exist only where no environmental factor is so unfavourable as to induce a long-term preponderance of deaths over births. Some environmental factors show well-defined spatial breaks (such as the supply of air on passing through the surface of water bodies), or temporal breaks such as the onset of plant growth in spring in temperate regions. More subtle effects are produced by factors that vary continuously along a gradient. Differences between regions in such things as mean temperatures and amounts of precipitation can gradually tip the balance of survival against species. Large ungulates like moose are limited by landform, climate, food supply, predation, and human activities (fig. 3). The distribution of such species will coincide with areas of the earth's surface where all these factors are sufficiently favourable to permit survival and reproduction. The impact of factors such as the severity of winter weather or human hunting need not result in total annihilation of a species to be effective limitations to distribution. Unremitting pressure at the edge of a species tolerance may cause a continued excess of deaths over births that eventually results in regional extirpation. Neither do limiting factors have to operate continuously to be effective. Animals in temperate regions may

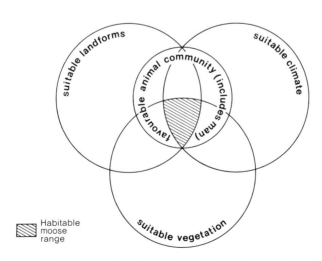

Fig. 3. Schematic diagram of environmental factors limiting moose distribution.

colonize northward during periods of years with mild winters, but lose all their territorial gains under the rigid limits imposed by one or more severe winters. Factors probably effective in limiting moose populations are discussed below.

LANDFORMS Moose range is centred on vast mid-continental plains in both Eurasia and North America. Much of the present-day range coincides with areas covered by Pleistocene glaciers and is characterized by hummocky till plains that have impeded drainage and many lakes. While moose use lakes for feeding on aquatic plants, for cooling themselves, and for escaping from flies in summer, they also inhabit vast regions where few lakes exist. Moose move routinely over rugged mountain ranges and through great differences in elevation. Although they seldom inhabit crags or other precipitous terrain, they readily invade high alpine shrub fields.

Of special importance to moose are the flood plains of large rivers and the inland deltas and marshy lake bottoms left by post-Pliestocene drainage changes (Berg and Phillips 1974). Periodic flooding and channel changes keep much of these areas in early stages of vegetation succession with heavy growths of forage plants such as willows (*Salix* spp.) and red-osier dogwood (*Cornus stolonifera*).

River valleys provide environments suitable for tree growth, extending stringers of forest cover far into arctic tundras. Moose have colonized these stringers both in North America (Kelsall 1972, LeResche et al. 1974) and in Eurasia (Egerov 1965, Kistchinski 1974).

In regions where elevation differences of more than 200 m occur within relatively short distances, moose move to sites where the wind exposure has lessened snow depth, or where the inclination toward the sun provides warmer daytime temperatures (Stelfox and Taber 1969, Prescott 1974, Telfer 1978*a*). In extensive mountain systems the orientation of the ranges may control ungulate distributions. For example, Stelfox and Taber (1969) pointed out that large ungulate populations, including moose, exist on the Rocky Mountains because the ranges have been thrust from the southwest toward the northeast, thus creating large areas of alpine shrub and grasslands that face the winter sun and have prevailing winds that blow away the snow. These mountains also have the advantage of being in the lee of more westerly ranges that catch snow from Pacific storms. The mountains that catch the snow—the Selkirks and Cascades—have few moose and a generally sparse ungulate fauna.

In New England, New Brunswick, and Nova Scotia the existence of elevated ridges that catch the snow has enabled moose to maintain their distribution in spite of large populations of disease-carrying white-tailed deer (*Odocoileus virginianus*) (Telfer 1967b).

Other landform characteristics beneficial to moose include the large areas of Alaska, Scandinavia, and elsewhere that have been at approximately the regional timberline during recent centuries. These regions support a shrub-dominated vegetation providing abundant forage. The occurrence of shrub or other forage-producing vegetation on mountains depends on the presence of substrates which are susceptible enough to weathering to produce an adequate depth of soil genetic material. On steep slopes underlain by resistant rocks, erosion may remove the products of weathering before vegetation can become widely established.

Perhaps the greatest impact that landforms have had on moose distribution and densities is through the limitation of human development. Large areas in North America and Europe that are climatically suited to agriculture are on glaciated shields where stoniness, lack of soil, and a low percentage of arable land have limited the conversion of forest to farms. The most intensive use applied to most of these areas has been timber production, which can be beneficial to moose.

CLIMATE   Two climatic factors have been thought to limit moose distribution: snow cover and extreme hot or cold temperatures. Almost all of the moose range coincides with regions having a persistent seasonal snow cover. However, snow cover varies greatly from region to region and from year to year. Key characteristics of snow cover are thickness, density, and hardness. Dry continental interiors are usually characterized by thin layers of light snow that do not form crusts, except in open areas such as extensive meadows and the frozen surfaces of large lakes, where the snow surface becomes hard-packed by the wind. In such places the snow cover is usually thin, seldom dense, and quite hard (Pruitt 1959).

In regions of maritime climate, the precipitation is greater and thick snow covers are the rule. Warmer temperatures lead to thawing and refreezing during the course of the winter, with the result that the snow settles and becomes quite dense with hard crusts forming at the surface (Kelsall and Prescott 1971, Skogland 1978). By March, maritime snow covers may contain many buried crusts, which prevent moose and other animals from sinking deeply.

Since moose are large, powerful animals they can travel through thick snow covers with surprising ease, but must pay a heavy bioenergetic toll (Stewart et al. 1977). Snow cover increases with elevation, so in mountainous areas, moose move to lower areas in winter (Edwards and Ritcey

1956, Krafft 1964). In the Canadian Rocky Mountains, however, moose may move up to timberline areas where open slopes may be blown clear of snow, as well as down into shallow snow in the valley montane zone (Stelfox and Taber 1969). In more level terrain moose may move long distances to areas of more shallow snow as reported by Knorre (1959) for northern European Russia. A more common behaviour is to concentrate in forest areas on southerly exposures in vegetation types where woody browse and warmer microclimates are available (Prescott 1974, Crête 1977).

In few places does snow accumulate to thicknesses sufficient to limit moose distribution. However, it may combine with other factors such as uncontrolled hunting and wolf predation to render moose populations vulnerable to local extirpation. In the past, moose were considered to be excluded from certain regions of deep snow such as western Yakutia, the adjoining Krasnoyarskiy Region of Siberia, and Ungava in northern Quebec. However, in recent years moose have either invaded these regions or increased from existing small population nuclei. Since moose are seasonally sedentary they may exist at low densities in remote country without being encountered by explorers and travellers (Kelsall and Telfer 1974).

Low temperatures are seldom limiting to moose. However, cold, in combination with high wind speeds, deep snow, and limited food may prevent a successful moose invasion of arctic tundra areas (Miller et al. 1972, Kelsall 1972). Knorre (1959) mentioned occasions when young moose froze to death in northern European Russia. Those deaths occurred during severe freezes that followed thawing and rain. More serious is the impact of a high bioenergetic costs on total regional populations, which may result in failures of the calf crop of the following spring (Stewart et al. 1977). Low temperatures and winds severe enough to cause subnormal development of embryos would either prevent moose from occupying a region or lead to dwindling numbers and local extirpation. Rather slight shifts in average winter severity along the northern boundaries of moose distribution may lead to substantial expansion or contraction of range boundaries.

High temperatures may be even more restrictive to moose than low temperatures (Kelsall and Telfer 1974). Intensive metabolic studies of moose at the University of Alberta's Department of Animal Science have established that moose experience difficulty in dissipating heat (Hudson pers. com.). An examination of the southern limits of moose distribution worldwide shows a close correspondence to the 20 C July isotherm. The more southerly extensions of the historic range are in cooler mountain ranges such as the Rocky Mountains in North America and the Caucasus Mountains in Eurasia. The historic range of moose in central and western France, a hilly but not mountainous region, extended south to about the 20 C July isotherm as well (Kurtén 1968, map, page 169). It is thus probable that although there was adequate browse in such places as the Midwest United States, the central Balkans, northern Italy, and the shrub deserts of western North America, moose could not inhabit them because high temperatures affected their reproductive performance.

VEGETATION Throughout the Holarctic region moose are associated with the boreal forest biome and are one of its characteristic animal forms. (The "spruce-moose" biome is a popular characterization.) Moose also occur in transition forests of mixed coniferous-deciduous stands to the south of the boreal forest from Manchuria in eastern China to Poland, Sweden, and, historically, Scotland. In North America northern parts of the transition zone dominated by northern hardwoods, from Nova Scotia to Minnesota, have been occupied by moose. In historic times moose occupied large areas of the deciduous biome in France, the Low Countries, and lower elevations in Germany. On the northern forest-tundra transition belt and higher elevation portions of the boreal forest, moose are distributed along major valleys (Mould 1977, Kelsall 1972, Kistchinski 1974) and occupy much of the near timberline shrublands in regions such as Norway, Newfoundland, and Alaska where large tracts of land occur at timberline elevations (Krafft 1964, LeResche et al. 1974). In alpine regions moose may move from subalpine forests that catch large accumulations of snow to the areas above the timberline that are blown free of it (Stelfox and Taber 1969).

Moose thus occupy areas containing a wide spectrum of vegetation, although their distribution is based on the northern coniferous forest formation. Since, however, they historically occupied the northern part of the deciduous forest formation of western Europe and even today may winter, in some areas, in shrub vegetation lacking coniferous stands (Peek 1974, Berg and Phillips 1974), it is a reasonable hypothesis that moose distribution is more strongly limited by other factors than by the occurrence of particular types of vegetation.

DISEASE Moose are subject to a variety of diseases. Parasites such as liver flukes (*Fascioloides magna*) carried by cattle and other livestock infect moose where livestock share their range (Wolfe 1974). Unnamed cattle diseases were thought to be partly responsible for moose declines in European Russia (Syroechkovskiy and Rogacheva 1974). However, the most striking example of moose perhaps having had their distribution limited by disease is their absence from the Appalachian Mountains south of the Allegheny Plateau in the eastern United States (Kelsall and Telfer 1974). The climate and vegetation of the higher parts of the Appalachians appear to be suitable for moose, but historically they have not occupied the area. Throughout eastern North America white-tailed deer (*Odocoileus virginianus*) carry the lungworm (*Parelaphostrongylus tenuis*). While this organism does little harm to the deer it is fatal to moose, and the high Appalachian deer populations may have created a situation in which colonizing moose have been quickly infected and have perished. In fact, observations of moose invading the Adirondack Mountains of New York suggest that even those not known to have been killed by poachers disappear within a year (Severinghaus and Jackson 1970). Moose once inhabited the Adirondacks but dwindled to local extirpation with a rise in white-tailed deer populations in the nineteenth century. While high summer temperatures sufficiently explain the limited moose penetration of the low-elevation deciduous forest farther west, *Parelaphostrongylus tenuis* seems to exclude moose from the Appalachians.

North American moose are hosts of the winter or moose tick (*Dermacentor albipictus*). Periodic heavy infestations of ticks have been reported from many areas of moose range. There has been much discussion as to whether the ticks are the cause of moose mortality. Recent work in Alberta indicates that they can be. In the winter of 1981–82 over 50% of the moose died in some districts of central Alberta (Lynch pers. com.). The extent to which ticks contributed to this die-off is unclear, but the winter was of moderate severity and tick loads averaged 63,000 individuals per moose in late winter at Elk Island National Park (Samuel pers. com.). Heavy tick loads seem related to short winters. Mild weather in late autumn lengthens the time when moose can become infested. In spring, early disappearance of snow cover allows engorged female ticks to drop off moose onto dry ground where conditions are suitable for reproduction. Ticks that drop on snow have a poorer chance of survival (Samuel pers. com.). Recent tick infestations in Alberta are localized geographically to the southern portion of the boreal mixedwood forest and adjacent aspen parkland ecotone. The high levels of mortality reported suggest that ticks have the capability to be a major factor in declines of moose populations and could conceivably limit moose densities or even distribution in areas particularly favourable for infestations.

HUNTING Moose have always been desirable game for human hunters, and drastic local reductions of moose numbers in Eurasia and parts of North America have been attributed to uncontrolled hunting. However, with rational management, moose populations can take considerable hunting pressure and maintain high numbers in close proximity to human activities, as in Scandinavia (Lykke and Cowan 1968). On the other hand, hunting combined with large-scale land clearance for agriculture and industry have extirpated moose in northwestern Europe, portions of the American Lake States, and in developed portions of their range in Ontario and Quebec. Perhaps the most significant aspect of the relationship of moose to man is the small proportion of historic moose distribution actually lost due to human activity.

PREDATION Other than man, the supreme predator of moose is the grey wolf (*Canis lupus*), which originally occupied the entire moose range (Frenzel 1974). Locally, predation by bears, both *Ursus arctos* and *U. americana*, can be important (Haglund 1974, Franzman et al. 1980). Predators almost never limit the distribution of their prey, as the extermination of a major prey species is about the worst thing that could happen to a predator. However, in unusual circumstances where heavy wolf predation is combined with a series of severe winters in a marginal range or with a sudden increase in human hunting, it might precipitate a population crash that could extirpate moose locally. Where the distribution of a species is reduced in such a manner, the key factor is difficult to identify.

Population Densities

The actual number of animals per unit area is an important measure of habitat selection. It is therefore useful to examine the range of reported population densities.

Moose can survive at very low densities. Estimates presented for Siberia by Syroechkovskiy and Rogacheva (1974) range from 0.0016 to 0.16 moose/km^2 in better-populated regions. Many regions were estimated to have in the vicinity of 0.03/km^2. A map presented by Filonov and Zykov (1974: 608) for European parts of the Soviet Union shows that most areas had 0.1 to 0.3/km^2. In northeast Siberia, Kistchinski (1974) reported moose numbers from 0.005 to 0.095/km^2. A map of densities in Quebec, based on aerial surveys made between 1964 and 1972 (Brassard et al. 1974: 72), showed most of the province with densities less than 0.04/km^2 while the most densely populated regions had more than 0.3/km^2.

Smaller areas of good moose range have much higher densities. For instance, surveys of blocks of range in Alberta varying in area from 91 to 413 km² made over the years 1958 to 1975, found densities from 0.39 to 1.39 and averaging 0.96/km² (Lynch 1975). Some of the counts were considered low due to unusually early moose movement to heavy cover in response to early snow. Riding Mountain National Park in Manitoba was reported to have densities of 1.0/km² in 1975–76 and 1.05/km² in 1976–77 (Carbyn pers. com.). Wintering densities in many areas can reach 8 to 15 moose/km², especially in early winter in flat terrain, but also in valley concentration areas in late winter.

When dispersal is prevented by fences moose herds can exist at considerably higher densities on good boreal forest range. Densities in a fenced reserve, Elk Island National Park, in Alberta, have averaged over 2.3/km² for 75 years and often were 3.0 to 4.5/km² before periodic reduction slaughters. On one occasion, densities were between 6 and 7.5/km² before a die-off occurred (Canadian Wildlife Service unpublished data).

In summary, vast subarctic areas with substantial mountain tundra and taiga areas show average regional densities of less than 0.1 moose/km². Better boreal and coniferous/deciduous transition ranges support regional averages of 0.1 to 0.3 moose/km² while excellent range in those areas with deciduous vegetation and soft, thin snow covers average 0.4 to 1.0 moose/km².

Habitat Use

Within any given area the density of moose varies markedly. Studies of habitat use by any animal must begin with a classification of the environment. The categories employed usually conform to the investigator's human perception of reality. The animals' own perception of their environment must ever remain mysterious, although it may be inferred to some extent by a correlation of habitat use with environmental variables. It has therefore been usual to classify moose habitat into categories within the broad variables of topography and vegetation and to measure the utilization within those categories.

As wild animals, moose are by definition not under the control of the researcher so many indirect methods have been employed to estimate their "use" of habitat categories. Because moose are large enough to be observed from aircraft, some of the largest-scale studies of habitat use have been

Circumpolar Distribution and Habitat Requirements of Moose 159

carried out by recording categories of habitats in which animals were observed during aerial census work. Other time-honoured techniques are counting fecal pellet-piles, counting tracks, and direct ground observation of moose. Counts of pellet piles have traditionally been viewed as an index of time spent by ungulates in various habitats. No information is available as to whether this is the case for moose. Recent studies of deer and wapiti (*Cervus elaphus*) defecations in relation to time spent in various habitats (Collins and Urness 1981) showed that in some cases data did not support the assumption that accumulation of pellet piles was proportional to time spent. However, the study involved tamed animals followed for sample time period, while most pellet-count surveys count accumulations for an entire season—often the fall, winter, and early spring combined. The actual relationship of such counts to time spent for any ungulate has been inadequately tested.

Track counts are an index of activity in various habitats and may yield different values than pellet-group counts (Cairns and Telfer 1980). Direct observation of moose is a valid technique in open vegetation types and hilly terrain, where a number of moose can often be seen from a single vantage point. Where evergreen vegetation makes moose difficult to observe, the method is unreliable. Observations made from a distance, be they ground or aerial, have the disadvantage of being difficult to combine with quantitative vegetation measurements at the actual observation site. In recent years the adaption of radio transmitters for attachment to animals, combined with refinement of capture techniques using drugs, has revolutionized all field biological studies including those of habitat use. Moose are large mammals, well-suited for instrumentation with telemetry transmitters; thus a new technique has been added, permitting estimates of habitat use to be made from radio locational fixes in various habitat categories (Phillips et al. 1973).

Most habitat use data, no matter how obtained, is presented as a comparison of actual use with that expected. Habitat categories that show actual use as being significantly greater than expected use are considered as "preferred" or "selected" (Neu et al. 1974). It is worth noting, however, that focusing on preferred categories may obscure the fact that a large proportion of the species is actually living in neutral or avoided habitats. This discrepancy occurs where preferred habitats cover a small proportion of the total land area. An example is presented by Kearney and Gilbert (1976), where the small white pine (*Pinus strobus*) type, which had the highest use (0.46 pellet groups/plot), contributed 229 moose-days of use compared with 1,020 moose-days for the white birch (*Betula papyrifera*) type which had only half the use intensity (0.23 pellet groups/plot). Habitat types can therefore be important although not especially preferred. The proportion of moose in low-preference types would be expected to increase if the populations are dense overall and the animals more likely to disperse into marginal habitat.

The phenomenon of habitat can be organized in the light of many concepts and especially on many scales. I propose that moose habitat be viewed on three scales: regional, local, and compartment or site scales.

1. *Regional scale.* Areas at this scale measure 1,000 km² or more. They could be comprised of political or physiographic regions such as counties, land holdings of large timber enterprises (or working circle subdivisions of such holdings), governmental forest administration districts, or natural features such as river basins, mountain ranges, and dead-ice moraines. A habitat region would contain a "population" of scores or hundreds of moose, usually on a year-round basis, and could serve as a separate moose management unit. Relevant information relating to habitat at this scale would include aerial survey results which determine seasonal concentration areas, proportions of the area in various major land classes, and forest types classified by deciduous/coniferous composition and broad height and density classes.

2. *Local scale.* This scale embraces areas measuring approximately 100 km² to 999 km², roughly one to several townships, or smaller watersheds, but only portions of major land systems. Habitat localities would normally contain at least seasonal ranges of ten or more moose, enough to constitute a self-sustaining breeding group. Relevant habitat information at this scale includes the mean stand size of more-detailed forest types, categorized by species composition data and 2 or 3-m height classes; measures of stand interspersion, biomass values for forage production in relation to forest type; and proportions of area with various aspects (the simplest ecological aspect classification for land units is into "sunny" aspects, from southeast through south to northwest, and "shaded" exposures from northwest to southeast). Other measurable characteristics at the scale of local habitat are proportions of the area with various topographic positions using such categories as flat, lower, middle- and upper-slope, and ridgetop in rolling country, while flat lowland, slope, and flat upland may be categories of ecological relevance even in quite level landscapes; pellet-group and track counts to determine moose use of various landform or vegetation categories; and comparative surveys of snow depth and condition on the landform-vegetation categories.

3. *Compartment or site scale.* Forest managers usually subdivide lands under their control into small fundamental units called compartments, which are in the order of a few hundred hectares in area. Other subdivisions at this scale are small watersheds, old land grants, or groupings of sections and quarter sections (in areas of North America covered by the co-ordinate system of public land surveys). Habitat compartments will comprise part or all of the seasonal home range of either individual moose or small groups. At the compartment scale, relevant moose habitat measurements include detailed forest stand analysis by tree diameter classes and stand basal areas, crown cover estimates, forage production by species, moose tracks, droppings, and browsing and bed sites in relation to stand parameters and local variations in snow conditions.

Topography

The wide range of topographic regions occupied by moose throughout their distribution is matched by their use of almost all landscape types in smaller-scale habitat units. At a regional scale, moose move from areas of thick snow cover to areas with thinner snow on valley floors and at timberline (Edwards and Ritcey 1956, Stelfox and Taber 1969, Peek 1974). Moose will use all terrain types except precipitous rocky crags. However, at certain times of the year, certain landscape features may be strongly favoured. In regions where moose are subject to wolf predation they seek island in rivers and lakes, and they resort to hummocks in musket for calving, as witnessed in Alaska by LeResche et al. (1974) and in Ontario by Peterson (1955).

For many years naturalists have considered ponds, lakes, and marshes to be essential to moose welfare (Leopold 1933). In fact, moose do make use of water bodies to escape flies and cool themselves, and these activities may affect overall survival. Moose also seek emergent vegetation as a source of sodium, which has been found to be in short supply in moose forage in at least one area (Isle Royale in Lake Superior, according to Botkin et al. 1973). It is also true that dense moose populations occur in regions with few bodies of water, such as the Porcupine Hills of southern Alberta and the Cobequid Hills of Nova Scotia (areas that have been studied by the author). The current lack of knowledge about the availability of nutrients such as sodium in the habitat in various regions prevents conclusions from being made about the influence of aquatic vegetation on the density of moose populations. The reported use of "licks" by moose in the regions of the uplands of Gaspasie in Quebec (Bouchard

1970), northern Ontario (Fraser and Reardon 1980), and northern Alberta (Best et al. 1977) are probably related to sodium deficiency. Licks constitute a special landscape feature that can draw ungulates into areas where they drink spring water and eat soil. Such water and soil may contain sodium and other minerals. In Sibley Provincial Park in Ontario, Fraser and Reardon (1980) found that springwater at lick sites contained significantly higher levels of sodium and chlorine than did streamwater in the area. They also found that moose preferred experimental mixtures of water laced with sodium compounds over those containing other minerals. In Alberta, analyses of licks in the Swan Hills, when compared with surrounding soils, showed high levels of sodium (Best et al. 1977).

Winter creates a bioenergetic drain on moose, as they must keep up their body temperature and travel through impeding snow. One behavioural response moose employ is to limit their activity to warmer sites. Moose become quite sedentary as winter progresses and are usually found on southerly facing slopes where they have high exposure to the sun (Brassard et al. 1974, Prescott 1974).

Aspect and drainage strongly affect vegetation, which in turn has a strong influence on moose distribution, but the benefit of radiant heat from the sun and the flow of cold air to the valley bottom (Geiger 1965) create a more comfortable environment for moose on sunny middle and upper slopes. This may explain why Prescott (1974) found moose in Nova Scotia preferring the upper third of slopes in sharply dissected terrain. Brassard et al. (1974) measured slopes in some Quebec moose-wintering areas and found they were gentle, averaging a 6.5% slope.

Vegetation

Moose derive two benefits from vegetation: food and cover. Cover serves as shelter from wind or hot sun in summer and as an interceptor of snow as well as protection from predators and human hunters. Most moose forage consists of the twigs and leaves of woody plants, although a great variety of plant groups are eaten in various parts of the moose range (Peterson 1955, Sablina 1970, Morow 1976). The greatest amounts of moose forage are produced in shrub or forest regeneration stands, where most of the stand biomass is within the height range where moose feed (fig. 3). Deciduous species provide most moose forage; however, in many localities, a significant proportion of winter forage is composed of conifers. Balsam fir (*Abies balsamea*) is a major diet item in Newfoundland (Bergerud and Manuel 1968), Quebec (Brassard et al. 1974), and Nova Scotia (Telfer 1976), while Scotch pine (*Pinus sylvestris*) and Norway

Table 1 A general forest classification for moose habitat studies

Non-arboreal vegetation

Marsh	Emergent vegetation
Fen	Wet meadow, largely *Carex* spp.
Bog	*Sphagnum* dominated with ericaceous shrubs and scattered conifers
Grassland	Upland grasses and forbs
Shrubland	>50% cover of plants of shrub growth form (generally less than 5m in height)

Arboreal vegetation

			Mature (12m)	
			Crown Cover	
	Saplings (5m)	Poles (5–12m)	Dense (>50% crown cover)	Open (<50% crown cover)
Deciduous (75% of basal area deciduous)	X			
Mixed wood (25–75% of basal area deciduous)	X	X	X	X
Coniferous (25% of basal area deciduous)	X	X	X	X

spruce (*Picea abies*) are browsed in Europe, sometimes causing heavy damage to regenerating commercial forest trees (Markgren 1974). As forest stands become older they grow out of the reach of moose and the shade of the tree crown layer discourages the growth of browse plants underneath. Insects, disease, and wind-throw may also kill trees and create an open crown cover. Similar effects are created by forest management regimes that prescribe partial cutting. Stands with less than 50% crown cover and with a basal area of less than approximately 17 m^2/ha may be considered "open" and will be quite productive of browse (Telfer 1972*a*).

A system of classification for forested moose habitat is suggested in table 1. Much of its information can be obtained by lumping together categories usually employed by foresters in timber inventories. Thus, those interested in moose can tap large banks of existing information on past and present range conditions. The suggested system subdivides vegetation into 17 categories. Some workers might favour a different and more complicated classification. In certain areas such efforts would be warranted. However, in my experience, the 17 categories presented are an adequate vegetative stratification for moose habitat studies.

Slow, long-term changes in vegetation may explain ungulate population trends (Telfer 1971). Geist (1971, 1974) has introduced the concept of *permanent* and *transient* moose habitats. The huge food supply created by regrowth following fire in boreal regions forms the key element in creating a transient habitat situation. Since severe burning occurs in periods of dry years there may be, on a regional and particularly a local scale, many years when transient habitat is limited. Moose then depend on permanent habitat such as alluvial flood plain areas where continuous disturbance by flooding and channel shifting keeps large areas in shrub stands or early-successional deciduous forest.

Moose Movements and Habitat

Movements of animals may be short and on a compartmental, or local, or on a regional scale. Moose make definite seasonal movements between habitats suitable to the season. In some cases these movements constitute migration (Pulliainen 1974). Altitudinal migrations are common in mountainous country, where moose summer at high elevations but move down to find thin snowcovers and/or riparian browse-producing forest in winter as reported from Montana by Peek (1974), from British Columbia by Edwards and Ritcey (1956), and from Norway by Krafft (1964). In less mountainous regions extensive horizontal migrations, some thought to be

Circumpolar Distribution and Habitat Requirements of Moose 165

as long as 300 km, have been documented, especially in northern Europe (Pulliainen 1974, Knorre 1959). Such movements are generally away from subarctic tundra-edge summer range to winter range farther south or in areas with thinner or softer snow.

LeResche (1974) reviewed North American migrations and movement patterns of moose and identified three kinds of movements, as follows:

Type A: Animals move short distances horizontally between seasonal ranges.

Type B: Moose move between traditional winter and summer-fall home ranges usually at different altitudinal levels.

Type C: Moose not only move between a summer-fall range and a winter range, but also move to a traditional spring range.

LeResche marshalled considerable evidence, especially from his own work on the Kenai Peninsula, Alaska, in favour of the view that moose involved in each of the three movement types may mingle on the same areas. A locality may thus be a year-round range for some individuals but only a winter range to others.

Moose move to take advantage of a different environment (LeResche 1974). Environments favoured at different seasons have certain characteristics as discussed below.

SPRING RANGE In spring and early summer, moose are calving and often move to lowland bogs (LeResche et al. 1974) or to islands and peninsulas (Peterson 1955), where snow melts first and vegetation greens up early. Such areas appear to be difficult for predators such as wolves and black bears to search thoroughly, and often provide opportunities for moose to escape by water if pursued. (It is noteworthy that moose calves can swim when not much more than a week old: Peterson 1955.) The spring and summer habitat of moose in the interior of Alaska was described by LeResche et al. (1974: 158) as "an intricate mosaic of black spruce forests, bogs, shrubs, and sub-climax hardwood communities, as well as numerous intermediate stages." Meandering streams and oxbow lakes were characteristic of this dynamic, permafrost-controlled landscape.

In spring and summer, moose are often observed in water, avoiding insects, keeping cool, and feeding on mineral-rich aquatic vegetation. The dense vegetation along water bodies and wetland edges provides protection against heat as well as providing super-abundant forage.

SUMMER-FALL RANGE In flat regions such as those found on much of the Canadian Shield, moose may remain close to water and indulge in heavy aquatic feeding (Peterson 1955, Joyal and Scherrer 1974, Belovsky 1978, Dodds 1974). They may also use relatively open upland sites presumably for feeding (Peek et al. 1976, Telfer 1968b). Peek et al. (1976) described "open" forest as having less than 990 trees/ha while Telfer (1968b) qualitatively described it as patchy forest interspersed with areas open to the sky. In some areas moose congregate near mineral licks for a few weeks in early summer (Best et al. 1977), later dispersing to summer ranges. Bulls and subadults usually arrive first at the licks, followed by cows with their newborn calves (Lynch pers. com.). As summer progresses, moose in regions of level terrain become more generally distributed across summer range habitat localities. In New Brunswick, moose were using 63% of a habitat locality of 93 km^2 in late June, compared with 13% under deep snow conditions in early April (Telfer 1968b). Studies by Peek et al. (1976) showed a broad distribution of moose activity over various forest cover types, with some preference for stands of trembling aspen (*Populus tremuloides*) and black spruce (*Picea mariana*) as summer progressed. Berg and Phillips (1974), working in northwestern Minnesota in a flat, poorly drained region supporting marsh and willow shrubland, found moose summering in open stands of short willows but spending much time in water.

In mountainous regions moose move from winter range either to calving areas or else directly toward their summer and fall range at higher elevations (Edwards and Ritcey 1956, LeResche 1974, Peek 1974). The vegetation used varies frm subalpine forest to timberline forest-shrubland mosaics and shrublands above timberline (LeResche 1974: table II). Edwards and Ritcey (1956) described subalpine summer range thus: "Preferred areas are wet from seepage with meadows among the trees or about high burns." Moose that summer in arctic tundra localities occupy riparian shrub stands and areas of sparse shrub growth on adjacent uplands (Kistchinski 1974, LeResche 1974).

EARLY WINTER HABITAT In northern Ontario, moose have been found to concentrate on traditional sites in the post-rut period where they remain at densities of 4 to 9 moose per km^2 (Thompson pers. com.). Similar early winter concentrations have been reported in the Peace-Athabasca Delta of Alberta (Allison pers. com.) where densities exceeded 12/km^2. In the Swan Hills of Alberta, moose densities of 4 to 6/km^2 were found in burns during early winter (Lynch pers. com.). The use of burns began about 5 years after the fire and coincided with development of substantial browse production. Peek et al. (1976) noted a shift to more intensive use of preferred forest types in early winter. Types used were open (less than 990 trees/ha with 50% of tracks counted in stands with less than 125 trees/ha). In Thompson's Ontario study area, the early winter range was also an open habitat with a dense shrub layer. Allison also found dense moose concentrations on deltaic shrub stands, mostly willows (*Salix* spp.).

Moose that spend the early winter on discrete ranges may be displaying a seasonal movement pattern additional to the three types described by LeResche (1974). This pattern is a variant of Type A, in which moose move short distances between summer, early winter, and late winter home range elements. This pattern may be designated Type D.

LATE WINTER HABITAT Because of the severity of the winters in the area of moose distribution and the obvious drain on the vitality of moose due to windchill and locomotion in deep snow, considerable effort has been expended on identifying and describing the winter habitat of moose, and in the management procedures needed to improve or maintain it. Wherever the movement of moose has been closely studied it has been found to shift in late winter to sites ecologically different from those used in the snow-free portions of the year.

Moose in mountainous regions such as the northern Rocky Mountains or the Scandinavian Mountains move downhill to riparian shrublands or aspen forests (Peek 1974, Markgren 1974, Stelfox and Taber 1969). Stelfox and Taber also reported moose moving *up* to timberline and alpine shrub areas to winter. They also found many moose wintering in the subalpine forest, possibly using small permanent habitat units like shrub-grown avalanche tracks, bogs, and riparian strips along small streams. In the Selkirk Mountains of British Columbia, where snow cover thickness often exceeds 2 m, moose utilize avalanche slopes where the shrub vegetation is exposed by periodic slides. Moose often move up such slopes to the timberline, walking on the packed snow of the slide tracks (Van Tighem pers. com.).

Where mountains adjoin broad lowlands and valleys, moose seek out burned areas, where available (studies cited by LeResche 1974: table II). One such area was described by Edwards and Ritcey (1956: 488) as "open, deciduous growth, rich in browse."

In regions of flatter terrain, moose cannot escape deep snow by altitudinal migration. However, dense coniferous forest catches part of the snowfall, and cold, calm conditions within the stands tend to keep the snow cover softer than it is in the open. Moose gradually move to areas that possess at least small stands of dense conifers. During a winter of deep snow in New Brunswick only 10% of moose tracks were found in dense conifer in January, compared with 85% at the end of March (Telfer 1970). In Minnesota, Peek et al. (1976) found most moose active in balsam fir (*Abies balsamea*), black spruce (*Picea mariana*), and pine (*Pinus banksiana, P. resinosa* and *P. strobus*) stands under snow with temperature conditions considered severe, while in milder periods (early winter and early spring), trembling aspen stands were used heavily in addition to black spruce and pine types. In severe periods Peek et al. found approximately 50% of the moose tracks in stands with more than 990 trees/ha. In contrast, in mild periods only 10 to 20% of moose tracks were found in such stands. In flat, willow-covered wetlands in Minnesota, moose moved into tall, dense willow and into available aspen and other deciduous forest during late winter (Berg and Phillips 1974).

Long-term studies of moose on Isle Royale in Lake Superior (Krefting 1974) have shown moose to use burns heavily in winter, but to move into mature mixed wood in winters with deeper snow. In the drier mid-continental areas of North America, frequent wildfires in the past have led to the creation of vast aspen stands. In central Alberta, moose concentrate in these areas by November and remain all winter. The other forest type selected is forested muskeg, also browse-producing, but one with patchy cover in black spruce clumps (Rolley and Keith 1980). The use of young aspen stands throughout the winter was observed in the similar environment of central Manitoba by Rounds (1981). Open, browse-producing shrublands were also reported to be the preferred winter habitat of moose in central Alberta and the Alberta foothills by Telfer (1978*a*) and Cairns and Telfer (1980).

Further east, in the North American deciduous-coniferous transition forest, moose concentrate in forest stands where outbreaks of insects such as spruce budworm (*Choristoneura fumiferana*) have opened the forest canopy (Krefting 1974, Brassard et al. 1974). The diversity of tree species in those forests has led to considerable selective cutting of particular species and sizes of timber. Even areas considered clearcut by foresters frequently have substantial patches of non-commercial trees remaining (Telfer 1972*b*). Thus logged forests exhibit the same structural features as do stands attacked by insects. In the most productive moose range in Quebec, Brassard et al. (1974) estimated that adequate winter cover consisted of 375 to 1,000 stems of evergreen conifers per ha with diameter breast high (DBH) greater than 17 cm, providing a basal area of 11 to 16 m^2/ha. Further north, more coniferous cover was required. The intensity of use in the Quebec concentration areas by moose was related to the amount of browse available (Brassard et al. 1974, Crête 1977). Further east, in the

Maritime Provinces, logging, insects, and tree diseases have created open, patchy forests in proper topographic positions for moose wintering. Forests have been further opened up by wind throw and by porcupine (*Erethizon dorsatum*) feeding there. The wintering areas exhibit the same characteristics of patchy tree cover, dense shrub, and sapling understory reported elsewhere (Dodds 1974, Prescott 1974, Telfer 1967a).

Considerable areas of forest are disturbed and set back to shrub stages of succession by flooding and by tree cutting of beaver (*Castor canadensis*). These areas provide small patches of browse plants interspersed with cover and become habitat for sodium-rich aquatic vegetation, thus benefiting the moose. Beaver also play a role in preventing wolf predation on moose by serving as an available buffer species. However, the presence of the beavers as alternate prey for wolves may help maintain higher wolf population densities, which in the long run may lead to heavier predation on moose.

In general, the vegetation of moose late winter range is characterized by

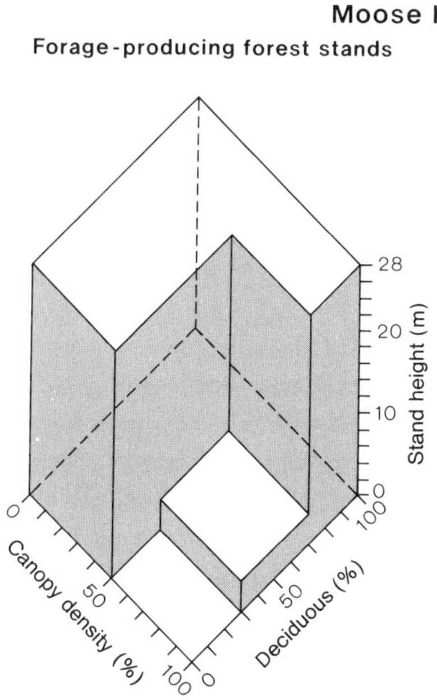

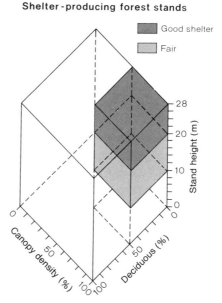

Fig. 4. Diagrammatic representation of relative value for moose winter-forage production of boreal forest stands of varying structure and composition.

Fig. 5. Diagrammatic representation of boreal forest conditions providing shelter for moose.

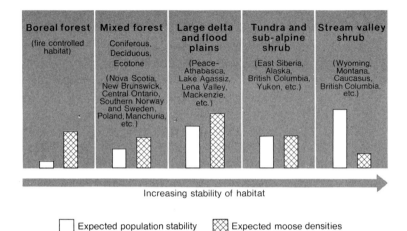

Fig. 6. Major circumpolar moose habitats, with diagrams hypothesizing relative moose-population densities and population stability in each habitat.

a dense shrub layer or "feeding stratum" (Crête 1977) under an open, patchy "cover stratum" of forest (fig. 4). Just as moose in mountainous areas retreat downhill as the snow becomes too deep for them, so moose in flatter terrain abandon open, food-rich forest for coniferous cover when forçed to by increasing snow accumulations. The forest structural condition providing adequate shelter for such situations is diagrammed in fig. 5.

Habitat Organization

Peek (1974) proposed a system of moose habitat classification based on the dynamic interactions of landform and vegetation and the influence of factors such as fluvial processes and fire frequency. He arranged boreal and mountain habitats on a continuum that ranged from most stable in time to least stable. He suggested that variations in behaviour, physiology, and morphology among moose populations were adaptations to different levels of habitat stability. Peek's concept has been expanded (fig. 6) in an attempt to organize the major landform-vegetation formation elements of the circumpolar moose habitat.

Stable habitats are not necessarily the most productive, nor are they highly correlated with expected moose densities or stability of populations. The expected densities of moose on a regional scale within the major habitats listed in fig. 6 correspond closely to the gross productivity of ecosystems in table 2 with the exception of mixed forest, which is considered more productive than boreal forest. The achievable densities of moose are mediated by factors that put vegetation at a height range where moose can eat it, in an edible species and plant structure, and within a snow environment that to varying degrees permits their access to existing forage. An additional factor affecting the ability of moose to utilize plant production is the availability of forest stands that provide adequate thermal cover near to forage.

Table 2   Estimated gross primary productivity of major moose habitats, based on Odum 1971

Habitat	Productivity (kcal/ha/year)
Flood plains	8,000
Mixed forests	5,000
Boreal forest	3,000
Stream valley shrubs	2,500
Tundra and alpine	200

Population stability (fig. 6) relates to fluctuations of numbers in time. Such fluctuations may be related to habitat changes, such as changes in the boreal forest brought about by fire, but are probably more the result of other factors. In the tundra and subalpine, the impact of extreme winter conditions and late spring creates instability in populations, although the habitat itself is stable and the full productivity of those ecosystems is within the reach of moose and is mostly edible. Stability in the mixed forests is predicted to be greater than in the boreal forest due to the lower frequency of burning. However, there are periodically major disturbances due to insects that cause mixed forests to provide enough forage to support great increases in moose numbers.

The stability of the stream-valley shrub habitat has been discussed by Peek (1974) and is based on a rather unchanging area of forage-producing vegetation used for winter range in snow conditions that seldom influence availability. Moose fan out over the surrounding mountains for spring, summer, and fall range.

Boreal Forest

Moose distribution largely coincides with the circumpolar belt dominated by coniferous, needle-leaved evergreens of the genera *Pinus, Picea,* and *Abies,* and of the needle-leaved deciduous genus *Larix.* Within these forests vast disturbances caused by fire are normal (Kelsall et al. 1977, Johnson 1980, Zackrisson 1980), but in some areas timber harvesting has become the major disturbance (Markgren 1974, Krefting 1974). Deciduous broad-leaved tree genera such as *Betula, Prunus,* and *Populus* follow disturbances in those regions, providing patches with a heavy yield of forage. The drainage in the flat, glaciated landscapes of much of the moose range, such as the Canadian Shield and Scandinavia, is impeded by glacial deposits, resulting in large areas of lakes and wetlands. Interspersed wetlands prevent fires from sweeping vast areas completely clean of older forest, but frequent ignitions ensure a variety of forest stand age classes. The result is a fine-grained mosaic of stands in various cover types and age classes. Miller (1976) described such a region in northern Manitoba where stands averaged 4 ha over a vast area.

Other parts of the moose range are either unglaciated (much of Siberia and Alaska) or well-drained due to steep topography (mountain ranges throughout). Burned and logged areas are much larger in such well-drained regions, with resulting forest stands providing a mosaic with a much larger grain. Since the burning rate of much of the boreal forest is of the order of 1% per year (Alexander 1980), regrowth to forest of the pole stage (table 1) usually requires 15 to 40 years, depending on site, seed, or sprout source availability and fire intensity. Thus, a substantial percentage of the vast boreal moose range was always in food-producing stands under primitive conditions.

Mixed Forest Regions

Much of the ecotone between the deciduous forest formation of eastern North America, east Asia, and Europe, and the conifer-dominated boreal forest formation to the north contains moose populations. In Europe the zone extends from southern Norway through south-central Sweden into Poland and across the centre of the Soviet Union, gradually narrowing between the dry steppes of west Asia and the boreal formation. Elements of this transition occur in mid-continental North America (in Minnesota) and extend through southern Ontario, Quebec, northern New England, and the Maritime Provinces. Much of the North American zone is hilly and chracterized by impeded drainage and many lakes. Stands of forest types resembling both boreal and deciduous formations occur side by side depending on slight temperature, soil moisture, and soil parent material differences (Eyre 1980). These are interspersed with other forest types composed of mixtures of boreal and deciduous species.

The impact of insect outbreaks and various pathological syndromes such as birch dieback in eastern North America have contributed to opening the forests. Fire frequency has been described as low in primitive times in the North American portion of this habitat (Wein and Moore 1979, Lorimer 1980), although occasional fires and additional windthrow have also created moose forage. The extensive deciduous and mixed forests provide a substantial browse supply from understory shrubs. In North America, shrubs such as *Acer spicatum, A. pensylvanicum, Viburnum cassinoides, Hamamalis virginiana,* and *Vaccinium* spp. provide a good understory of moose forage, along with saplings of tree genera like *Abies, Betula,* and *Populus.* The mixed forest contains basically permanent moose habitat with food-producing stands along bogs, lakeshores, and stream valleys, and in the upland deciduous forest. The occurrence of transient habitat was less pronounced under primitive conditions, but with the advent of timber harvesting the carrying capacity of these forests may have been raised.

Alluvial Habitat
Moose thrive on flood plains and deltas (Berg and Phillips 1974, Kistchinski 1974, Egorov 1965). These habitats may be described as "stable unstable." They are unstable in the short term, but stable in the long term. Continuous landscape change is occurring as shifting river channels and annual flooding maintain vegetation in an early successional stage and provide large disturbed areas that are invaded by shrubs. At the same time the spreading of fertile new sediment, the plentiful ground water and, in northern regions, the reduction of permafrost, create excellent growing conditions and promote high primary productivity. These habitats hold year-round moose populations of high density and provide important winter habitat (LeResche et al. 1974). Major examples of this type of habitat occur in the glacial Lake Agassiz area of Minnesota, and in the deltas such as the Saskatchewan in west-central Manitoba, the Peace-Athabasca in Alberta, and the Slave River in the Northwest Territories, as well as in the alluvial plains of the great northern rivers—the Athabasca, Slave, Mackenzie, Yukon, and Kuskokwim in North America, and the Kolyma, Indigirka, Yana, Lena, Enisei, Irtysh, Ob, and Pechora in Eurasia. On these landforms dense stands of the genus *Salix* are common along with *Betula, Populus, Prunus, Cornus, Corylus, Vaccinium,* and *Ribes.* All are good moose forage.

Tundra and Subalpine Shrub
These habitats occur along the courses of rivers flowing through the mainland tundras of arctic regions from Ungava to Norway and in northern mountainous areas, especially in the Northwest Territories, Yukon,

Alaska, and northeastern Siberia. Mould (1977) described typical tundra moose range on Alaska's Colville River. Considerable subalpine shrub areas also occur in Alaska, Norway, the Scandinavian Mountains, and in Lapland (Kelsall 1972, Markgren 1974, Kistchinski 1974, LeResche et al. 1974). These areas have low primary productivity, but all forage is within reach of moose and most is palatable. Severe winters cause losses in moose populations in these habitats, and many moose migrate to more sheltered localities. Among the permanent residents, population stability is low.

Stream Valley Habitats
Peek (1974) reviewed information on stream valley habitats in the northern Rocky Mountains. These areas also occur in British Columbia, the Yukon, Alaska, and the southern belt of Siberia, including the Altai mountain system. The principal feature of these moose ranges is the riparian growth, mostly of the genus *Salix* that provides excellent forage. These valleys often are rather dry, with open grasslands and open, dry forests of *Populus, Pinus* and, in the Rocky Mountains, *Pseudotsuga*. Some moose populations live year-round on these riparian linear habitats and others winter there, moving up into the mountains for the rest of the year.

Mechanisms Controlling Habitat Selection

Moose distribution, although extensive, has limits. Within those limits density varies and complex movement patterns have developed that put moose in certain habitats at certain seasons. What are the driving mechanisms that control this selection in time and space? Several aspects are discussed below.

Energy Relationships
Belovsky (1978, 1981) has argued persuasively that moose are "food maximizers." This strategy involves the animal in feeding at times and places when the maximum net energy intake per unit of time can be obtained. Since considerable time must be spent by ruminants like moose in masticating the cud, they feed at times of day and on sites where they can fill their rumens quickly with a minimum expenditure of energy. They then spend the requisite amount of time ruminating and then commence another feeding bout. Such a strategy is advantageous to moose because they are large animals with a substantial energy requirement, and in fall

Circumpolar Distribution and Habitat Requirements of Moose 175

and winter their diet consists of material with low digestibilities (usually in the 20 to 50% range). Gasaway and Coady (1974) use 40% digestibility in their calculations of energy balance for moose in Alaska.

On the other side of the balance, moose might be deterred from feeding in sites with high forage biomass on an optimum schedule by factors that either exact a high energy cost, or threaten the survival and reproduction of moose more than does sub-optimal foraging. High energy costs result from temperatures that are too low or too high, requiring increased metabolism to maintain homeostasis in body temperature. Locomotion in snow is also costly in energy terms. Moose move to areas with snowcovers that provide the easiest walking. Belovsky (1981) carried out an informative test of his hypothesis that moose are food maximizers by comparing their use of broad habitat groupings and the times of daily activity with predictions made from a model. The model integrated estimates of temperature regime and food availability, allowing for the constraints of rumen capacity, rumination time, and endurable limits of thermal stress. The data were collected during summer on Isle Royale, when the harassment of moose by wolves or reaction of moose to observers was apparently minimal. The model predictions generally agreed with the observed moose activity distribution data obtained by direct observation, suggesting that moose, in summer at least, use their habitat in a way that optimizes their energy balance.

As yet there are no studies tying moose winter habitat use to energy parameters. However, snow avoidance, the seeking of thermal cover, and behaviours such as bedding in deep snow for insulation (Knorre 1959) suggest a very close adjustment to requirements for optimum energy balance. Work with another cervid, white-tailed deer, show a strong correlation of winter habitat use with energy parameters (Moen 1968). One could predict that the use of habitat to maximize energy intake or to minimize energy loss might be most important in summer or winter, when weather conditions are at the extremes. Fall and spring should be thermoneutral periods, when moose could be active anywhere at any time of day. In fact, spring and fall are the times when moose are most active (Hauge and Keith 1981). Migrations and other movements, the heavy use of mineral licks, and the rut occur in thermoneutral seasons.

Factors liable to override energy optimization as a survival strategy involve predation by wolves and human disturbance. Wolves and moose have co-evolved over a long period. It would be largely counter-productive for wolves to drive moose into poor habitat that would lead to mortality other than at the jaws of a wolf. In fact, there is little in the literature to indicate that moose move from otherwise optimal habitat because of wolves. Human disturbance can have an impact on moose. One recent study at Elk Island National Park found moose avoiding cross-country ski trails, while wapiti and bison did not (Ferguson and Keith 1982). However, at Elk Island the habitat is highly interspersed (Cairns and Telfer 1980), so moose that avoided trails had the same landform and vegetation types available to them that they would have found near the ski trails.

Co-actions with other herbivores can also be responsible for shifts in moose habitat use. Such co-actions can take the form of *exploitive competition,* in which other species remove the moose food resource either at the same time or before moose get to it, or *interference competition,* where moose avoid areas containing other animals. There is some evidence for exploitive competition on winter ranges. Rounds (1981) noted that moose in Riding Mountain National Park in Manitoba made less use of shrubland that was heavily used by wapiti (*Cervus elaphus*) than reported elsewhere. Subjective reports by observers suggest that moose may move away from groups of wapiti and bison (*Bison bison*). While food appears to be super-abundant on moose ranges except in late winter, the presence of other browsers feeding in numbers cannot fail to reduce the availability of forage even in productive vegetation types like shrubland. Moose use is thus deflected to types with less forage production, but which might actually have more browse remaining per unit area by late winter. Throughout the boreal forest moose habitat in North America, snowshoe hares (*Lepus americanus*) have cyclic highs in population density that result in heavy impacts on available browse for two or three winters every decade (Keith and Windberg 1978, Wolff 1980). Heavy hare feeding can compete with moose feeding in certain circumstances (Dodds 1960). One could therefore predict shifts in moose habitat use to types less favoured by hares during and immediately after regional highs in the hare population cycle.

Shifts in habitat use due to weather, predators, or other disturbance or competition can all be modelled. The models might be expected to produce a general hypothesis of moose as a species whose distribution and habitat use is remarkably dependent on the optimization of energy intake and retention.

Summary

Moose first appear in the fossil record of the late Pliocene Epoch in Europe. Various moose forms spread throughout northern Eurasia and North America during the Pleistocene Epoch. Modern moose all belong to one species, *Alces alces*. Since Roman times they have lost relatively little of their world-wide distribution. Most lost range is in western Europe and central and eastern North America. Moose have been introduced to Newfoundland and adjacent areas, adding significantly to the distributional area.

Moose are distributed over vast mid-continental plains in Eurasia and North America, but also find habitats in mountains. Most of the circumpolar moose range is north of 50° N latitude. Moose are well adapted to cold but have low tolerance for hot weather. The southern boundary of their range roughly coincides with the 20 C isotherm. In eastern North America, disease may have limited moose distribution. Although moose may congregate in densities up to $15/km^2$, year-round densities seldom exceed the level of 0.5 to $1.5/km^2$.

Moose make seasonal use of a wide variety of habitats, including water bodies and marshes in some areas, all forms of forest stands, and arctic and alpine tundra. Early summer calving locations often are on islands, points, or muskeg hummocks where predation is minimized. Large stands of shrubs and saplings provide autumn and early winter habitats. Late winter habitat in flat terrain is usually dense evergreen forest, while in mountainous country valley bottoms with thin snow covers are preferred. On the basis of total year-round range, shrub lands, either produced by site and climate or by fire, logging, or insect disturbance, are most preferred.

Habitats used by moose may be ranked along a continuum from unstable to stable. Upland boreal forest has the lowest stability due to frequent burning. Mixed deciduous/coniferous forest and subalpine shrub have medium stability, while large deltas, flood plains, and stream valley shrub lands have high stability. Stable habitats provide a permanent refuge for moose populations, while disturbances in upland forest provide transient habitat supporting population increases.

The basic mechanism controlling moose movement and habitat selection appears to be optimization of energy status. However, distribution and habitat selection are also influenced by co-actions with other herbivores and disturbance by predators and man.

References

Alexander, M. E. 1980. Forest fire history research in Ontario: A problem analysis. pp. 96–109. In M. A. Stokes, and J. H. Dieterich, eds. Proceedings of the fire history workshop, October 20–24, 1980, Tucson, Arizona. USDA Forest Serv. Gen. Tech. Rep. RM-81. Rocky Mtn. Forest and Range Exp. Sta., Fort Collins, Colorado. 142 pp.

Allison, L. N.d. Personal Communication. Canadian Wildlife Service.

Belovsky, G. E. Diet optimization in a generalist herbivore: The moose. Theoret. Pop. Biol. 14: 105–134.

———. 1981. Optimal activity times and habitat choice of moose. Oecologia 48: 22–30.

Berg, W. E., and R. L. Phillips. 1974. Habitat use by moose in northwestern Minnesota with reference to other heavily willowed areas. Le natur. can. 101: 101–116.

Bergerud, A. T., and F. Manuel. 1968. Moose damage to balsam fir-white birch forests in central Newfoundland. J. Wildl. Manage. 32: 729–746.

Best, D. A., G. M. Lynch, and O. J. Rongstad. 1977. Annual spring movements of moose to mineral licks in Swan Hills, Alberta. Proc. North American Moose Conference and Workshop 13: 215–228.

Botkin, D., P. Jordan, A. Dominski, H. Lowendorf, and G. Hutchinson. 1973. Sodium dynamics in a northern forest ecosystem. Proc. Nat. Acad. Sci. 70: 2745–2748.

Bouchard, R. 1970. Etude préliminaire du comportement de l'orignal dans une vasiere de la réserve Matane, été 1965. Ministere du Tourisme, de la Chasse et de la Peche du Québec, Service de la Faune, Rapport 5: 235–253.

Brassard, J. M., E. Audy, M. Crête, and P. Grenier. 1974. Distribution and winter habitat of moose in Quebec. Le natur. can. 101: 67–80.

Cairns, A. L., and E. S. Telfer. 1980. Habitat use by 4 sympatric ungulates in boreal mixedwood forest. J. Wildl. Manage. 44: 849–857.

Carbyn, L. N.d. Personal Communication. Canadian Wildlife Service.

Crête, M. 1977. Importance de la coupe forestiere sur l'habitat hivernal del'orignal dans le sud-ouest du Quebec. Can. J. For. Res. 7: 241–257.

Collins, W. B., and P. J. Urness. 1981. Habitat preferences of mule deer as rated by pellet-group distributions. J. Wildl. Manage. 45: 969–972.

Dodds, D. G. 1960. Food competition and range relationships of moose and snowshoe hare in Newfoundland. J. Wildl. Manage. 24: 52–60.

———. 1974. Distribution, habitat and status of moose in the Atlantic provinces of Canada and northwestern United States. Le natur. can. 101: 51–65.

Edwards, R. Y., and R. W. Ritcey. 1956. The migrations of a moose herd. J. Mammal. 37: 487–494.

Egerov, O. V. 1965. Wild ungulates of Yakutia U. S. Dep. Interior and Nat. Sci. Found. Washington, D. C. 189 pp. [Translated from Russian by the Israel Program for Scientific Translation.]

Eyre, F. H. (ed.). 1980. Forest cover types of the United States and Canada. Soc. Amer. Foresters, Washington, D. C. 148 pp.

Ferguson, M., and L. B. Keith. 1982. Influence of nordic skiing on distribution of

moose and elk in Elk Island National Park, Alberta. Can. Field-Nat. 96: 69–78.
Filonov, C. P., and C. D. Zykov. 1974. Dynamics of moose populations in the forest zone of the European part of the USSR and in the Urals. Le natur. can. 101: 605–613.
Flint, R. F. 1971. Glacial and Pleistocene geology. John Wiley & Sons, Inc., New York. 553 pp. + plates.
Formazov, A. N. 1946. Snow cover as an integral factor of the environment and its importance in the ecology of mammals and birds. Occ. Pub. No. 1 Boreal Inst., Univ. Alberta, Edmonton.
Franzman, A. W., C. C. Schwartz, and R. O. Peterson. 1980. Moose calf mortality in the summer on the Kenai Peninsula, Alaska. J. Wildl. Manage. 44: 764–768.
Fraser, D., and E. Reardon. 1980. Attraction of wild ungulates to mineral-rich springs in central Canada. Holarctic Ecol. 3: 36–40.
Frenzel, B. 1968. The Pleistocene vegetation of northern Eurasia. Science 161: 637–649.
Frenzel, L. D. 1974. Occurrence of moose in food of wolves as revealed by scat analyses: A review of North American studies. Le natur. can. 101: 467–479.
Gasaway, W. A. and J. W. Coady. 1974. Review of energy requirements and rumen fermentation in moose and other ruminants. Le natur. can. 101: 227–262.
Geiger, R. 1965. The climate near the ground. Harvard Univ. Press, Cambridge. 611 pp.
Geist, V. 1971. Mountain sheep. Univ. Chicago Press, Chicago. 383 pp.
———. 1974. On the evolution of reproductive potential in moose. Le natur. can. 101: 527–537.
———. 1983. On the evolution of ice age mammals and its signficance to an understanding of specialations. Assoc. Southeastern Biol. Bull. 30: 109–133.
Haglund, B. 1974. Moose relations with predators in Sweden, with special reference to bear and wolverine. Le natur. can. 101: 457–466.
Hauge, T. M. and L. B. Keith. 1981. Dynamics of moose populations in northeastern Alberta. J. Wildl. Manage. 45: 573–597.
Hudson, R. N.d. Personal communication. Department of Animal Science, University of Alberta.
Johnson, E. A. 1980. Fire recurrence and vegetation in the lichen woodlands of the Northwest Territories, Canada. pp. 110–114. In M. A. Stokes and J. H. Dieterich, eds. Proc. Fire History Workshop, October 20–24, 1980, Tucson, Arizona. USDA Forest Serv. Gen. Tech. Rep. RM–81, Rocky Mtn. Forest and Range Exp. Sta., Fort Collins, Colorado. 142 pp.
Joyal, R., and B. Scherrer. 1974. Summer observations on moose activity in western Quebec. Proc. North Amer. Moose Conf. 10: 269–278.
Kearney, S. R., and F. W. Gilbert. 1976. Habitat use by white-tailed deer and moose on sympatric range. J. Wildl. Mange. 40: 645–657.
Keith, L. B., and L. A. Windberg. 1978. A demographic analysis of the snowshoe hare cycle. Wildl. Monogr. 58. 70 pp.
Kelsall, J. P. 1972. The northern limits of moose (*Alces alces*) in western North America. J. Mammal. 53: 129–138.
———, and W. Prescott. 1971. Moose and deer behavior in snow in Fundy National Park, New Brunswick. Can. Wildl. Serv. Rep. Ser. No. 15. 27 pp.

―――, and E. S. Telfer. 1974. Biogeography of moose with particular reference to western North America. Le natur. can. 101: 117–130.

―――, E. S. Telfer, and T. W. Wright. 1977. The effects of fire on the ecology of the boreal forest with particular reference to the Canadian north: a review and selected bibliography. Can. Wildl. Serv. Occ. Pap. No. 32.

Kistchinski, A. A. 1974. The moose in northeast Siberia. Le natur. can. 101: 179–184.

Knorre, E. P. 1959. Ecology of the moose. In G. A. Novikov, ed. Transactions of the Pechora-Ilych State Game Preserve, Issue VII, Komi Book Publishers, Syktyvkar. [Translated from Russian, Can. Wildl. Serv. Transl. No. TR-RUS-85. 240 pp. typescript.]

―――. 1961. Experimental moose farming. [Popular Science Essay.] Komi Pub. House, Syktyvkar Komi, A.S.S.R. 52 pp.

Krafft, A. 1964. Management of moose in a Norwegian forest. Pap. Norwegian State Game Res. Inst., Ser. 2(16): 5–61.

Krefting, L. W. 1974. The ecology of the Isle Royale moose with special reference to the habitat. Agr. Exp. Sta., Univ. of Minnesota, Tech. Bull., 297 For. Ser. 15. 75 pp.

Kurten, B. 1968. Pleistocene mammals of Europe. Aldine Pub. Co., Chicago. 317 pp.

Leopold, D. 1933. Game management. Charles Scribner's Son, New York. 48 pp.

LeResche, R. E. 1974. Moose migration in North America. Le natur. can. 101: 393–415.

―――, R. H. Bishop, and J. W. Coady. 1974. Distribution and habitats of moose in Alaska. Le natur. can. 101: 143–178.

Lorimer, C. G. 1980. The use of land survey records in estimating presettlement fire frequency. pp. 57–62. In M. A. Stokes and J. H. Dieterich, eds. Proc. Fire History Workshop, October 20–24, 1980. Tucson, Arizona. USDA Forest Serv. Gen. Tech. Rep. RM 81. 142 pp.

Lykke, J., and I. McT. Cowan. 1968. Moose management and population dynamics on the Scandinavian Peninsula, with special reference to Norway. Proc. 5th North Amer. Conf. and Workshop. Kenai, Alaska. Kenai Moose Res. Sta., Kenai, Alaska. 22 pp.

Lynch, G. M. 1975. Best timing of moose surveys in Alberta. Proc. North Amer. Moose Conf. and Workshop 11: 141–153.

―――. N.d. Personal communication. Fish and Wildlife Division, Alberta Department of Energy and Natural Resources.

Markgren, G. 1974. The moose in Fennoscandia. Le natur. can. 101: 185–194.

―――. N.d. Personal Communication. National Swedish Environment Protection Board.

Miller, D. R. 1976. Biology of the Kaminuriak population of barren-ground caribou. Part 3: Taiga winter range relationships and diet. Can. Wildl. Serv. Rep. Ser., No. 36. 41 pp.

Miller, F. L., E. Broughton, and E. Land. 1972. Moose fatality resulting from overextension of range. J. Wildl. Dis. 8: 95–98.

Moen, A. N. 1968. Energy exchange of white-tailed deer, western Minnesota. Ecology 49: 676–682.

Morow, K. 1976. Food habits of moose from Augustow Forest. Acta Theriol. 21: 101–116.
Mould, E. 1977. Habitat relationships of moose in northern Alaska. Proc. North Amer. Moose Conf. and Workshop 13: 144–156.
Nesbitt, W. H., and J. S. Parker (eds.). 1977. North American big game. The Boone and Crockett Club and the Nat. Rifle Assoc. Washington, D.C. 367 pp.
Neu, E. W., C. R. Byers, and J. M. Peek. 1974. A technique for analysis of utilization—availability data. J. Wildl. Manage. 38: 541–545.
Odum, E. P. 1971. Fundamentals of ecology, 3rd ed. W. B. Saunders and Co., Toronto. 574 pp.
Peek, J. M. 1974. On the nature of winter habits of Shiras moose. Le natur. can. 101: 131–141.
———, D. L. Urich, and R. J. Mackie. 1976. Moose habitat selection and relationships to forest management in northeastern Minnesota. Wildl. Monogr. No. 48. 65 pp.
Peterson, R. L. 1955. North American moose. Univ. Toronto Press, Toronto. 280 pp.
Phillips, R. L., W. E. Berg, and D. B. Siniff. 1973. Movement patterns and range use of moose in northwestern Minnesota. J. Wildl. Manage. 37: 266–278.
Prescott, W. H. 1974. Interrelationships of moose and deer of the genus *Odocoileus*. Le natur. can. 101: 493–504.
Pruitt, W. O. 1959. Snow as a factor in the winter ecology of the barren-ground caribou. Arctic 12: 159–179.
Pulliainen, E. 1974. Seasonal movements of moose in Europe. Le natur. can. 101: 379–392.
Reynolds, S. H. 1934. *Alces*. Part 7a. *In* Dawkins, W. B. and S. H. Reynolds, 1872–1962. A monograph of the British Pleistocene Mammalia Vol. III British Pleistocene Artodactylia. Paleontographical Soc., London.
Rolley, R. E., and L. B. Keith. 1980. Moose population dynamics and winter habitat use at Rochester, Alberta, 1965–1979. Can. Field-Natur. 94: 9–18.
Rounds, R. C. 1981. First approximation of habitat selectivity of ungulates on extensive winter ranges. J. Wildl. Manage. 45: 187–196.
Sablina, T. B. 1970. Evolyutsiya pishchevaritel'nor sistemy olenei [The evolution of the digestive system in deer.] [Translated from Russian for the Can. Wildl. Serv.] Nauka, Moscow. 338 pp. typescript.
Samuel, W. N.d. Personal communication. Department of Zoology, University of Alberta.
Severinghaus, C. W., and L. W. Jackson. 1970. Feasibility of stocking moose in the Adirondacks. N.Y. Fish and Game J. 17: 18–32.
Skogland, T. 1978. Characteristics of the snow cover and its relationship to wild mountain reindeer (*Rangifer tarandus tarandus*) feeding strategies. Arctic & Alpine Res. 10: 569–580.
Stelfox, J. G., and R. D. Taber. 1969. Big game in the northern Rocky Mountain coniferous forest. pp. 197–222. *In* Center for Natural Resource (ed.). Proc. Symposium on Coniferous Forests of the Northern Rocky Mountains. Missoula, 1968. 395 pp.
Steven, H. M., and A. Carlisle. 1959. The native pinewoods of Scotland. Oliver

and Boyd, Edinburgh. 368 pp.

Stewart, R. R., R. R. MacLennan, and J. D. Kinnear. 1977. The relationship of plant phenology to moose. Saskatchewan Department of Tourism and Renewable Resources, Tech. Bull. No. 3. 20 pp.

Syroechkovskiy, E. E., and E. V. Rogacheva. 1974. Moose in the Asiatic part of the USSR. Le natur. can. 101: 595–604.

Telfer, E. S. 1967a. Comparison of a deer yard and a moose yard in Nova Scotia. Can. J. Zool. 45: 485–490.

———. 1967b. Comparison of moose and deer winter range in Nova Scotia. J. Wildl. Manage. 31: 418–425.

———. 1968a. The status of moose in Nova Scotia. J. Mammal. 49: 325–326.

———. 1968b. Distribution and association of moose and deer in central New Brunswick. Paper presented at Northeast Fish and Wildl. Conf., Bedford, N.H. 17 pp. + append. (mimeo).

———. 1970. Winter habitat selection by moose and white-tailed deer. J. Wildl. Manage. 34: 553–559.

———. 1971. Changes in carrying capacity of deer range in western Nova Scotia. Can. Field-Natur. 85: 231–234.

———. 1972a. Forage yield in two forest zones of New Brunswick and Nova Scotia. J. Range. Manage. 25: 446–449.

———. 1972b. Forage yield and browse utilization on logged areas in New Brunswick. Can. J. Forest Res. 2: 346–350.

———. 1978a. Cervid distribution, browse and snow cover in Alberta. J. Wildl. Manage. 42: 352–361.

———. 1978b. Habitat requirements of moose—the principal taiga range animal. Proc. 1st Int. Rangeland Congress: 462–465.

Thompson, I. N.d. Personal communication. Canadian Wildlife Service, Ottawa.

Van Tighem, K. N.d. Personal communication. Canadian Wildlife Service, Ottawa.

Van Wormer, J. 1972. The world of the moose. J. B. Lippincott Co., New York. 160 pp.

Wein, R. W., and J. M. Moore. 1979. Fire history and recent fire rotation periods in the Nova Scotia Acadian forest. Can. J. Forest Res. 9: 116–178.

Wolfe, M. L. 1974. An overview of moose coactions with other animals. Le natur. can. 101: 437–456.

Wolff, J. O. 1980. The role of habitat patchiness in the population dynamics of snowshoe hares. Ecol. Monogr. 50: 111–130.

Zackrisson, O. 1980. Forest fire history: Ecological significance and dating problems in the North Swedish boreal forest. pp. 120–125. In M. A. Stokes and J. H. Dieterich, eds. Proc. Fire History Workshop, Tucson, Arizona, October 20–24, 1980: 120–125. Gen. Tech. Rep. RM-81. Rocky Mtn. Forest and Range Exp. Sta., Fort Collins, Colorado. 142 pp.

Population Growth in an Introduced Herd of Wood Bison (Bison bison athabascae)

George W. Calef

Abstract

In August 1963, 18 bison were transplanted from Wood Buffalo National Park to the Mackenzie Bison Sanctuary near Great Slave Lake, Northwest Territories. The animals were considered a pure strain of the wood bison (*Bison bison athabascae*) and were believed to be free of disease. The transplant was undertaken to preserve the wood bison genotype in the wild from loss through hybridization with plains bison (*B. b. bison*), which were introduced into the park between 1925 and 1928. The transplanted population of wood bison grew at an almost constant rate of 26% per year according to the growth equation $Y = 17.28e^{0.223X}$, ($r^2 = .99$), and numbered at least 560 animals in April 1979. The observed rate of increase ($r_i = 0.233$) is believed to be the maximum rate of increase (r_{max}) for bison under the natural conditions prevailing in the southwestern Northwest Territories, the environment from which the holotype of this subspecies was collected. The growth rate of the introduced population was much higher than has been observed for other bison populations in the Slave River region, primarily because of a low mortality rate. The high survival rate (greater than 95% in both calves and adults) in the Mackenzie Sanctuary bison population was attributed to a low incidence of disease and predation.

Introduction

When European explorers first visited the regions near the Slave River and Great Slave Lake in the Northwest Territories they found bison as part of the native fauna, living on the sedge and grass prairies in the boreal forest. Rhodes (1897) subsequently described the northern wood bison (from a bull killed near Fort Resolution, Northwest Territories) as a distinct subspecies (*Bison bison athabascae*) on the basis of its large size and dark woolly pelage (Banfield 1961).

In the late nineteenth century, the wood bison suffered much the same fate as their relatives on the plains (*B. b. bison*), being gradually extirpated from their ranges. By 1900 only a few hundred wood bison remained in the wilderness between Great Slave Lake and the Peace-Athabasca Delta. In 1893 the Canadian government acted to protect the wood bison from hunting, and in 1922, Wood Buffalo National Park was established to protect the recovering population and its habitat (Fuller 1966).

Unfortunately, in a short-sighted effort to hasten the recovery of bison in the park, the government transported over 6,600 plains bison from Wainwright Buffalo Park, Alberta and released them west of the Slave River, between the Peace-Athabasca Delta and Fort Fitzgerald. These introductions, while having the desired effect of increasing the population, resulted in hybridization between the subspecies and in "the virtual disappearance of the pure wood buffalo" (Fuller 1966).

However, in 1959 Novakowski located some bison in the almost inaccessible northwest corner of the park, isolated from the nearest hybrid bison by more than 75 miles (120 km). Five specimens were collected from a herd of about 200 animals along the Nyarling River in 1959. Banfield and Novakowski (1960) concluded that these animals indeed represented pure wood bison, a conclusion based on size, pelage, and skull measurements, and since corroborated for the descendants of these animals by Geist and Karsten (1977), who considered additional morphological features.

In February 1963, 77 bison were captured from the Nyarling River herd for transplanting to other areas where the wood bison genotype could be preserved from hybridization. The captured bison were held and tested for tuberculosis and brucellosis. Those testing positive were released back into the park. After a second testing for disease, 18 animals were released into the wild in the Mackenzie Bison Sanctuary on 14 August 1963, about 15 miles (24 km) northeast of Fort Providence, Northwest Territories (fig. 1). Another 23 bison, after similar screening for disease, were transported to Elk Island National Park in 1965. These were to be bred under controlled conditions to establish a source herd from which further transfers could be made. This paper describes and discusses the exponential increase of the free-ranging population in the Mackenzie Bison Sanctuary.

The Study Area

The Mackenzie Bison Sanctuary encompasses more than 7,000 square miles (18,000 km²) along the west shore of Great Slave Lake (fig. 1). The

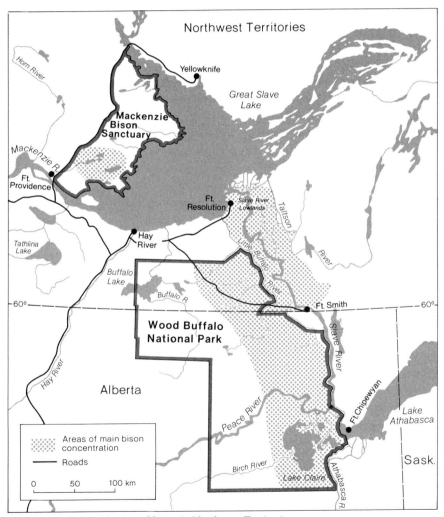

Fig. 1. Current distribution of bison in Northwest Territories.

area consists of extensive stands of jackpine (*Pinus banksiana*) and spruce (*Picea mariana*, *P. glauca*), which characterize the Upper Mackenzie section of the boreal forest region (Rowe 1972). The Upper Mackenzie section includes most of the range of the bison in the Northwest Territories and encompasses the Slave River Lowlands, the area from which the type specimen of the wood bison was collected (Banfield 1961). Thus, the Mackenzie Bison Sanctuary is similar in its major plant communities and climate to the historical ranges of the wood bison, and to the current ranges of hybrid bison in the Northwest Territories. Jacobson (1976) has described the geomorphology, plant associations and succession, and associated fauna of the sanctuary in more detail.

Within the area currently occupied by the bison in the sanctuary, numerous shallow lakes and large bogs break the forest, and sedge meadows and willow-sedge parklands surround the lakes. Many of the lakes appear to be drying up, and vegetation is recolonizing the lake bottoms. The bison feed most heavily on the wet meadows along the margins of the major lakes. Reynolds et al. (1978) found that bison preferred to feed in the wet meadows on the Slave River Lowlands, and Peden and Reynolds (1981) showed that the productivity of wet meadows in the Mackenzie Sanctuary was similar to that of wet meadows on the Slave River Lowlands. Jacobson (1976) concluded that the plant associations of the sanctuary were quantitatively similar to those of Wood Buffalo National Park. Although the food habits of the wood bison in the sanctuary have not been studied, presumably they will be similar to those of the Slave River Lowlands bison, mainly sedges (*Carex* spp.) and reed grasses (*Calamagrostis* spp.) (Reynolds et al. 1978).

Methods

Total counts of the bison in the Mackenzie Sanctuary were made by aerial survey in 1967, 1969, and at least once each year from 1972 through 1981 inclusive. The counts were conducted by various employees of the Northwest Territories Wildlife Service and the Canadian Wildlife Service. Unpublished reports detailing the results of individual surveys are on file with the Northwest Territories Wildlife Service in Yellowknife.

The survey procedure was to search all of the meadows in the area of bison concentration and the major trails connecting them. The lakes were approached at an altitude of 700 to 1,000 ft (210 to 300 m) above ground level, usually from the north, as the bison tend to feed along the north shores, especially in winter. Most bison, except mature males in small groups, react to aircraft by fleeing toward the forest. The high altitude approach allowed the observers to see the entire meadow and note the location of the herds. Most of the meadows were small enough to see at a glance; the half-dozen or so larger meadows were circumnavigated.

By circling with the aircraft each group of bison could be counted before the animals ran into the forest. When the observers felt that any group was too large to count accurately during the survey, a 35 mm colour photograph was taken to obtain a correct count. The location of each group was plotted on a 1:250,000 scale topographic map to record the distribution and to avoid recounting any groups.

Although undoubtedly a few animals were missed during some aerial surveys, the surveys probably located a high proportion of the bison in the sanctuary, and the counts were in all likelihood accurate. The bison were always located in the same area, and the majority of them were always found in the open meadows around the half-dozen or so major lakes. Most of the surveys were carried out in late winter (January to March); the massive, dark bison show up well against the snow in the open meadows, and they move when the aircraft approaches. When observers located tracks in an area they searched until the animals were found. Most of the groups were small enough to be easily counted; usually two observers verified the total for each group. If any doubt existed as to whether an accurate count had been obtained, a photograph was taken for later confirmation. Finally, as will be shown later, the rate of population increase revealed by successive surveys was close to the physiological maximum for the species; it is not possible that many animals were missed.

Estimation of Calf Production

Counts of calves were made on three surveys. Each herd was circled at a low altitude until the observers were satisfied that an accurate count had been obtained. The small, reddish calves are easily recognizable during the summer months, but care must be taken that calves who are hidden behind adults are not missed. For this reason, aerial photographs cannot be taken to confirm calf counts (Calef 1976).

Results and Discussion

Population Growth Rate

The population of 18 bison released in the Mackenzie Bison Sanctuary in August 1963 has grown exponentially to over 700 animals prior to calving in 1981. This represents a yearly average increase of almost 26% (table 1).

Table 1   Observed increase of the wood bison population in the Mackenzie Bison Sanctuary, 1963–80

	Number of bison observed	Population expected if growth rate = 26%/year
1963	18	18
1964	—	23
1965	—	30
1966	—	38
1967	42	48
1968	56	60
1969	—	76
1970	—	96
1971	111	121
1972	122	152
1973	171	192
1974	237	242
1975	299	305
1976	356	384
1977	450	484
1978	580	610
1979	645	768
1980	701	967

The rate of population increase over the 15-year period from 1963 to 1979 was remarkably constant; the least squares regression that best fits the change in the number of bison counted over time, $Y = 17.28e^{0.233X}$, has a correlation coefficient (r^2) of 0.99 (fig. 2); thus the instantaneous rate of increase (r_i) = 0.233 (Caughley and Birch 1971). The results of the 1980 and 1981 censuses are not included in calculating the rate of increase because these counts may have underestimated the population. However, including the 1980–1981 results changes the calculation only slightly.

Any population growing at a constant rate of increase soon reaches a stable age distribution. In the case of the bison population in the Mackenzie Sanctuary, calves account for approximately 20% of the total number of animals, yearlings for 16%, 2- to 3-year-olds for 13%, and 4-year-olds and older for about 51% of the population (table 2).

The observed 26% annual rate of population growth requires a minimum calf crop (calves as a percentage of total animals, including calves) of at least 20.6% $\left(\dfrac{26}{100 + 26}\right)$, assuming that all calves survive and no adults die. The observed calf percentage during the 3 years for which it was estimated averaged just slightly higher than this required minimum (table 3).

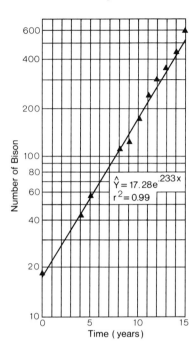

Fig. 2. Population growth of wood bison in Mackenzie Bison Sanctuary, 1963–1979.

$\hat{Y} = 17.28 e^{.233x}$
$r^2 = 0.99$

Since the observed calf percentage is only slightly higher than that required to account for the observed population increase, survival must be greater than 95% among both calves and adults.*

Although age-specific reproductive rates have not been measured for the wood bison of the Mackenzie Sanctuary, a reasonable estimate may be made based on a few known facts about bison reproduction. The primary sex ratio in bison is close to 1:1, with a slight preponderance of males (Fuller 1966, Meagher 1973, Haugen 1974). Adult sex ratios are more difficult to determine. Fuller (1960) felt that females somewhat outnum-

*Calculation for calf mortality is as follows: average observed calf crop = 21.5%, which is equivalent to 27.38 calves per 100 adults (i.e, $\frac{27.38}{100 + 27.38} \times 100 = 21.5\%$). Calf production required for a 26% annual increase = $\frac{26}{100 + 26} \times 100 = 20.6\%$. Thus, the greatest possible mortality = 27.38 − 26. Therefore maximum calf mortality = $\frac{1.38}{27.38} \times 100 = 5\%$. These calculations assume no adult mortality. If some adults are dying, then calf mortality must be less than 5% to compensate, since the observed calf crop is so close to the required calf crop required for a 26% increment that the maximum total mortality for the herd (calf and adult) equals 5%.

Table 2   The establishment of a stable age distribution by a population of 18 animals growing at 26% per year

Population size	\multicolumn{15}{c}{Age class (yrs)}

Population size	1	2	3	4	5	6	7	8	9	10	11	12	13	14	15
18															
23	5														
30	7	5													
38	8	7	5												
48	10	8	7	5											
60	12	10	8	7	5										
	(20)	(16.6)	(13.3)												
76	16	12	10	8	7	5									
96	20	16	12	10	8	7	5								
121	25	20	16	12	10	8	7	5							
152	31	25	20	16	12	10	8	7	5						
192	40	31	25	20	16	12	10	8	7	5					
242	50	40	31	25	20	16	12	10	8	7	5				
305	63	50	40	31	25	20	16	12	10	8	7	5			
384	79	63	50	40	31	25	20	16	12	10	8	7	5		
484	100	79	63	50	40	31	25	20	16	12	10	8	7	5	
610	126	100	79	63	50	40	31	25	20	16	12	10	8	7	5
	(20.6)	(16.4)	(13.0)												

Note

Numbers in parenthesis indicate percentage of total population. Note that after 5 years an essentially stable age distribution has been achieved. Population size for any year is the previous year's population multiplied by 1.26. The number of animals in age class 0–1 in any year is 0.26 times previous year's population. All age classes are moved into the next higher age class each year. The total population (column 1) is always the sum of all age classes + 18 — the original population.

George Calef's colour photographs of northern landscapes and wildlife illustrate the fragile and sensitive ecology of the North.
Caribou trail, winter range north of Yellowknife, Northwest Territories.

Rutting caribou, Blackstone River Valley, Yukon Territory (top); Mountain caribou bull, Mackenzie Mountains, Northwest Territories (bottom).

Bull caribou crossing Blackstone River, Yukon Territory (top); Post-calving herd, Bathurst caribou herd, near Bathurst Inlet, Northwest Territories (bottom).

Bearberry and glacial boulders, Big Lake, Northwest Territories (top); Breakup on the barren-lands, Bathurst caribou calving ground, Northwest Territories (bottom).

Autumn tundra, Big Lake, Northwest Territories.

Bull moose in early autumn blizzard, Mount McKinley National Park, Alaska.

Bull moose, Mount McKinley National Park, Alaska (top); Bull bison, Slave River Lowlands, Northwest Territories (middle); Bison herd, Slave River Lowlands, Northwest Territories (bottom).

Dall sheep rams, Mount McKinley National Park, Alaska.

Colour photographs by George Calef.

Table 3 Observed calf production by wood bison in the Mackenzie Bison Sanctuary

Date of survey	No. of bison classified	Sample as percentage of estimated population	No. of calves	Percentage of calves
2 August 1975	299	100	62	20.7
12 July 1976	239	67	51	21.3
11 June 1977	359	80	80	22.3
Total	897	–	193	21.5

Table 4 Pregnancy rates of selected bison populations (sample sizes in parentheses)

| Population | Age of females | | | | All females |
	1–2	2–3	3–4	>4	
Wood Buffalo Park[1]	3%(NA)	39%(NA)	66%(NA)	68%(NA)	57% (733)
Ft. Niobrara Nat. Wildlife Refuge[2]	0%(17)	87%(39)	65%(23)	87%(81)	75% (160)
Yellowstone Park[3] (1940–44)	0%(15)	0% (2)	50% (4)	>90% (80)	86% (101)
Yellowstone Park[3] (1964–66)	–	14% (7)	27%(11)	>57% (53)	52% (71)

Notes
1. Fuller, 1966.
2. Haugen, 1974.
3. Meagher, 1973.

bered males in Wood Buffalo Park, but Novakowski (1958) presented results of a slaughter in the park that indicated a sex ratio close to 1:1. However, because the mortality of both calves and adults has been so low in the Mackenzie Sanctuary, the adult sex ratio must be very similar to the primary ratio, and for all practical purposes it can be assumed to be 1:1.

Bison virtually always produce a single offspring and rarely give birth before 3 years of age (Fuller 1966, Haugen 1974). Pregnancy rates vary considerably from herd to herd and from year to year, but under favourable conditions they are as high as 90% for females 3 years old and older (table 4).

Table 5 Probable age-specific reproduction of wood bison in the Mackenzie Bison Sanctuary

	Age class				
	0–1	1–2	2–3	4+	Total
Age structure of population[1]	20	16	13	51	100
Number of cows assuming 1:1 sex-ratio	10	8	6	26	50
Pregnancy rate (%)	0	0	90	90	
Calves produced	0	0	5	23	28
Expected calf crop as percentage of total population					21.9 ($^{28}/_{128}$)
Observed calf crop[2]					21.5

Notes
1. See Table 2.
2. See Table 3.

Using these facts I have constructed the most probable reproductive schedule for the bison of the Mackenzie Sanctuary (table 5). The observed calf crops will be produced if 90% of the females aged 3 years and older give birth each year. Possibly some 2-year-old animals give birth, as has been observed in semi-domestic populations (Meagher 1973), but that is at odds with what has been observed in other free-ranging populations and is not required in theory to produce either the observed population growth or the observed calf crops. To postulate yearling pregnancy in the Mackenzie Sanctuary population seems an unnecessary assumption.

Because the reproductive rate is near the maximum for the species under favourable natural conditions, and because survival of both juveniles and adults is greater than 95%, the rate of increase $r_i = 0.233$ must be considered the maximum rate of increase (r_{max}) for wood bison in the Upper Mackenzie section of the boreal forest the holotype environment for the subspecies.

Causes of the Irruption of the Wood Bison Population

Although the reproduction of the introduced wood bison has on average been somewhat higher than that of other bison herds in the Northwest Territories, the main difference lies in the much lower death rate in the introduced population, especially among calves (table 6). The close agree-

Table 6 Comparison of productivity and mortality statistics for three bison populations in the Northwest Territories

Population	Period	Population trend	Initial calves (%)	Calf mortality (%)	Adult mortality (%)	Source
Mackenzie Bison Sanctuary	1963–79	Increasing at 26%/year	21.5	<5	<5	Present paper
Slave River Lowlands	1974–76	Decline of 35% in 1975; decline of 27% in 1976	18	94	33/1975 26/1976	Calef, 1976
Slave River Lowlands	1976–78[1]	Decline of 16.7 in 1977; no decline in 1978	17–19	70	21/1977 5/1978	Van Camp, 1978a
Wood Buffalo National Park	1950–53	Stable	18–23[2]	72	8	Fuller, 1966

Notes
1. The wolf population was reduced substantially between 1976 and 1978 by a control program and high natural losses.
2. The 23% calf crop is unlikely from reproductive data given by Fuller, 1966. This figure comes from classifications made in July when bulls are segregated from cows thus increasing the apparent percentage of calves: I have used an initial calf crop of 20% in computing calf mortality.

ment between the estimated number of calves and the increment required to produce the observed population increase means that very few animals die each year. Indeed, some of the original 18 bison may still be alive today, as bison may live for 40 years. Observations made during the surveys provide corroborating evidence of low mortality; no dead bison or bison skeletons have ever been reported from the sanctuary. During the surveys in the Slave River Lowlands and in Wood Buffalo National Park, observers regularly see remains of bison that have been killed by wolves (*Canis lupus*) and have died of various diseases. For example, on a single survey of the Slave River Lowlands in March 1975, I saw nine bison carcasses (Calef 1975). Bison skulls and other bones are easily found in these areas. In contrast, I have walked along and flown over the north shore of Falaise Lake, the area most heavily used by bison in the sanctuary, without finding any skeletal remains.

Predation and disease constitute the major causes of death among the bison of the Slave River Lowlands and Wood Buffalo National Park. Many authors have reported wolves preying on bison (Soper 1941, Fuller 1966, Calef 1976, Van Camp 1978a). Bison form the bulk of wolves' diet in these areas, especially in winter (Fuller and Novakowski 1955, Oosenbrug and Carbyn in press). Calef (1976) and Van Camp (1978a) believed that predation by wolves played a major role in the decline of the bison population of the Slave River Lowlands from 1974 to 1978.

Brucellosis, tuberculosis, and anthrax have all been documented in the bison of Wood Buffalo Park and the Slave River Lowlands (Fuller 1966, Choquette and Broughton 1967, Coupland 1975). Fuller (1966) reported that approximately 40% of the bison examined in Wood Buffalo Park had either tuberculosis or brucellosis, and estimated that 4 to 6% of the bison died from tuberculosis each year. Anthrax outbreaks on the bison ranges south of Great Slave Lake have occurred regularly during the past 20 years, claiming the lives of hundreds of bison during some summers (Coupland 1975).

The wood bison that were captured on the Nyarling River were screened twice for diseases. Those released into the sanctuary were believed to be uncontaminated (Novakowski 1963). However, several cases of tuberculosis and one case of brucellosis have turned up in the wood bison of Elk Island National Park (Babbage undated). These wood bison are descendants of animals captured from the same herd as the Mackenzie Sanctuary bison and were screened for tuberculosis and brucellosis in the same way. The bison were transferred to Elk Island National Park in 1965, and tuberculosis was discovered in 1969. Thus it is not clear whether the wood bison brought tuberculosis with them from Wood Buffalo National Park, or contracted the disease at Elk Island National Park. Although subsequent testing has not been carried out in the sanctuary, it seems unlikely that disease could be either causing significant mortality or lowering the reproductive rate. A collection of ten animals to test for disease is planned for the spring of 1982.

Wolves appear to be absent from the Mackenzie Bison Sanctuary. Only once have wolves been observed during aerial surveys, and no record of kills exists. Twice I have camped for a week on Falaise Lake (once in September and once in late November) and walked throughout the areas used by the bison without seeing any wolves or their tracks, or hearing any howling.

Apparently wolves were once abundant in the region, but they were poisoned during the late 1950s (Art Look, former Northwest Territories game officer and local trapper, pers. com.). Why they have not returned is

Population Growth in an Introduced Herd of Wood Bison   195

puzzling, for wolves are common on the caribou ranges north of the sanctuary, and in addition to bison, moose (*Alces alces*) and woodland caribou (*Rangifer tarandus caribou*) are common in the sanctuary, offering suitable prey. Even if wolves do return to the area they might not immediately begin to prey on bison. Apparently wolves must learn to kill certain prey. The wild bison herd near Delta Junction in Alaska has not lost any of its members to wolves, although the predators are common in the region and prey on moose and caribou (Van Camp pers. com.). On the Slave River Lowlands the situation is reversed; the wolves prey almost exclusively on bison and seldom take moose.

Comparison with Other Ungulate Population Irruptions
The exponential increase of the Mackenzie wood bison represents a particularly well-documented example of the irruption of an ungulate population. Caughley (1970) states that such irruptions are most commonly observed in introduced populations, but may also occur in established populations whenever there is "a large discrepancy between the number of animals the environment can carry, and the number of animals actually present" (Riney 1964). Caughley (1970) stated that irruptions can occur in the presence of predators, although most of the examples he cited involved areas in which predators were absent or reduced. He also provided evidence that in an irruption of thar (*Hemitragus jemlahicus*) in New Zealand, the lowering of the death rate was far more significant than the increase in the birth rate, a conclusion echoed by Bergerud (1980) for several populations of North American caribou (*Rangifer tarandus*). Bergerud argued, however, that the absence or reduction of predators initiated caribou irruptions and there is similar evidence from studies of other North American ungulates (Peek 1980). Thus a low predator population is probably a necessary if not a sufficient requirement for an irruption of ungulates.

Wood bison are enjoying two advantages that stimulate a rapid population increase: an underexploited habitat and a scarcity of predators. Like the irrupting populations of caribou and thar mentioned above, the wood bison currently have a much reduced death rate compared with other bison populations in the Northwest Territories.

The Future of the Wood Bison Population
Van Camp (1978c) and others have suggested that the increase rate of the wood bison population is slowing. This suggestion is based on observations of decreased yearling percentages during late-winter surveys. In my

opinion yearlings in this population cannot be identified accurately from the air in late winter and are usually underestimated. The fact that the surveys for 1979 and 1980 recorded lower population estimates (645 and 701 respectively) than would be expected, based on the growth rate of 26% per year, is also inconclusive. The observers on both those surveys reported that they probably missed some animals. Perhaps the population has now reached a size at which completely accurate total counts are no longer feasible.

I believe that the wood bison population of the Mackenzie Sanctuary will continue its rapid growth for many more years. The carrying capacity of the sanctuary has been conservatively estimated at 14,000 head, based on standing crop and productivity of the forage (Peden and Reynolds 1981). Moreover, most expanding populations go through a period of dispersal, during which emigration occurs as the population continues to expand its range. Dispersal has not yet occurred in the wood bison population, despite the presence of large tracts of unoccupied habitat adjacent to the currently used range.

However, the time must come when the increase of the Mackenzie bison population will halt. In many ungulate populations the period of increase is followed by a period of stability in numbers, and then a population crash (Caughley 1970). Whether such a decline can be forestalled and high productivity maintained through cropping by either humans or other predators is not clear.

The transplanting of the threatened wood bison to the Mackenzie Bison Sanctuary stands as a remarkable success in the conservation of the subspecies. Let us hope that the new transplant of 28 wood bison that took place near Nahanni Butte in the summer of 1980 will prove equally successful, and that thriving populations of these magnificent animals, the largest terrestrial grazers in North America, will ultimately be of great benefit to the residents of the Northwest Territories.

Afterword

Almost 100 years have passed since the last herds of wild buffalo vanished from the Great Plains. The extermination of the buffalo was merely the most spectacular and important loss (in human terms) among the scores of species of wild animals and plants destroyed by the pioneers' invasion of the North American frontier.

I have shown how close the wild wood bison came to following their plains counterparts into oblivion. Yet today, in the Northwest Territories, truly wild bison are living where they have lived from time immemorial, chewed by flies, battered by blizzards, trembling in the dark at the hunting pack's howl. It is ironic but perhaps fitting that the "last frontier," as the northern wilderness is often called, has provided sanctuary for these refugees from the first frontier.

But have we learned our lesson? I fear not. Proposals are afoot to turn the Slave River Lowlands into a game ranching area; to make the bison semi-domestic creatures. The Northwest Territories government is regularly confronted with schemes for farming, for ranching musk-oxen, for introducing reindeer to replace caribou. The Yukon government, feeling the same pressure, is at work on an agricultural policy. Hidden in all these proposals are the seeds of the destruction of the wilderness: predator control, road building, land clearing, fencing, and competition for good grazing lands.

Virtually all of the world's best agricultural land is already under intensive cultivation, the once-rich biological communities shredded by massive monocultures. Northern agriculture promises to be, at best, a marginal enterprise. Most of its proponents are attracted by the lure of free land and the prospect of subsidies from fledgling governments that want "development." The wilderness that northern agriculture would destroy has a value far beyond the few head of skinny cattle (or beefalos) or bales of hay it might produce. As Henry Beston said, "The world today is sick to its thin blood for lack of elemental things. . . . " We do not need a few more acres of marginal farm land. We need whole environments, so that we may feel like whole men again; we seek, like Thoreau, "an entire heaven and an entire earth."

Aldo Leopold put it beautifully:

"Man always kills the things he loves, and so we the pioneers have killed our wilderness. Some say we had to. Be that as it may, I am glad I shall never be young without wild country to be young in. Of what avail are forty freedoms without a blank spot on the map?"

Perhaps we, like the wood bison, have our last chance in the lands of the North.

In June 1982, 1002 bison were counted in the Mackenzie Bison Sanctuary, indicating that the population is still expanding rapidly. The dispersal phase seems to be starting for this population; in 1983 several dozen bison were observed at Mink Lake, about 40 km to the northwest of the centre of concentration of the population.

Ten adult bison were killed and tested for disease in March 1982. No evidence of tuberculosis or brucellosis was detected, and the animals were all in excellent condition. However, only 2 of 5 females were pregnant. Aerial surveys conducted since 1979 also have indicated lower calf ratios in the herd now than those reported from 1975 to 1977.

Three dead bison, apparently killed by wolves, have been observed in the past two years.

Acknowledgments

I thank the Northwest Territories Wildlife Service for providing me with the opportunity to study the bison on the Slave River Lowlands and the Mackenzie Bison Sanctuary from 1975 to 1977, and for their co-operation during the preparation of this paper. I am grateful to Frank Geddes for providing a helpful critical review of the manuscript and for encouraging me to include an afterword. Hal Reynolds and Val Geist also offered suggestions for improving the manuscript.

References

Babbage, G. W. Undated. Wood bison (*athabascae*) management isolation area. Unpub. rep., Elk Island Nat. Park. 5 pp. typescript.

Banfield, A. W. F. 1961. The wood bison type specimen. J. Mammal. 42: 553–554.

———, and N. S. Novakowski. 1960. The survival of the wood bison (*Bison bison athabascae* Rhoads) in the Northwest Territories. Nat. Mus. Can. Nat. Hist. Pap. No. 8. 6 pp.

Bergerud, A. T. 1980. A review of the dynamics of caribou and wild reindeer in North America. *In* E. Reimers, E. Garre, and S. Skjenneberg, eds. Proc. 2nd Int. Reindeer/Caribou Symp., Røros, Norway, 1979: 556–581. Direktoratet for vilt og ferskvannsfisk, Trondheim.

Blood, D. A. 1963. Structure of confined bison herds with regard to prevention of excessive inbreeding. Unpub. rep., Can. Wildl. Serv. 12 pp. typescript.

Calef, G. W. 1975. Bison census: Game management zones 3 and 5. March 31–April 2, 1975. Unpub. rep., N.W.T. Wildl. Serv. 9 pp. typescript.

———. 1976. Status of bison in the N.W.T.: A progress report. Unpub. rep., N.W.T. Wildl. Serv. 68 pp. + append., typescript.

Caughley, G. 1970. Eruption of ungulate populations, with emphasis on Himalayan Thar in New Zealand. Ecology 51: 53–71.

———, and L. C. Birch. 1971. Rate of increase. J. Wildl. Manage. 35: 658–663.

Choquette, L. P. E., and E. Broughton. 1967. Anthrax in bison: Wood Buffalo National Park and the Northwest Territories Report for the year 1967. Unpub. rep., Can. Wildl. Serv. 23 pp. + append., typescript.

Coupland, R. W. 1975. Progress report: Comparisons of cattle and bison for meat production on prairie ranges in the Mackenzie District: The disease aspect. Unpub. rep., Can. Wildl. Serv. 11 pp. typescript.

Fuller, W. A. 1960. Behavior and social organization of the wild bison of Wood Buffalo National Park, Canada. Arctic 13: 3–19.

———. 1966. The biology and management of the bison of Wood Buffalo National Park. Can. Wildl. Serv. Manage. Bull. Ser. 1, No. 16. 52 pp.

———, and N. S. Novakowski. 1955. Wolf control operations, Wood Buffalo National Park, 1951–52. Can. Wildl. Serv. Wildl. Manage. Bull. Ser. 1, No. 11. Ottawa. 23 pp.

Geist, V., and P. Karsten. 1977. The wood bison in relation to hypotheses on the origin of the American bison. Z. fur Saugetierkunde 42: 119–127.

Haugen, A. O. 1974. Reproduction in the plains bison. Iowa St. J. Res. 49: 1–8.

Jacobson, R. 1976. Preliminary habitat assessment: Mackenzie Bison Sanctuary, N.W.T. Unpub. rep., Can. Wildl. Serv. 49 pp. typescript.

Meagher, M. M. 1973. The bison of Yellowstone National Park. U.S. Nat. Park. Serv. Sci. Monogr. Ser. No. 1. 176 pp.

Novakowski, N. S. 1958. Report on the tagging, testing and slaughtering of bison in the Lake Claire area, Wood Buffalo National Park, October, November 1957. Unpub. rep., Can. Wildl. Serv. 14 pp. typescript.

———. 1963. Wood bison transfer: Completion report. Can. Wildl. Serv. Rep. 4 pp. typescript.

Oosenbrug, S., and L. Carbyn. In press. Winter predation and activity pattern of a pack of wolves in Wood Buffalo National Park. *In* F. Harrington and P. Paquet, eds. Proc. Portland Wolf Symp., Portland, Oregon, 1979. Noyes.

Peden, D. G., and H. W. Reynolds. 1981. Estimates of annual herbage production in the Falaise Lake area of the Mackenzie Bison Sanctuary, Northwest Territories, 1979. Unpub. rep., Can. Wildl. Serv. 123 pp. multilith.

Peek, J. M. 1980. Natural regulation of ungulates (What constitutes a real wilderness?). Wildl. Soc. Bull. 8: 217–227.

Reynolds, H. W., R. M. Hanson, and D. G. Peden. 1978. Diets of the Slave River Lowland bison herd, Northwest Territories, Canada. J. Wildl. Manage. 42: 581–590.

Rhodes, S. N. 1897. Notes on living and extinct species of North American Bovidae. Proc. Acad. Nat. Sci. Phil. 483–502.

Riney, T. 1964. The impact of introductions of large herbivores on the tropical environment. IUCN Pub., New Ser. 4: 261–273.

Rowe, J. S. 1972. Forest regions of Canada. Can. Forest. Serv. Pub. No. 1300. 172 pp.

Soper, J. D. 1941. History, range, and home life of the northern bison. Ecol. Monogr. 11: 347–412.

Van Camp, J. 1978a. Summary of progress wolf-bison project, Slave River Lowlands, N.W.T. Unpub. rep., N.W.T. Wildl. Serv. 22 pp. typescript.

———. 1978b. Summary of progress wolf-bison project, Hook Lake area, N.W.T. Unpub. rep., N.W.T. Wild. Serv. 29 pp. typescript.

———. 1978c. Wood bison population census, Mackenzie Bison Sanctuary July 12, 1976–March 18, 1978. Unpub. rep., N.W.T. Wildl. Serv. 9 pp. typescript.

Polar Bear (*Ursus maritimus*) Ecology and Environmental Considerations in the Canadian High Arctic

Ian Stirling
Wendy Calvert
Dennis Andriashek

Introduction

In 1970 the Canadian Wildlife Service initiated studies of the population ecology of the polar bear (*Ursus maritimus*) in the Canadian High Arctic (fig. 1). The objective was to provide management recommendations, with particular reference to the review of Inuit quotas in the area, and to conserve a viable polar bear population.

The discovery of natural gas in the High Arctic resulted in a proposal to build a pipeline to transport the resource to southern markets. It was proposed that the pipeline in the first phase would run from the discoveries on Melville Island (west of Bathurst Island) across Bathurst, Cornwallis, and Somerset islands and their associated inter-island channels, to the mainland and thence to southern Canada. In response we collected ecological data with which to assess ahead of the event the possible detrimental effects of pipeline construction on polar bears, and to establish a baseline against which the effects of environmental damage could be evaluated. Specifically, data were sought on population size, age-specific reproductive parameters, distribution, seasonal movements, the number of relatively discrete subpopulations affected, and the location of important maternity denning, feeding, and summer retreat areas for polar bears.

From 1970 through 1974 the research was of a preliminary nature. From 1975 through 1977 field work was more intensive because of additional funds received from the Arctic Islands Pipeline Project (AIPP). The

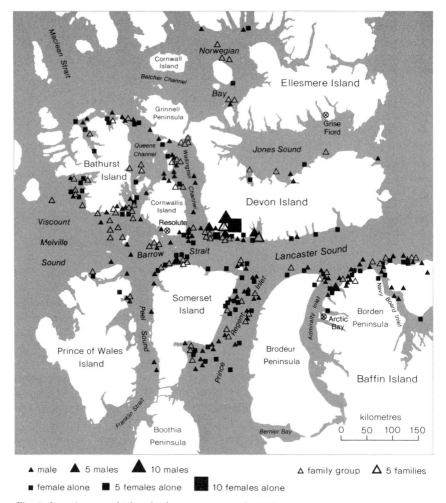

Fig. 1. Locations at which polar bears were tagged, 1970-77.

main study area consisted of western Lancaster Sound, Barrow Strait, eastern Viscount Melville Sound, and north of Cornwallis and Bathurst islands (fig. 1). A limited amount of field work was conducted in eastern Lancaster Sound.

The objectives of this paper are to summarize the results of population ecology studies of polar bears in the High Arctic, to comment on the possible effects on polar bears of man-made environmental disturbance, and to identify gaps in our present knowledge.

Although the research described in this paper was conducted in the High

Arctic, the approach is representative of management and environmental assessment studies of polar bears conducted throughout their range in Canada. For comparison, reference will be made to similar studies done in other areas.

Life History of the Polar Bear

The polar bear is circumpolar in distribution. In Canada its range extends from the pack ice of the Arctic Ocean and High Arctic Islands to southern James Bay. Tagging and recapture programs, particularly in Canada, Norway, and Alaska have shown that most populations are fairly localized (DeMaster and Stirling 1981).

Pregnant female polar bears occupy maternity dens from about early November to late March or early April. Unlike black bears (*U. americanus*) or brown bears (*U. arctos*), the rest of the population does not den for the winter, although any polar bear may dig and use one for a few days during a period of bad weather. Van de Velde (1957, 1971) and Harington (1968) have reported instances of females with older cubs denning as well, but to an unknown extent.

Maternity dens are usually dug in deep snowbanks on the leeward side of steep slopes or on stream banks located near the sea. The entrance slopes upward into the main chamber and may be several metres long. Most dens have one or sometimes two rooms, often with alcoves dug into the walls and a ventilation hole dug through the roof. An average den is about 2 m long and 1.5 m wide by 1 m high (Harington 1968). Lentfer (1975) reported one instance of maternity denning north of the Alaskan coast on the drifting pack ice of the Beaufort Sea. However, we have no evidence that polar bears den on the sea ice near the Canadian Arctic Islands.

Polar bears, like several other groups of mammals, have delayed implantation. This means that the fertilized egg does not begin to grow immediately, but remains in the uterus in a dormant state. Thus although polar bears mate in May, the fertilized egg does not implant and begin to grow until about September. In captivity the young, normally two, are born anywhere from late November to January. In the wild there may be a variation in birth dates because of latitudinal differences. Polar bears are born with their eyes closed and weigh less then 1 kg at birth. The females metabolize their fat reserves to produce milk to nurse the cubs. By the time they leave the den in March or April the cubs weigh approximately 10 kg.

For up to 10 days after breaking out of the den, the female and cubs appear to play near the den and return to it often. Females sometimes dig out and eat vegetation at this time. The family then returns to the sea ice to hunt seals. Most polar bear cubs in the High Arctic stay with their mothers until they are 2⅓ years old, although a few may remain with the female into their fourth or fifth year. In comparison, a large proportion of the cubs in Hudson Bay remain with their mothers for only 1⅓ years (Ramsay and Stirling 1982).

When fully grown, adult male polar bears in Canada weigh about 450 to 550 kg; adult females weigh about 160 to 270 kg. The average life expectancy is 12 to 15 years, although some bears live past 20, and the occasional individual exceeds 30 years of age.

Ringed seals (*Phoca hispida*) and bearded seals (*Erignathus barbatus*) are the main food species of polar bears. Walruses (*Odobenus rosmarus*) are occasionally killed (Kiliaan and Stirling 1978), but it appears unlikely that they form a significant part of the bears' diet. Similarly, Freeman (1973) and Heyland and Hay (1976) recorded belugas (*Delphinapterus leucas*) that had apparently been attacked by polar bears. Observations have also been made of bears catching sea birds by diving and coming up beneath them, and of bears diving for and eating kelp (Lønø 1970, Stirling 1974, Russell 1975).

During the winter and spring, polar bears are dispersed offshore from the mainland and over the ice-covered inter-island channels of the Central and High Arctic Islands. They tend to concentrate along the pressure ice that parallels the coastlines and lies across the mouths of bays and in the vicinity of the floe edge. As break-up proceeds through the spring and early summer, the bears move with the ice in order to be able to hunt seals more effectively. In areas such as Hudson Bay, where the ice melts completely during the summer, the bears are forced onto the land where they wait for freeze-up. As soon as the sea freezes in the fall they move back onto the ice to hunt seals. The pattern of seasonal movements and distribution probably varies, depending on the location and consistency of ice formation, dispersal, and distribution over the years.

Ringed and bearded seals maintain their breathing holes from freeze-up to break-up by abrading the ice with the heavy claws on their foreflippers. Their breathing holes are located on the last cracks to close over after freeze-up (Smith and Stirling 1975). In areas where wind, water currents, or tidal action cause the ice to crack and subsequently refreeze, seals are apparently more accessible to polar bears, and the bears are able to hunt more successfully (Stirling and McEwan 1975, Stirling et al. 1975). In summer, seals are captured by stalking or waiting for the animal to surface at a breathing hole (Stirling 1974). In winter and spring, bears also dig seal pups, and sometimes adults, from lairs beneath the snow (Stirling and Latour 1978). Captured seals are not always completely eaten by the bear that caught them (Stirling 1974, Stirling and McEwan 1975).

Materials and Methods

The methods and rationale for the marking and recapture of individual polar bears, location of maternity dens, calculation of productivity, collection of specimens from Inuit hunters, location of summer retreats, age determination, calculation of mortality rates, and estimation of population size are all explained in Stirling et al. (1980).

Results and Discussion

Distribution and Movements

From 1970 through 1977, 478 polar bears were tagged in the High Arctic. Figs. 2 to 4 summarize the recorded movements polar bears made between their original capture sites and where they were recaptured, or killed by Inuit hunters. In general, tagged polar bears showed a high degree of fidelity to the areas in which they were first captured. However, a few longer movements have been recorded (Kiliaan et al. 1978, Stirling et al. 1978, Schweinsburg et al. 1982).

Table 1 Mean straight-line distances moved by polar bears in different seasons and years

Period of comparison	Sex	Sample size	Mean distance moved (km) ($\pm$ 95% confidence limit)
Spring to spring	male	68	98.55 $\pm$ 22.85
	female	40	113.67 $\pm$ 62.13
Spring to summer	male	14	148.93 $\pm$ 45.36
	female	18	144.78 $\pm$ 66.88
Summer to spring	male	7	190.71 $\pm$ 102.58
	female	13	231.54 $\pm$ 89.29
Two or more years between captures	male	29	144.07 $\pm$ 41.45
	female	25	183.08 $\pm$ 71.77

The straight-line distances that males and females move between capture and recapture or kill locations are summarized in table 1. Of the 214 movements recorded, 15 were over 300 km. There were no significant differences between the distances moved by males and by females in any of the time periods.

An additional 436 polar bears were tagged to the south of the study area, in the Central Artic. In these two areas combined, 140 recaptures were made on 124 individual bears, 48 were shot by Inuit hunters, and 6 were killed by hunters outside the two areas, up to the end of 1977 (Stir-

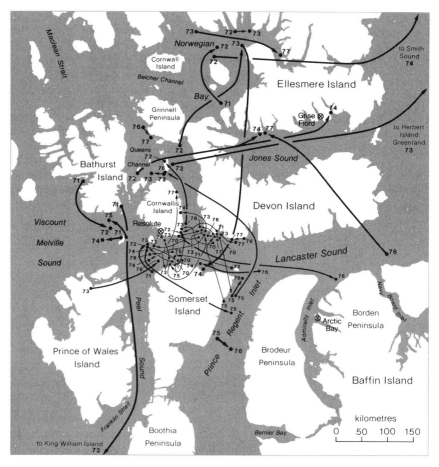

Fig. 2. The recorded movements of polar bears captured between mid-March and mid-June and recaptured or shot during the same season in one or more subsequent years. Numbers indicate years in which bears were captured, recaptured, or killed.

ling et al. 1978). Only 3 tagged bears had moved between the study area and the Central Arctic. No exchange was recorded between the study area and Hadley Bay on northeastern Victoria Island, where 84 polar bears were tagged between 1972 and 1975 (Stirling et al. 1978). Four polar bears from the study area were reported to have been shot by Greenland Inuit (fig. 2), and two from the eastern Beaufort Sea were recaptured in the study area (Stirling et al. 1975). These data suggest that for environmental impact assessment or management purposes the polar bears within the study area form a relatively discrete subpopulation. Although the western

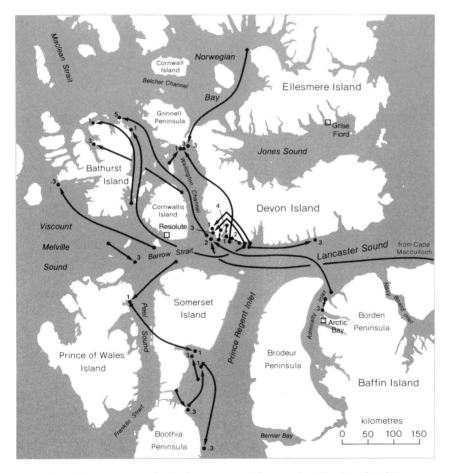

Fig. 3. Recorded movements of polar bears captured between late March and mid-June and then recaptured or shot between mid-June and the end of August of the same year or subsequent years. Numbers indicate years between being captured and recaptured or killed.

and southern boundaries seem to be defined clearly enough, the northern and eastern boundaries are not.

There are not enough data from the mark and recapture studies to make conclusive statements about seasonal movements. However, in general the recorded movements of polar bears between winter and summer (fig. 3) go from western Lancaster Sound and Barrow Strait to the bays of southern Devon Island, or to the north and west of Cornwallis Island. The recorded movements of polar bears in those areas between summer and winter tend to do the reverse (fig. 4). Schweinsburg et al. (1982) found a similar

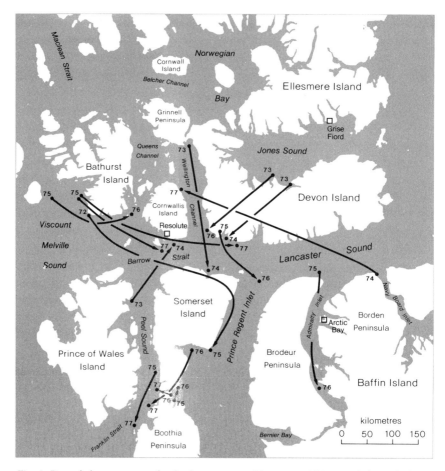

Fig. 4. Recorded movements of polar bears captured between mid-June and the end of August and then recaptured or shot between late March and the end of May of the following year or subsequent years. Numbers indicate years in which bears were captured, recaptured, or killed.

pattern in eastern Lancaster Sound. Coupled with the subjective observations we made while working in the field, we interpret this to mean that during the winter and spring bears concentrate in eastern Barrow Strait and Lancaster Sound, hunting seals where there is continuous ice cover. During the summer, as break-up proceeds westward through Lancaster Sound and Barrow Strait, many bears either move into the bays on the south coast of Devon Island to hunt seals where the ice remains a few weeks longer, or continue to move west or northwest as circumstances dictate, in order to remain on the ice. This extended hunting period is probably especially important for females with cubs of any age and pregnant females that need to gain additional fat before giving birth in the coming winter. Individual polar bears show fidelity to these bays, which further emphasizes the bays' ecological significance.

Less is known about the distribution of bears during late summer in their relatively ice-free retreats, such as the Borden and Brodeur peninsulas on northern Baffin Island and the southeastern coast of Devon Island. Many bears occur in these areas, but it is not known what proportion of the population they represent. Some bears go inland, where they dig and rest in snow dens (Schweinsburg 1979).

A few additional comments are appropriate in relation to the movement data discussed above and the delineation of subpopulations. In general, some clumping of marked bears has been recorded adjacent to settlements or field camps such as Resolute or southwestern Devon Island (figs. 2 to 4), where logistic facilities are established. Had field work been concentrated at other locations within the study area, similar localized movement patterns probably would have resulted. Thus instead of clearly discrete "subpopulations" in the study area, it appears that there is a continuum of polar bears whose home ranges overlap to varying degrees. The size of these home ranges may vary in relation to local environmental factors, especially ice conditions. Similarly the density of overlapping home ranges probably differs from area to area because of geographic variations in the productivity of the sea. Activity in any area may affect the polar bears whose home ranges include all or part of the affected area. This also means that polar bears further away from a disturbed area than the distance of the mean home-range diameter may not be affected. Although we do not know the size of the home ranges of male and female polar bears, we have records of straight-line distances moved between seasons. Movement data suggest that the average distance over which any activity may have an effect on polar bears is probably 150 to 200 km from the edge of the area being influenced.

Maternity Denning Areas

Fig. 5 summarizes information gathered from maternity denning surveys in the study area. Because of the limitations on the kind of data it is possible to collect (Kiliaan et al. 1978), we are able to discuss relative densities only. Nowhere in the High Arctic have we found high numbers of dens or cubs-of-the-year similar to those reported from Wrangel Island, USSR (Uspenski and Kistchinski 1972), Svalbard (Larsen 1976), or the Manitoba coast of Hudson Bay (Stirling et al. 1977a). This may have been influenced by inadequacies in our survey techniques, or we simply may have missed the high-density productivity areas. Another explanation might be that suitable habitat for denning in the area under consideration is vir-

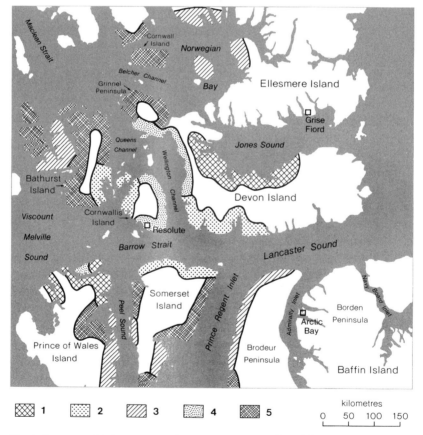

Fig. 5. Summary of information on denning areas: (1) confirmed denning areas; (2) denning recorded, but apparently at relatively lower densities; (3) possible denning areas; (4) surveyed, no positive data; (5) inadequately surveyed.

tually unlimited, so that the density is low even in the most suitable areas. In comparison, relative to the vast expanses of apparently adequate feeding habitat, there are few islands available where maternity denning could take place north of the Siberian coast or Svalbard, and none off the coast of northern Manitoba. Consequently it seems likely that polar bear maternity denning would be more concentrated in those areas than it is in the Canadian Arctic Archipelago.

Reproductive Parameters

The mean litter sizes of females accompanied by cubs-of-the-year, yearlings, and 2-year-olds (calculated from Stirling et al. 1978: table 6) was 1.63 ($n = 38$), 1.70 ($n = 27$), and 1.44 ($n = 9$) respectively, which is comparable to other areas of the Arctic (Stirling et al. 1975, 1980, Lentfer et al. 1980, Schweinsburg et al. 1981). Only in James and Hudson bays does it appear that the litter size may be consistently larger than in the more northerly parts of polar bear range (Jonkel et al. 1976, Stirling et al. 1977a, Ramsay and Stirling 1982).

The age of first conception for the majority of female polar bears in the High Arctic is 4 years. Thus, they are 5 years old when they have their first cubs. Since cubs remain with their mothers at least 2⅓ years, the minimum breeding interval is 3 years. The reproductive value of females rises with age to a maximum in 15- to 19-year-old females (table 2). Although the sample size of older bears is small, it appears that reproductive senility begins to develop after 20 years of age. The mean natality rate of females aged 5 to 29 years was 0.421 cubs per year.

Table 2   Age-specific litter-produced rates, litter sizes, and natality rates of female polar bears in the High Arctic study area (sample sizes in parentheses)

Age class of females (yrs)	Age-specific litter-produced rate (litters/yr)	Age-specific mean litter size (cubs/litter)	Age-specific natality rates (cubs/yr)
3	0.034 (29)	1.00 (1)	0.034
4	0.375 (24)	1.44 (9)	0.540
5	0.235 (17)	1.60 (5)	0.376
5–9	0.273 (110)	1.67 (33)	0.456
10–14	0.229 (70)	1.94 (18)	0.444
15–19	0.429 (21)	1.50 (8)	0.644
20–29	0.063 (16)	2.00 (1)	0.126
5-29	0.249 (217)	1.69 (62)	0.421

Table 3 Number of polar bears of each age and sex class captured or killed in the High Arctic between 1970 and 1977[1]

	Killed bears		Captured bears	
Age	Male	Female	Male	Female
0	0	1	34	37
1	10	7	17	20
2	11	17	17	21
3	30	9	15	8
4	18	13	19	16
5	23	8	11	13
6	10	7	17	11
7	14	4	13	6
8	8	9	4	16
9	10	1	7	13
10	2	3	5	9
11	6	4	10	11
12	3	1	2	4
13	1	0	1	10
14	1	1	6	7
15	3	2	4	6
16	3	1	0	5
17	1	0	0	2
18	2	1	0	2
19	2	0	0	3
20	1	1	2	1
21	0	0	0	0
22	0	0	0	1
23	1	1	0	2
24	0	0	1	0
25	1	0	0	2
26	0	0	0	1
27	0	0	0	0
28	0	0	0	1
29	0	0	0	1
Total	161	91	185	229

Note
1. The samples are not complete because specimens were not turned in from all the bears shot, nor collected from all bears captured.

Age Structure and Mortality Rates

Most adult polar bears in the study area live for 12 to 15 years, but a small number survive for over 20 (table 3). In the sample of polar bears handled during our studies females were more abundant than males in the older age classes, but the difference was not statistically significant ($X^2 = 3.8$, $df = 4$, $p^6.05$). The annual mortality rate of the captured females, based on a sample of 172 animals, was 12.1%. The capture sample for males from the study area was not adequate for the calculation of an annual mortality rate, but a value of 16.2% was calculated from a combined sample of 326 male polar bears from the Central and High Arctic (Stirling et al. 1978). However, the samples may have been biased by the fact that from 1970 to 1974 preference in sampling was given to females, family groups, and previously tagged animals.

Estimates of Population Size

Population size was estimated from mark and recapture data (table 4) using a method developed by DeMaster et al. (1980) that is similar to the Lincoln-Peterson method. It was not possible to attempt an estimate prior to 1975 because the sample was too small and because an unknown portion of the sample was taken selectively. For such estimates the quality improves with both time and sample size. From 1975 to 1977 the population size was estimated at 1,758, 1,224, and 1,675. It is possible that these estimates are low because of the enormous area covered and because mark

Table 4 Summary of mark and recapture data from the High Arctic and estimates of population size ($\hat{N}$) at time i (from DeMaster et al. 1980)

i	1	2	3	4	5	6	7	8
Year	1970	1971	1972	1973	1974	1975	1976	1977
n_i	18	19	57	45	48	79	102	110
m_i	0	0	6	9	6	5	13	14
R_i	18	19	57	44	48	73	102	108
$\hat{p}_i$	—	—	0.105	0.020	0.125	0.063	0.127	0.127
$\hat{M}_i$	—	15.66	21.45	63.03	85.29	110.74	155.51	212.72
$se\hat{M}_i$	—	1.79	2.76	5.79	8.52	11.40	15.42	20.74
$\hat{N}_i$	—	—	—	—	—	1757.82	1224.45	1674.96
$se\hat{N}_i$	—	—	—	—	—	788.63	341.74	451.22
p(cap)	—	—	—	—	—	0.045	0.084	0.066

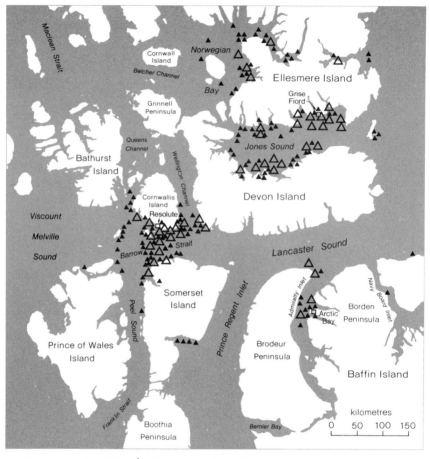

▲ 1 known polar bear kill   △ 5 known polar bear kills

Fig. 6. Known locations in the High Arctic where polar bears were killed by Inuit hunters, 1968–77.

and recapture efforts tended to be concentrated in localized areas. Therefore not all bears would have an equal probability of being captured, and the estimates would be biased too low. However, the data to date suggest that the minimum size of the population in the study area is about 1,700 polar bears. Schweinsburg et al. (1982) estimated the population of polar bears in eastern Lancaster Sound in 1979 to be 1031 ± 236.

Inuit Hunting Patterns and Utilization of Polar Bears

In prehistoric times Inuit hunting camps were widespread throughout the High Arctic; as a result polar bears were hunted on almost all coastal areas at one time or another. However, in recent years the Inuit have moved into the much more centralized arctic communities we know today. This has resulted in significant changes in hunting patterns (Freeman 1976a, 1976b).

The known locations where Inuit hunters killed polar bears from 1968 to 1977 are illustrated in fig. 6. The records are incomplete for Pond Inlet, the eastern entrance to Pond Inlet, and along the coast of Bylot Island. At present little hunting of polar bears occurs along the south coast of Devon Island, probably because travel over the ice is difficult for the Inuit during the winter and spring and because bears are available nearer to the settlements.

From 1972 to 1978 the sale of polar bear hides from the study area annually grossed between about $40,000 when the market was down, to about $150,000 in 1973–74 when prices were high (table 5, based on Stirling et al. 1978, Smith 1979). It is clear from the values given above that, despite the price fluctuations inherent in the fur market, polar bear hides will probably continue to form an important part of the cash income of Inuit hunters in the High Arctic.

Under the Northwest Territories Game Ordinance a limited sport-hunt at the request of particular settlements has been permitted since January 1970. The tags allotted to the sport-hunt are taken from the settlement quota and cannot be used at a later date, even if the sport-hunt is un-

Table 5 Average known prices (in dollars) paid to hunters and estimated total value for polar bear hides (number of skins in parentheses) from the four settlements in the Lancaster Sound region, 1972–78

Settlement	1972–73	1973–74	1974–75	1975–76	1976–77	1977–78
Arctic Bay	840(12)	1686(10)	959(11)	620 (5)	717 (3)	929 (7)
Grise Fiord	1824(13)	–	579 (7)	352(19)	–	–
Pond Inlet	–	1590 (5)	566(11)	536 (9)	855(10)	1081(18)
Resolute	1041(34)	–	573 (8)	412 (5)	247 (5)	496(13)
Average price ($)	1173(59)	1654(15)	687(37)	439(38)	663(18)	853(38)
Total quota	92	92	92	92	92	94
Estimated total value to above settlements ($)	107,916	152,168	63,204	40,388	60,996	80,182

Table 6 The number and cost (in dollars) of sport-hunts by Inuit settlements hunting in the Lancaster Sound region, 1970–75

Year	Settlement	No. applicants	No. successful sport-hunts	Cost per hunt ($)
1970	Resolute	3	3	2,500
1971	Resolute	4	4	2,500
1972	Resolute	4	3	2,500
1973	Pond Inlet	4	3	2,500
1974	Pond Inlet	1	1	3,500
1975	Pond Inlet	1	0	4,500
Total		17	14	

successful. After 1973, when the price of polar bear hides rose considerably, Inuit interest in the sport-hunt declined (Smith and Jonkel 1975). For many, the effort involved in servicing a sport-hunt did not justify the financial gain. Table 6 lists the settlements in the study area that offered sport-hunts from 1970 to 1975, the number of applicants, the number of successful hunts, and the cost per hunt. Because there are few countries where sport-hunters can legally take polar bears, guided hunts continue to represent a potential source of revenue for Inuit settlements throughout the Canadian Arctic.

Any significant disruption of polar bear distribution, survival, or natality rates could have significant economic effects on adjacent communities, and important cultural effects as well.

Vulnerability to Environmental Disruptions

It is difficult to predict the effects of future environmental damage on polar bears because there is little data on how polar bears have been affected in the past by known environmental changes, natural or man-made. However, the results from two studies indicate that polar bears are vulnerable to some changes.

Preliminary laboratory studies indicate that oil on the fur seriously affects the ability of a polar bear to thermoregulate (Øritsland et al. 1982). This is probably more significant for young bears, which use a greater proportion of their metabolic energy for thermoregulation. Two of the three

Polar Bear Ecology and Environmental Considerations   217

bears that ingested oil while grooming themselves died. It is not known whether polar bears would avoid leads covered with oil, but the effects could be quite detrimental if the bears do not avoid them. Bears that ingest oil can probably be saved if cleaning and internal treatment begin promptly. However, the problems associated with trying to clean bears in the field after widespread contamination would be difficult and expensive, and efforts may have to concentrate on the most valuable bears in the population, such as adult females.

During our long-term studies in the eastern Beaufort Sea, we recorded large-scale changes in the distribution and abundance of both seals and polar bears as a result of natural causes (Stirling et al. 1975, 1976, 1977b, 1982, DeMaster et al. 1980). In a period of only 1 year the number of ringed and bearded seals dropped by 50% and their productivity by 90%. In about 2 years polar bear numbers dropped by a third and their productivity by a half. It took about 5 years for the population to recover. It seems likely that it would take a population even longer to recover from a major decline caused by man-made disturbances and the attendant environmental damage.

Although the mechanisms can only be speculated upon at this stage, there appeared to be an association between heavy ice conditions and the lowered physical condition, reproduction, and survival of seals and polar bears. Presumably, in years when ice conditions are heavier there is a greater risk of problems such as blowouts or oil spills. Consequently the extent of the potential damage to marine mammals resulting from man-made environmental disasters will be greatest in years when ice conditons are heavy and the populations already weakened. Detailed studies of the free-ranging behaviour, hunting abilities, and habitat utilization of polar bears should increase our understanding of the relationship between bears, seals, and sea ice. This in turn should increase our ability to predict how man-made changes to the environment might influence polar bears.

There are several ways in which man-made disruptions might affect polar bears. They include the disruption of spring feeding and breeding areas, the disturbance of maternity denning areas, the disturbance of summer feeding and refuge areas, man-bear conflicts, changes in Inuit hunting patterns, and other considerations. Each of these is discussed in detail below.

Disruption of Spring Feeding and Breeding Areas
Polar bears show a high degree of fidelity to their spring feeding and breeding areas. Large-scale, regular disruptions might result from icebreaker

traffic through these areas during the winter and spring, prior to the time that break-up would normally occur. It is not known whether icebreaker traffic could influence the pattern of freeze-up or break-up, or what the consequences would be for seals and bears. The significance of a disturbance would probably depend on when and where it occurred.

Disturbance of Maternity Denning Areas
Despite the fact that maternity denning has been recorded only at low densities in the Arctic Islands, some areas appear to be more important than others (fig. 5). Within the study area the more important areas include the southwestern coast of Devon Island, the north coast of Devon Island along the southern edge of Jones Sound, the northern tip of Brodeur Peninsula of Baffin Island, and the northern coasts of Prince of Wales and Somerset islands. If the earlier conclusion of widespread maternity denning at low density is correct, then even a high level of disturbance at localized sites such as airstrips or camps would probably have only a slight or negligible effect. Alternatively, sources of disturbances such as roads, seismic exploration, or pipelines that run parallel to long sections of coastline might cause extensive disruption to bears in maternity dens or the displacement of pregnant females looking for potential denning sites. The degree of females' fidelity to specific denning areas is unknown, so we cannot predict whether displaced females would be able to relocate. One way to minimize the detrimental effects of large-scale activities over the extensive areas in which maternity denning occurs might be to not permit these activities between 1 October and 30 April to a distance of 10 km inland, to minimize the disturbance of pregnant or postparturient females.

Disturbance of Summer Feeding and Refuge Areas
Polar bears concentrate in and show a high degree of fidelity to summer feeding and sanctuary areas. The bays in these areas are ice-covered late into the summer, extending the period during which polar bears can feed on seals (Stirling 1974); this is of particular importance to family groups and subadults. Schweinsburg et al. (1977) reported that those age and sex classes are more highly represented in such areas during the summer. The localized nature of these bays makes them extremely important, possibly to the point of being critical.

Large-scale intensive construction operations in areas such as Maxwell and Radstock bays on southwest Devon Island might have a significant impact by displacing family groups and subadult bears, in particular, to areas where they could not feed as successfully. This would result in higher mortality rates for younger animals. It seems appropriate to afford these areas the maximum amount of protection possible.

Some polar bears are found along the coastline or inland during the open-water period. Such bears are usually transient because of the lack of food; therefore their displacement would probably not have a significant impact.

Man-Bear Conflicts
In several areas land- or ice-based human activities may occur in areas where polar bears concentrate. This applies particularly to activities on the sea ice along the coastline of the inter-island channels during the winter and early spring, and along the beaches of the islands during the summer open-water period. Bears may enter camps throughout the year, but especially during the summer, when more of them are on the land and away from their normal food, seals. This situation will result in a greater but unpredictable number of conflicts between men and bears, in which several bears and possibly some men may be killed (Schweinsburg and Stirling 1976). Even when all reasonable precautions are taken, conflicts may occur simply because camps have odours that attract bears.

Changes in Inuit Hunting Patterns
There may be significant geographic changes in hunting patterns and more requests for increased quotas as a result of Inuit hunters moving to work and live at new industrial sites or at outpost camps. The possible consequences of redistributed or increased hunting pressure will vary with each area and should be monitored closely.

Other Considerations
Vehicle and aircraft movements would most likely have only minor or negligible local impact. The possible effects of chronic pollution or local contamination of areas because of events such as ruptured fuel storage units are unknown. So far only preliminary baseline toxicological studies have been undertaken (e.g., Bowes and Jonkel 1975).

On a more general note, it is clear that polynyas (areas of open water surrounded by ice) are extremely important to marine mammals for overwintering, feeding, and migration (Stirling and Cleator 1981). Almost all of these areas throughout the Arctic are threatened with some degree of disturbance or damage as a result of proposals such as offshore drilling in Lancaster Sound and Davis Strait, AIPP, Arctic Pilot Project, Polar Gas-Y line, and Beaufort Sea drilling and shipment of oil (Stirling and Calvert

1983). A common denominator of many of the environmental studies being done and proposed is that they are site-specific. This is acceptable and even necessary up to a point, but unless such research is fitted into a more meaningful ecological framework the benefits are (and have been) much reduced. Therefore we should place the highest priority on long-term (5 to 10 years) studies of ecological interrelationships, with two objectives in mind: the development of a predictive capability in terms of potential problems resulting from man's activities, and the establishment of sound baseline data, so that when the inevitable problems arise in the future we will be able to assess the effects.

Acknowledgments

This research was supported by the Canadian Wildlife Service, the Polar Continental Shelf Project, and the Arctic Islands Pipeline Project. The help of the following in the laboratory and in the field is gratefully acknowledged: W. R. Archibald, T. Chowns, D. DeMaster, C. Jonkel, H. P. L. Kiliaan, P. B. Latour, R. E. Schweinsburg, O. Shannon, Pauline Smith, Kathy Smyth, and A. Viedeman.

The Inuit Hunters and Trappers associations from all the settlements within the study area gave much valuable advice and assistance.

References

Bowes, G. W., and C. J. Jonkel. 1975. Presence and distribution of polychlorinated biphenyls (PCB) in arctic and subarctic marine food chains. J. Fish. Res. Bd. Can. 32: 2111–2123.

DeMaster, D., M. C. S. Kingsley, and I. Stirling. 1980. A multiple mark and recapture estimate applied to polar bears. Can. J. Zool. 58: 633–638.

———, and I. Stirling. 1981. *Ursus maritimus*. Mammal. Species 145: 1–7.

Freeman, M. M. R. 1973. Polar bear predation on beluga in the Canadian Arctic. Arctic 26: 163–164.

———, ed. 1976a. Inuit land use and occupancy project. Volume one: Land use and occupancy. Supply and Services Can., Ottawa. 263 pp.

———, ed. 1976b. Inuit land use and occupancy project. Volume three: Land use atlas. Supply and Services Can., Ottawa. xvi + 153 maps.

Harington, C. R. 1968. Denning habits of the polar bear (*Ursus maritimus* Phipps). Can. Wildl. Serv. Rep. Ser. 5. 30 pp.

Heyland, J. D., and K. Hay. 1976. An attack by a polar bear on a juvenile beluga. Arctic 29: 56–57.

Jonkel, C., P. Smith, I. Stirling, and G. B. Kolenosky. 1976. Notes on the present status of the polar bear in James Bay and the Belcher Islands. Can. Wildl. Serv. Occ. Pap. 26. 40 pp.

Kiliaan, H. P. L., and I. Stirling. 1978. Observations on overwintering walruses in the eastern Canadian High Arctic. J. Mammal. 59: 197–200.

———, I. Stirling, and C. Jonkel. 1978. Notes on polar bears in the area of Jones Sound and Norwegian Bay. Can. Wildl. Serv. Prog. Notes, No. 88. 21 pp.

Larsen, T. 1976. Polar bear dens in Svalbard, 1972 and 1973. pp. 199–208. In M. R. Pelton, J. W. Lentfer, and G. W. Folk, eds. Bears: Their biology and management. IUCN, New Ser. 40.

Lentfer, J. W. 1975. Polar bear denning on drifting sea ice. J. Mammal. 56: 716.

———, R. J. Hensel, J. R. Gilbert, and F. E. Sorensen. 1980. Population characteristics of Alaskan polar bears. pp. 109–116. In C. J. Martinka and K. L. MacArthur, eds. Bears: Their biology and management. Bear Biology Assoc. Conf. Ser. No 3. U.S. Gov. Printing Office, Washington, D.C.

Lønø, O. 1970. The polar bear (Ursus maritimus Phipps) in the Svalbard area. Nor. Polarinst. Skr. Nr. 149. 130 pp.

Øritsland, N. A., F. R. Engelhardt, F. A. Juck, R. J. Hurst, and P. D. Watts. 1982. Effect of crude oil on polar bears. Env. Stud. No. 24. Dep. Indian and Northern Affairs, Ottawa. 268 pp.

Ramsay, M. A., and I. Stirling. 1982. Reproductive biology and ecology of female polar bears in western Hudson Bay. Natur. Can. (Rev. Ecol. Syst.) 109: 941–946.

Russell, R. H. 1975. The food habits of polar bears of James Bay and southwest Hudson Bay in summer and autumn. Arctic 28: 117–139.

Schweinsburg, R. E. 1979. Snow dens used by polar bears in the Canadian High Arctic. Arctic 32: 165–169.

———, D. J. Furnell, and S. J. Miller. 1981. Abundance, distribution and population structure of polar bears in the lower central Arctic islands. N.W.T. Wildl. Serv. Completion Rep. 2. 80 pp.

———, L. J. Lee, and P. B. Latour. 1982. Distribution, movement and abundance of polar bears in Lancaster Sound, Northwest Territories. Arctic 35: 159–169.

———, and I. Stirling. 1976. More research needed to minimize conflicts between men and polar bears. Oilweek 27: 54–55.

———, I. Stirling, S. Oosenbrug, and H. Kiliaan. 1977. A status report on polar bear studies in Lancaster Sound. N.W.T. Fish and Wildl. Serv. Rep. to Norlands Petroleum Ltd. 55 pp.

Smith, P. A. 1979. Résumé of the trade in polar bear hides in Canada, 1977–78. Can. Wildl. Serv. Prog. Notes, No. 103. 6 pp.

———, and C. J. Jonkel. 1975. Résumé of the trade in polar bear hides in Canada, 1973–74. Can. Wildl. Serv. Prog. Notes, No. 48. 5 pp.

Smith, T. G., and I. Stirling. 1975. The breeding habitat of the ringed seal (Phoca hispida): The birth lair and associated structures. Can. J. Zool. 53: 1297–1305.

Stirling, I. 1974. Mid-summer observations on the behavior of wild polar bears (Ursus maritimus). Can. J. Zool. 52: 1191–1198.

———, D. Andriashek, P. Latour, and W. Calvert. 1975. The distribution and abundance of polar bears in the eastern Beaufort Sea. A final report to the Beaufort Sea Project. Fish. and Mar. Serv., Dep. Env., Victoria, B.C. 59 pp.

———, W. R. Archibald, and D. DeMaster. 1977b. Distribution and abundance of seals in the eastern Beaufort Sea. J. Fish. Res. Bd. Can. 34: 976–988.

———, and W. Calvert. 1983. Environmental threats to marine mammals in the Canadian Arctic. Polar Record 21 : 433–449.

———, W. Calvert, and D. Andriashek. 1980. Population ecology studies of the polar bear in the area of southeastern Baffin Island. Can. Wildl. Serv. Occ. Pap. 44. 31 pp.

———, and H. Cleator, eds. 1981. Polynyas in the Canadian Arctic. Can. Wildl. Serv. Occ. Pap. 45. 70 pp.

———, C. Jonkel, P. Smith, R. Robertson, and D. Cross. 1977a. The ecology of the polar bear (*Ursus maritimus*) along the western coast of Hudson Bay. Can. Wildl. Serv. Occ. Pap. 33. 64 pp.

———, M. C. S. Kingsley, and W. Calvert. 1982. The distribution and abundance of seals in the eastern Beaufort Sea, 1974–79. Can. Wildl. Serv. Occ. Pap. 47. 23 pp.

———, and P. B. Latour. 1978. Comparative hunting abilities of polar bear cubs of different ages. Can. J. Zool. 56: 1768–1772.

———, and E. H. McEwan. 1975. The caloric value of whole ringed seals (*Phoca hispida*) in relation to polar bear (*Ursus maritimus*) ecology and hunting behavior. Can. J. Zool. 53: 1021–1027.

———, A. M. Pearson, and F. L. Bunnell. 1976. Population ecology studies of polar and grizzly bears in northern Canada. Trans. 41st North Amer. Wildl. Conf. 41: 421–430.

———, R. E. Schweinsburg, W. Calvert, and H. P. L. Kiliaan. 1978. Population ecology of the polar bear along the proposed Arctic Islands gas pipeline route. Final rep. to the Env. Manage. Serv., Dep. Env., Edmonton. 93 pp.

Uspenski, S. M., and A. A. Kistchinski. 1972. New data on the winter ecology of the polar bear (*Ursus maritimus* Phipps) on Wrangel Island. pp. 181–197. *In* S. Hererro, ed. Bears: Their biology and management. IUCN, New Ser. 23.

Van de Velde, F. 1957. Nanuk, king of the arctic beasts. Eskimo 45: 4–15.

———. 1971. Bear stories. Eskimo (New Ser.) 1: 7–11.

Plant Communities

Lichen Woodland in Northern Canada

J. Stan Rowe

Introduction

The characteristic vegetation of the northern half of the boreal zone or taiga in Canada, occupying about one million square kilometres, is a woodland of widely spaced spruce (accompanied in some places by pine, larch, birch, or fir) with a ground-cover layer of light-reflectant lichens through which dwarf shrubs are scattered. In the apt words of Hare (1964), this aesthetically appealing landscape constitutes "the celebrated lichen woodlands with their open crown and handsome lichen floor."

Although the appropriateness of calling the northern subzone of the boreal zone "the Subarctic" has long been debated (in recent years by Löve 1970, Blüthgen 1970, Hustich 1979), the term is a useful descriptor of the open boreal woodland plus the adjacent forest-tundra ecotone or Hemiarctic (Rousseau 1952), and defined as such it has recently been endorsed by Ahti (1980). Here I briefly review some of the varieties of subarctic woodland ecosystems and their obvious or suspected causes.

Described as a savanna by Dansereau (1951), as "park-like" by Raup (1946), and as a parkland by Rousseau (1952), the lichen woodlands are structurally unique and readily recognizable from the air and on aerial photographs. On the ground they display certain typical features such as a usual predominance in the openings between the trees of the fruticose *Cladina* "reindeer lichen" species (less commonly, *Stereocaulon* and *Cetraria*), a dispersed pattern of low ericoid shrubs (particularly species of

Fig. 1. Black spruce–lichen woodland (*Picea mariana–Cladina mitis*) near Porter Lake, Northwest Territories.

Vaccinium, Ledum, and *Empetrum*), and the virtual confinement of bryophytes (such as *Pleurozium, Dicranum,* and *Ptilidium*) to the shelter of standing trees, stumps, and fallen trunks. Herbaceous plants are rare or only sparsely represented, although the hemiparasitic half-shrub *Geocaulon lividum* is often prominent. Descriptions of the vegetation are given in recent papers by Johnson and Rowe (1975), Maikawa and Kershaw (1976), Rencz and Auclair (1978), and Bradley et al. (1982).

Names are classes, and "lichen woodland" encompasses a compositional range of landscape ecosystems. Only at the structural or physiognomic level of generalization does it portray a unit "thing," namely the northern subformation of the boreal zone that roughly coincides with the discontinuous permafrost zone of Brown (1967) between the mean annual isotherms of -4 C and -7 C. Close inspection reveals differences from place to place, such as a prominence of *Picea glauca* toward the forest-tundra edge (Gubbe 1976), on relatively deep soils (Rencz and Auclair 1978), especially those derived from sedimentary rocks (Zoltai et al. 1979), and the usual dominance of *P. mariana* elsewhere. As to the ground cover, *Cladina mitis* and *C. amaurocraea* are most evident on the drier and more exposed sites; *C. stellaris, C. rangiferina, Cetraria,* and *Stereocaulon* species on snow collection and moister sites; and *Cladonia coccifera, C. gracilis,* and *C. crispata* in younger stands.

Lichen woodlands clothe a variety of landforms. They occur occasionally in mountainous terrain (Steere et al. 1977), but according to Viereck and Schandelmeier (1980) they are not as widely distributed in Alaska as in Canada, where they are common on bedrock uplands, on both mineral soils and peatlands that harbour permafrost, and on many kinds of nonpermafrost terrain. Islands of the type are not uncommon even in the middle and southern boreal subzones associated with closed-crown conifer forests (Ahti 1959, Lambert and Maycock 1968), constituting "edaphic climaxes" on such excessively drained materials as dune sands and coarse stratified drift. Again, in the sporadic permafrost belt of northern Alberta and central and northern Saskatchewan and Manitoba, occasional patches of peatland display the whitish or yellowish lichen-rich surfaces between open-canopied black spruce. These southern forms, as well as the more prevalent ones of the North, show that more than one environment can produce the type. Is there, then, any single explanation for it?

Adaptations of Terricolous Lichens

Behind the community concept is the reasonable idea of obligate symbiotic relationships. Behind this in turn is the idea of coevolution, and the development over time of mutually beneficial interconnections between organisms in a particular environment. Following this line of reasoning, and assuming that lichen woodlands are of ancient, preglacial origin, it is logical to suppose that the cohabiting lichens, trees, shrubs, and soil organisms have evolved obligate ties and together form an organically interdependent community. These assumptions are implicit in hypotheses that explain the maintenance of lichen woodland ecosystems by reference to the functions and processes involving the interaction of their component parts, including climate in the air layer, the soil layer, and the biota of both (see, for example, the hypothesis of Kershaw 1978, as later discussed).

It is also reasonable to suppose that communities are assemblages of organisms whose common denominator is a tolerance of the particular environment that they share. The relationships that exist between the juxtaposed organisms may be facultative rather than obligate. The relatively short time since the deglaciation of most northern terrain, allowing only 8,000 years or so of intercourse between organisms there, may not be long enough for appreciable symbiotic coevolution to have occurred. Taking this view—consistent with the "individualistic community" concept of Gleason (1939)—perhaps an explanation of lichen woodland may be found by examining the ecologies of the component species and by seeking the common denominator for their coexistence and structural relationships. Is it perhaps the ability to withstand stress in a severe and unpredictable environment? On this proposition Larsen (1980) has explained the depauperate characteristics of the subartic flora, invoking random incursions of arctic air during the growing season as the destabilizing influence.

Lichens are typical stress-tolerant organisms (Grime 1979). They are perennial ascomycetes that have attained autotrophic status by capturing algae, giving them the ability to live above ground. Smith (1962) suggested that many are adapted to alternating wet and dry environmental phases. Some genera such as *Stereocaulon, Lecidea,* and *Peltigera* are symbiotic with blue-green algae that can fix atmospheric nitrogen—an obvious advantage on nutrient-poor sites (Crittenden and Kershaw 1978). Because they lack vascular tissue, lichens cannot wick-up moisture from the substratum; they must get the water and minerals they need from aerial fallout and from the snowmelt of spring, when for a brief period they may be flushed by nutrient-rich water. They are nourished more by the atmosphere than by the soil, and are ombrotrophic (as are bog species) rather than minerotrophic (as are fen species). In the North, lichen-rich bogs are common, but lichen-rich fens are unknown. Hare (1950) remarked that lichen woodlands conspicuously avoid steep, rocky slopes, and the reason may be that slopes tend to be subsidized by nutrient inflow from upslope (as in the mountains).

It follows that the fruticose ground lichens should prosper in those northern environments where potential competitors fail; that is, in those non-subsidized habitats where inhospitable regimes of moisture, nutrients, and temperature render the growth and survival of leafy plants impossible or at least precarious. The idea that lichen woodlands identify stressful environments is also supported by characteristics of the nonlichen component: the wide spacing and slow growth rates of the xeromorphic, woody perennials, many of which are evergreen (Ingestad 1973, Chapin 1980, Mattson 1980); the prevalence of mycorrhizal ericaceous shrubs (Stribley and Read 1974) with high root/shoot ratios (Rencz and Auclair 1978); and the depauperate vascular flora that is noticeably lacking in herbs (Larsen 1980).

Wherever environments are such that a closed canopy of trees, shrubs, herbs, and mosses is prevented, opportunities exist for the photophilous but otherwise undemanding lichens. Three examples illustrate how lichens flourish in the Subarctic, where the substratum, for one reason or another, imposes wide spacing on the vascular species and the bryophytes.

Woodland Structure Imposed by Substrate

Northeast of Fort Smith in the Northwest Territories, in a band extending southward from Great Slave Lake between 110° and 111° longitude, we mapped a forest area within which large patches of lichen woodland occur on granitic bedrock (Bradley et al. 1982). Between the widely spaced jackpine and black spruce—rooted in rock crevices or in dispersed pockets of thin drift veneer—lichens grow profusely, especially *Cladina mitis* and *C. rangiferina,* along with the saxicolous *Actinogyra muhlenbergii, Rhizocarpon geographicum, Parmelia centrifuga,* and *Peltigera polydactyla.* The common dwarf shrub is *Vaccinium myrtilloides.* This same kind of lichen woodland is common north of Great Slave Lake, where surface outcroppings of both limestone and granite limit the survival of vascular species and thereby favour the *Cladina* lichens (Thieret 1964). A similar type was described by Lindsey (1952) near Great Bear Lake, where bouldery beach materials impose wide spacing on *Picea glauca* to create a lichen-rich type.

Westward in the Mackenzie Valley a different kind of lichen woodland occurs on the mounded mineral soils described by Zoltai (1975). The fine-textured sediments seem an unlikely substrate for *Cladina* lichens, but the soils provide the key prerequisite for lichen survival: the confinement of rooted competitors to spaced loci. Frost heaving (cryoturbation) has produced a pattern of mounds on whose raised centres the lichens thrive. Black spruce, larch, ericoid shrubs, and bryophytes survive at the more stable perimeters of these hummocky sorted circles. Personal observations confirmed that the lichen woodlands visible on aerial photographs near Canol, across the river from Norman Wells, are the expression of *Cladina* species (particularly *C. mitis* and *C. stellaris*), with appreciable *Cetraria nivalis.* In marked contrast to its commonness eastward, *Stereocaulon paschale* is virtually absent (Bird et al. 1980).

A third kind of lichen woodland is common on frozen organic terrain—on palsas and peat plateaus—throughout the Canadian Low Subarctic (see, for example, Ahti and Hepburn 1967). In contrast to the vegetation pattern on cryoturbated mineral soils where lichens occupy the tops of the mounds with shrubs and trees between them, on raised permafrost bogs the woody species tend to occupy the rises, with the *Cladinae* in the lower peripheral positions. Apparently the rigours of spring flooding in bog-surface depressions, followed by summer drying, can be tolerated only by the lichens. The tops of the mounds are a less hazardous microhabitat for rooted plants, and here, on the stable frozen peat, the trees grow straight. Both shade and the bryophytes under the spruce crowns effectively exclude the lichens. But whenever environmental changes reduce either the tree canopy or the vigour of the peat-forming mosses, the lichens invade from the hollows to the mounds. Thus, after crown fires, whose intense back-radiation kills the mound-building *Sphagnum fuscum*, succession leads to lichen-rich vegetation (Rowe et al. 1974). Even without the incidence of fire, the elevation of bogs during the normal aggradation of peat, accentuated by ice segregation in palsas and peat plateaus, eventually initiates a surface-drying cycle that favours the expansion of lichens as the *Sphagnum* declines in vigour. In northern Ontario Railton and Sparling (1973) noted a predominance of lichen vegetation on palsa mounds higher than 60 cm, with *S. fuscum* dominating lower features. In the upper Mackenzie Valley I found decomposed gelatinous layers of white *Cladina* between strata of fibric peat, evidence that environmental fluctuations have alternately favoured lichen woodland and moss woodland on peat plateaus.

Lichen Woodland on Deep Drift

Most of the detailed studies of lichen woodland have been made on the Precambrian Shield, where on either side of Hudson Bay the type occupies enormous areas. In contrast to the sites just described, where visible characteristics of the substrate (rock fissures, congeliturbated mounds, and peat mounds) impose the wide spacing of trees that allows lichens to prosper, the reasons for the woodland structure on deep glacial drift are not immediately apparent. Because stand openness seems to be the key characteristic, various explanations for it have been advanced. These include radiation deficiencies at high latitudes (Vowinckel et al. 1975), cold wet soils (Hare 1950, Hustich 1951, Lucarotti 1976), nutrient-poor soils (Moore 1980, 1981), allelopathic effects of lichens on tree seedlings (Brown and Mikola 1974, Fisher 1979), and recurrent fire that prevents succession to closed-crown forest (Kershaw 1977, 1978). None is entirely convincing in explaining the genesis of the open woodland structure and its perpetuation.

The fire hypothesis (Kershaw 1977, 1978) is ingenious, explaining the

Fig. 2. Lichen woodland on deep till, Abitau Lake, Northwest Territories. The open *Picea mariana-Cladina-Stereocaulon* type on the well-drained, drumlin breaks abruptly into a permafrost bog-forest type on lower slopes.

woodland structure as resulting from sparse postfire regeneration of spruce. Kershaw agrees with Quebec workers (Hare 1950, Lucarotti 1976, Rencz and Auclair 1978) that the well-developed lichen mat mulches the soil, prevents evaporation, and thus maintains moisture near field capacity throughout the growing season. But he considers the latter soil condition to be favourable (rather than unfavourable, as most of the Quebec studies suggest) because it allows a filling-in by spruce seedlings and layers to "the full development of lichen woodland with an average intertree distance approaching 10 metres" (Kershaw 1977: 408). According to Kershaw, normal succession to closed-crown forest would continue were it not for fire, which kills the encroaching trees and restores the open structure in which the lichens flourish. In Clementsian terms, lichen woodlands are a persistent fire subclimax, dominated by *Cladina stellaris* in the eastern Subarctic and by *Stereocaulon paschale* in the western Subarctic.

These ideas of close structural and compositional obligate ties between lichens, trees, and fire in the community are appealing. The evergreen woody species and the lichens are inflammable and invite fire, which then creates conditions that perpetuate the woodland. As in the classic facilitating model of succession, the lichen mat improves the environment and allows full development of spaced-tree stands. The latter in turn are effective windbreaks and snow traps, and ameliorate the surface temperature regime for the lichens. At the same time there is negative feedback, such as the inhibition of mycorrhizae by the lichen leachate (Fisher 1979), which stabilizes the open structure by curtailing the ability of trees to absorb essential nutrients.

The model, though intriguing, is doubtful. When one surveys lichen woodland on the Precambrian Shield from the air, and sees it stretching on for hundreds of kilometres, noting the regularity with which the type appears on the well-drained uplands, and noting too that it characterizes "islands" of coarse drift in wetlands where fire is excluded, then intuitively it seems that the structure is stable and not just subclimax to closed-crown forest. Nor does the recurrent-burning hypothesis explain why woodland becomes more and more open from south to north, with tree densities and sizes steadily decreasing toward the forest-tundra border. Undoubtedly fires rejuvenate lichen woodland (Johnson and Rowe 1975), but it is difficult to believe that fires are a sufficiently dependable zonal influence to account for the widespread prevalence of the type.

In fact there is evidence that tree density stabilizes in the woodland stage, with or without preceding fire. Auclair (1982) reported spruce regeneration in old woodlands in quantities sufficient only to maintain the open structure. Within his area in northern Quebec he did not find evidence of succession to closed forest. Carroll (1978) also reported evidence of the stabilization of woodland in the open lichen phase near Lake Athabasca, as did Johnson (1981) farther north at Porter Lake in the Northwest Territories. Within the study areas of the last two workers I measured plots in very old black spruce-lichen stands and found an all-aged structure, evidence of a continuous replacement of overmature dying trees by seedlings and layers, though not in sufficient numbers to close the canopy. Thus the type is maintained without fire on well-drained soils to ages in excess of 250 years.

The Soil Drought Hypothesis

A principle of ecology is that in all the spectrum of ecosystem components and outside influences (of which the measurable ones are called "factors"), some are more important than others in their triggering role. An explanation of complex systems is therefore sought by analysing, simplifying, and picking out the various parts that may prove to be the key climatic, edaphic, or biotic factors.

The subarctic lichen woodlands seem to be polygenetic; a variety of environmental circumstances limit the density of trees and their leafy associates, allowing or encouraging the terricolous lichens to flourish on the uncanopied surfaces. Nevertheless there may be a common thread that runs through many of the variations of the type. In northern Europe the consensus seems to be that droughty soils are the key (Ahti 1961, Sjörs 1963). In northern Canada this particular hypothesis has not been seriously considered.

That soil drought in soil surface layers where roots are concentrated may be important is indicated both by the prevalence of fires (requiring a dry surface) and by the well-drained characteristics of the soils of the Shield. The latter are podzolic and brunisolic (Moore 1980, Bunting in Kershaw et al. 1975, Tarnocai in Bradley et al. 1982). The water retention capacity of such soils is usually low; even moderate droughts could cause a rapid decline from field capacity to wilting point in shallow rooting layers. Black and Bliss (1980) noted water-stress-related reduction of photosynthesis in black spruce seedlings, with high mortality at low levels of water loss even when trees appeared to be unaffected. Thus soil drought can be expected to exert a significant effect on stand density early in succession.

Fig. 3. Eluviated dystric brunisol profile under *Picea mariana–Cladina stellaris–Vaccinium myrtilloides* in northern Saskatchewan. Such soils are nutrient-poor, with a low water-retention capacity.

Moreover, evidence is increasing that in nutrient-poor ecosystems the roots of vascular plants tend to concentrate in zones of maximum biological activity, particulary in soil near-surface zones and at the organic-mineral interface, where decomposition and nutrient release are most active (Chapin and Van Cleve 1981). Sirén (1955) showed that spruce roots concentrated in and near the organic surface mat as stands aged, a condition likely to increase their susceptibility to drought. If it be remembered that the chief dominant—*Picea mariana*—is a shallow-rooted species, and that most research workers have characterized lichen woodland as having a *xeric surface*, then regardless of the amount of moisture deeper in the soil, a mechanism for periodic mortality and thinning of subarctic tree stands emerges. Finally, extremes are more important than means, and despite short-term measurements that have reported lichen woodland soils at field capacity throughout the growing season, one drought year in ten would surely be sufficient to drastically influence the woodland trees and shrubs.

Conclusions

Lichen woodland is the typical vegetation in the Subarctic, the northern subzone of the boreal zone. The term includes a variety of ecological types developed on rock, on frost-mounded mineral soils, on frozen peatland, and especially on well-drained uplands. One or both of the spruces, *Picea mariana* and *P. glauca*, usually dominates the woodland, although a component of *Larix laricina, Betula papyrifera, Pinus banksiana*, and in Quebec, *Abies balsamea*, is often present.

Explanations for the well-drained types on deep soils invoke the effects of low levels of solar radiation on tree photosynthesis, or the effects of cold soils, nutrient-deficient substrata, lichen-tree interactions, and fire. None is entirely convincing.

Field ecologists have suggested that after fire the early successional stands are open because of regeneration failure, and that they increase in density with age, forming first woodland and eventually closed-crown forest. Others believe that the stands begin well-stocked, and thin with age. It is likely that both hypotheses are correct, but in different parts of the Subarctic; the second one is certainly true for the Low Subarctic. Evidence from Quebec and the south-central Northwest Territories suggests that the woodland structure, however initially formed, can persist as a stable form over long periods even in the absence of fire.

The hypothesis that intermittent drying of the soil surface just under the lichen mat may be an important triggering factor in forming and maintaining the open woodland structure has not yet been carefully examined. Soil drought may be a key factor in lichen woodlands, especially on the Precambrian Shield uplands.

References

Ahti, T. 1959. Studies on the caribou lichen stands of Newfoundland. Ann. Bot. Soc. Vanamo 30: 1–44.
———. 1961. The open boreal woodland subzone and its relation to reindeer husbandry. Arch. Soc. Vanamo 16 (Supp.): 91–93.
———. 1980. Definition and subdivision of the Subarctic: A circumpolar view. In G. R. Brassard, ed. Canadian Botanical Association (Supp. to Vol. 13, No. 2): 3–10. Dep. Biology, Univ. Waterloo, Waterloo, Ontario.
———, and R. L. Hepburn. 1967. Preliminary studies on woodland caribou range, especially on lichen stands, in Ontario. Wildl. Res. Rep. No. 74. Ont. Dep. Lands and Forests, Downsview, Canada. 134 pp.
Auclair, A. N. D. 1982. The role of fire in lichen-dominated tundra and forest-tundra. In R. W. Wein and D. A. MacLean, eds. The role of fire in northern circumpolar ecosystems. SCOPE 18, John Wiley & Sons, Toronto.
Bird, C. D., J. W. Thomson, A. H. Marsh, G. W. Scotter, and P. Y. Wong. 1980. Lichens from the area drained by the Peel and Mackenzie rivers, Yukon and Northwest Territories, Canada. I. Macrolichens. Can. J. Bot. 58: 1947–1985.
Black, R. A., and L. C. Bliss. 1980. Reproductive biology of black spruce at treeline. Ecol. Monogr. 5: 331–354.
Blüthgen, J. 1970. Problems of definition and geographical differentiation of the Subarctic with special regard to northern Europe: Ecology and conservation. 1. Ecology of the subarctic regions. Proc. Helsinki Symp.: 11–33. UNESCO, Paris.
Bradley, S. W., J. S. Rowe, and C. Tarnocai. 1982. An ecological land survey of the Lockhart River map area. Ecol. Land Cl. Ser., Lands Directorate, Env. Can. Ottawa.
Brown, R. J. E. 1967. Permafrost in Canada. Map 1246A, Geol. Serv. Can. Energy, Mines and Resources, Ottawa.
Brown, R. T., and P. Mikola. 1974. The influence of fruticose lichens upon the mycorrhizae and seedling growth of forest trees. Acta Forest. Fennica 141: 1–22.
Carroll, S. 1978. The role of fire in the jack pine–lichen woodlands of the Athabasca Plains region of Canada. pp. 47–49. In Fire ecology in resource management. Information Rep. NOR-X-210, Northern Forest Res. Centre, Env. Can., Edmonton, Alberta.
Chapin, F. S. 1980. The mineral nutrition of wild plants. Annu. Rev. Ecol. and Syst. 11: 233–260.
———, and K. Van Cleve. 1981. Plant nutrient absorption and retention under differing fire regimes. pp. 301–321. In H. A. Mooney, T. M. Bonickson, N. L.

Christensen, J. E. Lotan, and W. A. Reiners, eds. Fire regimes and ecosystem properties. Gen. Tech. Rep. WO-26. USDA Forest. Serv., Washington, D.C.

Crittenden, P. D., and K. A. Kershaw. 1978. Discovering the role of lichens in the nitrogen cycle in boreal-Arctic ecosystems. Bryologist 18(2): 258–267.

Dansereau, P. 1951. Description and recording of vegetation upon a structural basis. Ecology 32: 172–229.

Fisher, R. F. 1979. Possible allelopathic effects of reindeer-moss (*Cladonia*) on jackpine and white spruce. Forest Sci. 25: 256–260.

Gleason, H. A. 1939. The individualistic concept of the plant association. Amer. Midland Natur. 21: 92–110.

Grime, J. P. 1979. Plant strategies and vegetation processes. John Wiley & Sons, Toronto. 222 pp.

Gubbe, D. M. 1976. Vegetation. pp. 90–244. *In* D. M. Gubbe, ed. Landscape survey, District of Keewatin, N.W.T., 1975. Rep. for Polar Gas Project, Toronto, by R. M. Hardy and Assoc. Ltd.

Hare, F. K. 1950. Climate and zonal divisions of the boreal forest formation in eastern Canada. Geogr. Rev. 40: 615–635.

———. 1964. New light from Labrador-Ungava. Ann. Assoc. Amer. Geogr. 54: 459–476.

Hustich, I. 1951. The lichen woodlands in Labrador and their importance as winter pastures for domesticated reindeer. Acta Geogr. 12: 1–48.

———. 1979. Ecological concepts and biographical zonation in the North: The need for a generally accepted terminology. Holarctic Ecol. 2: 208–217.

Ingestad, T. 1973. Nitrogen and cation nutrition of three ecologically different plant species. Physiol. Plant. 38: 29–34.

Johnson, E. A. 1981. Vegetation organization and dynamics of lichen woodland communities in the Northwest Territories, Canada. Ecology 62: 200–215.

———, and J. S. Rowe. 1975. Fire in the subarctic wintering ground of the Beverley caribou herd. Amer. Midland Natur. 94: 1–14.

Kershaw, K. A. 1977. Studies on lichen-dominated systems. XX. An examination of some aspects of the northern boreal lichen woodlands in Canada. Can. J. Bot. 55: 393–410.

———. 1978. The role of lichens in boreal tundra transition areas. Bryologist 81: 294–306.

———, W. R. Rouse, and B. T. Bunting. 1975. The impact of fire on forest and tundra ecosystems. Indian and Northern Affairs Pub. No. QS-8038-000-EE-A1. Ottawa. 81 pp.

Lambert, J. D. H., and P. F. Maycock. 1968. The ecology of terricolous lichens of the northern conifer-hardwood forests of central eastern Canada. Can. J. Bot. 46: 1043–1078.

Larsen, J. A. 1980. The boreal ecosystem. Academic Press, Toronto. 500 pp.

Lindsey, A. A. 1952. Vegetation of the ancient beaches above Great Bear and Great Slave lakes. Ecology 33: 535–549.

Löve, D. 1970. Subarctic and Subalpine: Where and what? Arctic and Alpine Res. 2: 63–73.

Lucarotti, C. 1976. Post-fire change in mycofloral species and mesofauna populations in lichen-woodland soils, Schefferville, Quebec. M.Sc. thesis. McGill Univ., Montreal.

Maikawa, E., and K. A. Kershaw. 1976. Studies on lichen-dominated ecosystems. XIX. The postfire recovery sequence of black spruce–lichen woodlands in the Abitau Lake Region, N.W.T. Can. J. Bot. 54: 2679–2687.

Mattson, W. J. 1980. Herbivory in relation to plant nitrogen content. Annu. Rev. Ecol. and Syst. 11: 119–161.

Moore, T. R. 1980. The nutrient status of subarctic soils. Arctic and Alpine Res. 12: 147–160.

———. 1981. Controls on the decomposition of organic matter in subarctic spruce–lichen woodland soils. Soil Sci. 131: 107–113.

Railton, J. B., and J. H. Sparling. 1973. Preliminary studies on the ecology of palsa mounds in northern Ontario. Can. J. Bot. 51: 1037–1044.

Raup, H. M. 1946. Phytogeographic studies in the Athabas Ka-Great Slave Lake region. II. J. Arnold Arboretum 27: 1–85.

Rencz, A. N., and A. N. D. Auclair. 1978. Biomass distribution in a subarctic *Picea mariana–Cladonia alpestris* woodland. Can. J. Forest Res. 8: 168–176.

Rousseau, J. 1952. Les zones biologiques de la péninsule Québec-Labrador et l'hémiarctique. Can. J. Bot. 30: 436–474.

Rowe, J. S., J. S. Bergsteinsson, G. A. Padbury, and R. Hermesh. 1974. Fire studies in the Mackenzie Valley. Arctic Land Use Research Program. Dep. Indian Affairs and Northern Dev., Ottawa. ALUR Rep. 73-74-61. 123 pp.

Sirén, G. 1955. The development of spruce forest on raw humus sites in northern Finland and its ecology. Acta Forest. Fennica 62: 4. 408 pp.

Sjörs, H. 1963. Amphiatlantic zonation: Nemoral to Arctic. pp. 109–125. *In* A. Löve and D. Löve, eds. North Atlantic biota and their history. Pergamon, New York.

Smith, D. C. 1962. The biology of lichen thalli. Biol. Rev. 37: 537–570.

Steere, W. C., G. W. Scotter, and K. Holmen. 1977. Bryophytes of Nahanni National Park and vicinity, Northwest Territories, Canada. Can. J. Bot. 55: 1764–1767.

Stribley, D. P., and D. J. Read. 1974. The biology of mycorrhizae in the ericaceae. IV. New Phytol. 73: 1149–1155.

Thieret, J. W. 1964. Botanical survey along the Yellowknife Highway, N.W.T., Canada. II. Vegetation. Sida 1: 187–239.

Viereck, L. A., and L. A. Schandelmeier. 1980. Effects of fire in Alaska and adjacent Canada: A literature review. BLM-Alaska Tech. Rep. 6. U.S. Dep. Interior, Bureau of Land Manage., Alaska State Office. 124 pp.

Vowinckel, T., W. C. Oechel, and W. G. Boll. 1975. The effect of climate on the photosynthesis of *Picea mariana* at the subarctic tree line. I. Field Measurements. Can. J. Bot. 53: 604–620.

Zoltai, S. C. 1975. Structure of subarctic forests on hummocky permafrost terrain in northwestern Canada. Can. J. Forest Res. 5: 1–9.

———, D. J. Karasiuk, and G. W. Scotter. 1979. A natural resource survey of Horton–Anderson River area, Northwest Territories. Parks Can., Ottawa. 160 pp.

Tundra Plant Communities of the Mackenzie Mountains, Northwest Territories; Floristic Characteristics of Long-Term Surface Disturbances

G. P. Kershaw

Introduction and Overview of the CANOL Project

The CANOL Project was a World War II venture commissioned by the United States Army but engineered, designed, constructed, and operated by civilian American contractors, employing a work force consisting of both American and Canadian nationals. The project was designed to transport crude oil produced at Norman Wells on the Mackenzie River (at that time the northernmost producing oilfield on the continent) to Whitehorse in the Yukon Territory (fig. 1). The oil was refined in Whitehorse, and the finished products were transported by pipelines to Carcross and Watson Lake in the Yukon and to Fairbanks and Skagway in Alaska. These products were provided for the defence of Alaska and Canada against the Japanese and for the fueling of aircraft being flown to the USSR as part of the lend-lease agreements (Richardson 1944b).

The CANOL Project entailed the construction of a 10.2-cm-diameter, 737-km-long, crude oil pipeline from Norman Wells to the Alaska Highway at Johnson's Crossing, and a 15.2-cm-diameter, 192-km-long pipeline from that point to Whitehorse. From Whitehorse, 5.1-cm and 7.6-cm pipelines distributed products to the locations mentioned above (Finnie 1945, Hemstock 1945, U.S. Army 1950) (fig. 1). A total of 2,575 km of pipelines in four separate systems eventually were constructed, as well as 828 km of gravel-surfaced tote road, 829 km of telephone system, 2,415 km of primarily new winter roads, 10 landing strips along the Mackenzie

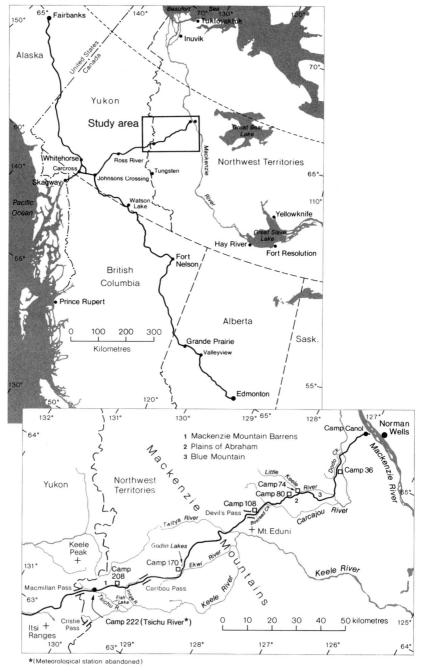

Fig. 1. The CANOL Projects encompassed a large portion of northwestern Canada. The area of this study included approximately 31% of the Northwest Territories section of CANOL No. 1 and the adjacent undisturbed plant communities.

River (U.S. Army 1950), and 58 wells, of which 55 produced commercial quantities of oil (U.S. Army 1950). In addition to this, 2,736 km of water routes were upgraded (Finnie 1945) (fig. 1). Construction began in October 1942 and the Whitehorse refinery commenced operations on 30 April 1944. The estimated total project cost was $138 million U.S. (Truman Committee 1944), exceeding that of the Alaska Highway. Several projects were carried out within CANOL; most of these involved air routes and winter roads.

Only the Northwest Territories sections of the CANOL No. 1 pipeline, road, and associated telephone system were considered in this study. This section of the project, referred to as CANOL No. 1 East by historical sources, was constructed between 10 October 1942 and 12 March 1944 and abandoned by 31 May 1945. In the postwar years oil and other developments in Alberta encouraged recovery of some of the materials along the route. In much of the area salvage operations were completed by the fall of 1953.

Only areas above timberline (i.e., alpine tundra) were studied. Timberline was defined by the upper (elevational) limit of trees (Tranquillini 1979). Any member of a tree species (i.e., *Picea glauca, P. mariana*, or *Abies lasiocarpa* in the study area) that possessed a single central trunk growing at least 2 m above the mean winter snow depth was considered a tree (Pruitt 1978)[1] The height of layered branches at the base of the tree indicate the mean winter snow depth. Timberline occurs at successively lower elevations from the southwest to the northeast in the study area. Six separate sectors of tundra were studied between RMP 56.5 near Canol Lake and RMP 231 in Macmillan Pass (fig. 1). These extend over 117 km, or 31% of the total route east of the continental divide at Macmillan Pass on the Yukon–Northwest Territories border.

The elevations of the tundra affected by the CANOL Project range from 775 m asl near the Mackenzie River to 1,740 m asl on the Plains of Abraham. This alpine area includes portions of the Selwyn and Mackenzie mountain chains. Dramatic changes in elevation occur along the route. For example, on the ascent to the Plains of Abraham the pipeline gradient is 55 m/km over a distance of 13.7 km, occasionally achieving grades of 17%.

The CANOL Project was one of the largest projects ever undertaken in northern Canada. It was rapidly conceived and executed with little understanding of northern environmental limitations. As a result of inadequate

---

1. Vascular plant nomenclature follows Porsild and Cody (1980).

planning and no environmentally sensitive regulation of the project, its potential for negative environmental impacts was great. For example, the only restriction that was placed on activities during the operation of the project was a moratorium on hunting; apparently this was ignored (Flood 1946). No mitigative measures designed to reduce environmental impacts were undertaken except as they resulted from normal engineering practices. No attempt was made to bypass ecologically sensitive areas in an effort to reduce potential negative biological consequences. Finally, no rehabilitation program was instituted to enhance recovery on CANOL disturbances that at the time of this study were 34 to 36 years old.

The choice of areas for field investigation was based mainly on the initial aerial photographic interpretation of 1944 and 1974 photography. Three main considerations governed the choice of study sites in the field:
(1) Each site had to be representative of the conditions found in that area.
(2) Sites with the largest number of disturbance types were selected in an effort to maximize the amount of information gained.
(3) Disturbances had to be a short distance from one another and close to a suitable undisturbed area that could be used as a control site.

Fieldwork was conducted from 15 June to 29 August in 1977, from 13 July to 3 October in 1978, and from 23 June to 27 August in 1979.

Five contiguous 0.5 x 1 m quadrats were used for sampling vegetation in areas dominated by erect shrub communities. In other communities five contiguous 0.5 x 0.5 m quadrats were sampled. For each quadrat, plant cover was estimated for each species[2] using a modified Braun–Blanquet cover classification, and verbal descriptions of physiognomy, structure, and vigour were made.

Voucher collections were identified by L. J. Kershaw (vascular plants), C. LaFarge-England (bryophytes), and J. Marsh (lichens). Vascular plant identifications were verified by W. J. Cody and G. A. Mulligan at the Vascular Plant Herbarium, Biosystematics Research Institute, Canada Department of Agriculture (DAO), and by G. W. Argus at the National Herbarium of Canada, National Museums of Canada (CAN). Vascular plant collections were placed in both CAN and DAO, and one was retained in the

---

2. Due to taxonomic difficulties with many of the nonvascular plant taxa, it was often necessary to refer to plants by genus (e.g., *Cladonia* sp., *Polytrichum* sp.) or by broader groups (e.g., crustose lichens, moss spp.). Through plant collections, it was later possible to compile relatively complete species lists for each site. However, percentage cover estimates for each taxon could not always be calculated, as each of these categories could include several species. Cover by vascular plants was recorded for each species present.

Department of Geography, the University of Alberta. Unless otherwise stated, nomenclature follows that of Porsild and Cody (1980) and Argus (1973) for vascular plants; that of Crum et al. (1973), Hale and Culberson (1970), and Steere (1978) for bryophytes; and that of Thomson (1967, 1979) for lichens.

Several phases were involved in the analysis of the vegetation data. These included standard arithmetic and statistical treatments of the aboveground phytomass and shrub age data, derivation of species cover means in each stand (Honsaker 1981), and the application of COMPCLUS (Gauch 1979) to the quadrat data.

The cover values from each of the five contiguous quadrats were averaged for each plant or plant group, and these data were then made available for the COMPCLUS program. COMPCLUS is a computer program that identifies stands on the basis of species composition and cover similarities. The clusters identified represent plant communities. In the analysis of control sites, a "percentage distance" (Gauch 1979) radius of 82.66 was used with no data transformations. The results emphasize the dominant components for each site based upon percent cover values.

The number of species present at each site and those held in common with the control were noted, and the results of this species richness analysis were then subjected to Sorensen's treatment in order to develop an Index of Similarity (Mueller-Dombois and Ellenberg 1974) comparing the plant communities of the disturbances with those of their controls.

The mean cover value for each species or plant group at a site was compared with that of its control by subtracting the average cover in the control from the average value recorded in an associated disturbance. This produced control-corrected data for all disturbances at each sample location. Mean differences from controls were then compiled by deriving the average of all of the relevant control-corrected values for that species. A positive value indicated an increase in cover, and a negative value reflected a decrease for that species. When a species was not present in a disturbance it was recorded as missing from a disturbance while present on the associated control in one case, but present on the disturbance at another site.

Results and Discussion

A total of 311 stands (i.e., 1,555 samples) were sampled to determine their botanical characteristics. In all, 169 lichen, 115 bryophyte, and 303 vascular plant taxa were collected—a total of 587 plant taxa.

Table 1 A general summary of the relative differences in substrate characteristics between CANOL disturbances and undisturbed conditions, Northwest Territories.

Characteristics[1]	Road	False start road	Bladed trail	Camp yard	Bulldozer track	Gravel pit	Gravel pit access road	Oil spill
Attitude								
macro-relief	elevated and terraced 0–3 m	elevated, depressed and terraced 0–2 m	depressed 10–100 cm	elevated 10–100 cm	in organics, 2 parallel ruts 5–25 cm deep	depressed 0.5–3 m	elevated and depressed 0.5–1 m	similar to control
micro-relief	level 5–10 cm	very irregular 20–100 cm	irregular 10–50 cm	level 0–15 cm	level 0–5 cm	irregular 10–75 cm	level to irregular 10–25 cm	similar to control
ditching	none to both sides	generally none	none	randomly located	none			
Compaction	extreme	slight to none	slight	extreme to none	none to moderate	slight to none	slight to moderate	slight
Surface composition	mineral	mineral-organic	mineral-organic	mineral-organic	similar to control	mineral	mineral	oil saturated organics and residual tar
gravel content	high (75–100%)	none to high (0–100%)	none to high (0–75%)	little to high (20–100%)	similar to control	high (80–100%)	little to high (20–100%)	none
Predisturbance soil	buried or removed	intact and patchy or removed	intact and patchy to removed	intact and patchy to removed	intact	removed	buried or removed to discontinuous	intact
Substrate climate								
winter	colder, drier	colder-warmer, wetter-drier	warmer, moister-wetter	colder-warmer, moister-drier	warmer, moister	warmer, moister	cooler-warmer, moister-drier	similar to control

summer	warmer, drier	warmer, wetter-drier	cooler, moister-wetter	warmer, moister-drier	cooler, moister-wetter	warmer, wetter-drier	warmer, wetter-drier	warmer-hotter, drier
Drainage surface	unrestricted runoff	restricted runoff and retarded	collecting area	unrestricted runoff	collecting area	collecting area	unrestricted runoff-collecting area	similar to control
internal	freer	unimpeded-freer	unimpeded-freer	freer	similar to control	unimpeded-freer	unimpeded-freer	unimpeded
Predisturbance vegetation	eliminated	some changes-eliminated	partially-completely eliminated	discontinuous	similar to control-some changes	eliminated	partially-completely eliminated	eliminated
Timing of disturbance	all seasons	all seasons	mainly fall and winter	all seasons	thaw season	all seasons	all seasons	all seasons
Post-1945 disturbance	limited re-disturbance during salvage	some sections redisturbed	probably not redisturbed	limited re-disturbance during salvage	no redisturbance	no redisturbance	no redisturbance	some spills created during salvage
Other remarks	road often submerged in areas of thermal erosion	most stretches never used, 1 section used in winter only	may channelize surface runoff and cause rill erosion	permafrost aggradation in the shade of some buildings	most resulted from 1 to few vehicle passes	a few used for road maintenance	often only an accumulation of fill lost from trucks	areas affected vary from few to 200m²

Note
1. These characteristics are relative to the undisturbed condition.

Eight distinct disturbances were studied (table 1). The condition of each at initiation varied considerably, and after 34 to 36 years these differences in the degree of surface disruption have resulted in variations in their plant community characteristics.

Seven major physiognomically defined plant communities were identified within the study area. Their definition was based primarily on structure and consequently there is often more than one floristically defined plant community in each. For example, areas of decumbent shrub tundra could be dominated by several species, including *Salix reticulata, S. arctica,* and *Dryas integrifolia.* Conversely, plant communities dominated by *Betula glandulosa* could fall within the category of erect deciduous shrub tundra when plants were robust or within the decumbent shrub tundra category when adverse conditions had resulted in shrubs of reduced size. Each floristic community will be discussed within the structural plant community that best typifies it. This was done to ensure that the detail of species composition would not be obscured by comparisons including the several floristically dissimilar plant communities that have been lumped into one physiognomically defined plant community.

The COMPCLUS analysis served to identify 15 plant communities from the 89 control stands. The plant community titles include the plants with the highest mean cover values in that cluster.

Only species with mean disturbance cover values at least 2% greater or less than those recorded on the control sites will be discussed. The many rare plants in the various communities are not discussed on an individual basis. A plant with less than 2% may be rare because it requires special microhabitat characteristics in order to survive, or it may have limited reproductive capability. These plants may prove most sensitive to disturbances and as such could be significant indicators of change. To use them in this way requires a detailed understanding of the autecology of each species; few such studies exist. However, rare plants are considered when discussing species richness and similarity coefficients at a site.

Species richness is equivalent to the number of plant taxa in the control or disturbance stand. When determining species richness, even those taxa occurring only once in several samples were included in the total.

Erect Deciduous Shrub Tundra

Erect deciduous shrub tundra includes those plant communities with a shrub layer over 30 cm tall and covering at least 50% of the area under plants. In the study area such communities were dominated by *Salix* spp. and *Betula glandulosa.* Variation in floristic composition occurred, with a

Fig. 2. Erect deciduous shrub tundra is common at lower elevations throughout the western portion of the Mackenzie Mountains. Shrubs vary in height from 30 to 200 cm.

number of distinct, floristically defined plant communities present. These included *Hedysarum alpinum*–Moss, *Salix reticulata*–*Salix lanata*–Moss, and *Betula glandulosa*–*Cladonia stellaris*–Moss communities.

These plant communities were abundant in the Joker Ridge and Bolstead Creek–Devil's Pass ecosections. Detailed descriptions of these ecosections are presented in Kershaw (1983). Generally, erect deciduous shrub tundra dominated lower elevations on imperfectly drained to well-drained soils (fig. 2).

HEDYSARUM ALPINUM – MOSS COMMUNITY This plant community can be classified as riparian since it dominates on flood plains where annual flooding occurs. Robust *Salix alaxensis* generally exceed 1.5 m in height and commonly reach 4 m. The understory is dominated by *Hedysarum alpinum*, mosses such as *Campylium stellatum*, *Bryum bimum*, *Distichium inclinatum*, *Drepanocladus sendtneri*, *D. uncinatus*, *Hypnum lindbergii*, and *Tomenthypnum nitens*, as well as the prostrate *Salix reticulata* (table 2).

Hedysarum alpinum had less cover on all disturbances than in controls. *Salix alaxensis*, which commonly comprises a high percentage of the cover

Table 2 Floristic characteristics of tundra plant communities in the Mackenzie Mountains, Northwest Territories.

	Erect deciduous shrub tundra			Decumbent shrub tundra		Sedge meadow tundra				Lichen heath tundra	Fruticose lichen tundra	Cushion plant tundra		Saxicolous Lichens	Crustose lichen tundra
Plant species or group	1 *Hedysarum alpinum*-Moss	2 *Salix reticulata-Salix lanata*	3 *Betula glandulosa-Cladonia stellaris*-Moss	4 *Salix barrattiana*-Moss	5 *Salix polaris-Dactylina beringica*	6 *Dryas integrifolia-Carex* spp.	7 *Carex membranacea-Sphagnum* spp.	8 Misc. Mosses-*Carex* spp.	9 *Carex* spp.-*Kobresia simpliciuscula*	10 *Dryas integrifolia-Cassiope tetragona*	11 *Cladonia stellaris-Alectoria ochroleuca*	12 *Dryas integrifolia-Cetraria tilesii*	13 *Dryas integrifolia-Rhizocarpon umbilicatum*	14 *Lecanora epibryon*	15 *Rhizocarpon inarense-Umbilicaria proboscidea*
LICHENS															
Buelliaceae															
Buellia papillata															
Rinodina roscida											*				
Cladoniaceae															
Cladonia sp.			0.3					0.4		0.4	0.8				
Cladonia acuminata										+					
Cladonia amaurocraea			1.1								0.4				
Cladonia arbuscula			4.1				+	1.0		0.8	0.3	0.6			

Species				
Cladonia bellidiflora	+		*	
Cladonia carneola	+		*	
Cladonia cenotea	*		+	
Cladonia chlorophaea		0.1		*	+
Cladonia coccifera	0.4	0.2		*	0.4
Cladonia cornuta	0.7		+		
Cladonia crispata	*		1.0		
Cladonia deformis	*		*		+
Cladonia ecmocyna		0.8	0.6		
Cladonia fimbriata	+		+		
Cladonia gonecha	0.1	0.2	*		+
Cladonia gracilis	1.6		+	0.1	4.0
Cladonia macrophylla	*				+
Cladonia major	+			*	
Cladonia meta-corallifera					
Cladonia mitis	36.3	0.5	*	*	9.5
Cladonia phyllophora					* +
Cladonia pleurota	*			*	* +
Cladonia pocillum	0.7		+		+
Cladonia pyxidata				0.2	1.4
Cladonia rangiferina	4.2		2.1 2.4	0.1 0.8	2.6
Cladonia stellaris	25.8		10.9	0.6	21.6 +
Cladonia subfurcata	+				
Cladonia subulata	0.1		+		+
Cladonia uncialis	+		+	0.3	0.4
Cladonia verticillata	+		*		0.1
Collemataceae					
Collema sp.				*	

Table 2 Continued

Plant species or group	1	2	3	4	5	6	7	8	9	10	11	12	13	14	15
Lecanoraceae															
Icmadophila ericetorum			+				+								
Lecanora epibyron										0.5		1.8	1.3	23.5	
Lecanora urceolaria										1.5					
Ochrolechia sp.												*			
Ochrolechia androgyna			0.1									1.6			
Ochrolechia geminipara							+				+				
Ochrolechia inaequatula										+					
Ochrolechia uliginosa										1.9					
Lecideaceae															
Bacidia alpina															+
Lecidea sp.															+
Lecidea demissa											0.6				
Lecidea glaucophaea															*
Lecidea granulosa			0.1					0.2							
Lecidea pantherina															*
Lecidella stigmatea										0.1					
Rhizocarpon sp.											4.2				2.6
Rhizocarpon eupetraeoides											1.4				
Rhizocarpon geographicum											*				
Rhizocarpon inarense											+				37.0
Rhizocarpon riparium							+				1.4				
Rhizocarpon umbilicatum													12.4		
Toninia lobulata												5.6			
Leprariaceae															
Lepraria neglecta								1.0							

Species	C1	C2	C3	C4	C5	C6	C7	C8
Nephromataceae								
Nephroma arcticum	+							
Nephroma expallidum		0.8	0.1		0.3		0.1	
Parmeliaceae								
Asahinea chrysantha	0.3					0.6		
Cetraria commixta				0.8	2.2			
Cetraria cucullata					+	3.5	1.5	0.8
Cetraria delisei						0.7		
Cetraria ericetorum						1.2	0.1	
Cetraria islandica	0.9		0.2		1.0	2.2	5.9	0.9
Cetraria laevigata	0.2				*	0.5	12.5	+ 0.1 +
Cetraria nivalis	2.6		0.1		0.2	2.4	5.9	0.2 0.1 +
Cetraria pinastri	0.4	0.1				+		
Cetraria richardsonii	0.4		0.1		0.1	0.4	0.6	0.1
Cetraria sepinicola	1.9							
Cetraria tilesii				0.1	+	1.9		3.9 0.3 + +
Parmelia centifuga	0.2							
Parmelia separata	0.2						0.2	
Parmelia septentrionalis	0.7							
Parmeliopsis ambigua							+	
Parmeliopsis hyperota								
Peltigeraceae								
Peltigera aphthosa	0.3	+	0.5		0.7	+	1.2	
Peltigera canina	0.1					+		
Peltigera malacea	+				+			
Peltigera pulverulenta					0.8			
Solarina bispora					0.1	0.1		0.1 0.1 0.1
Solarina crocea	0.1		0.1		0.6			
Pertusariaceae								
Pertusaria dactylina						0.3		1.4

Table 2 Continued

Plant species or group	1	2	3	4	5	6	7	8	9	10	11	12	13	14	15
Physciaceae															
Physcia caesia															1.4
Sphaerophoraceae															
Sphaerophorous globosus											0.8				
Stereocaulaceae															
Stereocaulon alpinum			+		0.3						+				
Stereocaulon paschale			2.7					0.9							
Stereocaulon tomentosum								0.1							
Umbilicariaceae															
Agyrophora rigida											0.3				0.3
Omphalodiscus decussatus															1.4
Omphalodiscus krascheninnikovii															1.4
Omphalodiscus viginus															1.4
Umbilicaria hyperborea			+					+			1.2				5.0
Umbilicaria proboscidea			+								0.9				35.5
Usneaceae															
Alectoria ochroleuca			+							3.2	10.8	1.9	0.1	+	+
Cornicularia aculeata										+					
Cornicularia divergens										+	+				
Dactylina arctica			+					0.1		0.8	0.2	0.2	+	+	+
Dactylina beringica			0.8		0.9			1.3		0.4	1.0	0.2	+		
Dactylina ramulosa						*				0.1		0.1	+		
Evernia perfragilis													+		
Thamnolia subuliformis						0.4				1.6		0.6	0.5	0.1	+
Thamnolia vermicularis										+	0.3	1.7			+

Taxon							
Verrucaceae							
Polyblastia gothica							+
Polyblastia hyperborea							+
Polyblastia integrascens							+
Polyblastia sendtneri			*	1.5		0.2	
Verrucaria muralis			*	*		1.4	2.6
Soil lichens	0.6	1.4	1.1	2.7	0.6	4.9	3.2
Crustose lichens		0.1		9.7	9.0	10.8	
Saxicolous lichens		3.0		6.6	3.5	5.4	30.0
BRYOPHYTES							
Amblystegiaceae							
Calliergon stramineum	*		+				*
Campylium stellatum				*		*	
Drepanocladus sp.				+			
Drepanocladus revolvens				+			
Drepanocladus sendtneri						*	
Drepanolcadus uncinatus	+		*				
Andreaceae							
Andrea rupestris							
Aulacomniaceae							
Aulacomnium sp.				*			
Aulacomnium acuminatum				*			
Aulacomnium palustre	8.0				+		
Aulacomnium turgidum						*	
Bartramiaceae							
Conostomum tetragonum							
Philonotis tomentella	*						

Table 2 Continued

Plant species or group	1	2	3	4	5	6	7	8	9	10	11	12	13	14	15
Brachytheciaceae															
Brachythecium sp.								*							
Brachythecium erythrorrhizon								*							
Brachythecium starkei				*											
Brachythecium turgidum													*		
Cirriphyllum cirrosum								*		+					
Tomenthypnum sp.								*		+				*	
Tomenthypnum nitens	*											*		*	
Bryaceae															
Bryum sp.				*				*				*		*	
Bryum bimum		*													
Bryum pseudotriquetrum				*									*		
Bryum tortifolium				*											
Pohlia cruda															
Catascopiaceae															
Catascopium nigritum						*		*						*	
Dicranaceae															
Dicranum sp.			+							*	*		*		
Dicranum elongatum			+							*	*				
Ditrichaceae															
Ceratodon purpureus							+			*					
Distichium capillaceum									+	*			*		
Distichium inclinatum			*			*				*			*		
Ditrichum sp.												+	*	*	
Ditrichum flexicaule														*	

Encalyptaceae
Encalypta sp.
Encalypta alpina
Encalypta procera

Entodontaceae
Orthothecium cryseum
Pleurozium schreberi

Fissidentaceae
Fissidens osmundoides

Grimmiaceae
Grimmia sp.
Rhacomitrium canescens
Rhacomitrium lanuginosum
Schistidium sp.

Hylocomniaceae
Hylocomnium splendens

Hypnaceae
Hypnum sp.
Hypnum bambergii
Hypnum cupressiforme
Hypnum lindbergii
Hypnum procerrimum
Hypnum revolutum
Ptilium crista-castrensis

Mniumaceae
Cinclidium stygium
Cyrtomnium sp.

Table 2 Continued

Plant species or group	1	2	3	4	5	6	7	8	9	10	11	12	13	14	15
Cyrtomnium hymenophyllum										+					
Plagiomnium cuspidatum								*							
Plagiomnium ellipticum								*							
Rhizomnium pseudopunctatum								*							
Orthotrichaceae															
Orthotrichum sp.															+
Polytrichaceae															
Pogonatum alpinum								*							
Polytrichum sp.			10.6		*		0.1	*							
Polytrichum commune			1.0				+	*							
Polytrichum juniperinum			+								2.3	*			
Polytrichum piliferum			0.3		*			*		*					
Polytrichum strictum															
Pottiaceae															
Tortella arctica										+					
Tortella fragilis										+		+	*	*	
Tortella tortuosa										*		*	*	*	
Tortula latifolia				*											
Tortula norvegica										+		+			
Tortula ruralis						*									
Ptilidiaceae															
Blepharostoma trichophylla						*									

Species	1	2	3	4	5	6	7	8	9	10	11	12	13	14
Rhytidiaceae														
Rhytidium rugosum								+						
Sphagnaceae														
Sphagnum sp.						22.0	*							
Theliaceae														
Myurella julacea					+		+					*	+	
Moss species	24.0	7.0	8.7	16.8	0.6	2.7	41.7	1.9	5.3	0.5	0.4	2.8	0.1	1.4
Hepatics	*					+	+	0.1	+		+	+		+
VASCULAR PLANTS														
Betulaceae														
Betula glandulosa	+	16.3				0.6	+		0.5	4.8	2.1			
Boraginaceae														
Mertensia paniculata		0.2	7.0				+							
Myosotis alpestris			0.2				+							
Caprifoliaceae														
Linnaea borealis										+				
Caryophyllaceae														
Cerastium beeringianum						0.1								
Melandrium apetalum					0.2			+	+	+	0.2	+	0.5	0.1
Minuartia arctica				+		0.1		+						0.2
Minuartia biflora								+				0.3		
Minuartia rossii				0.2					0.4		0.3	+		
Silene acaulis		+		0.6					1.5	0.6	1.8			0.7
Stellaria laeta			0.2							0.5	+	0.3		
Stellaria sp.			+				0.1			+		0.2		

Table 2 Continued

Plant species or group	1	2	3	4	5	6	7	8	9	10	11	12	13	14	15
Compositae															
Antennaria densifolia				0.9								0.2	+		
Antennaria ekmaniana						+				0.1		0.1			
Antennaria monocephala			0.5		0.2			0.2			0.2				
Arnica alpina			0.2			+				0.4		+	+		
Arnica lessingii	+														
Arnica louiseana											+				
Artemisia arctica			5.3		0.2			1.7			0.8				
Artemisia tilesii				0.6											
Aster sibiricus	0.8	0.1													
Chrysanthemum integrifolium						0.4									
Erigeron eriocephalus	+									0.1		+	0.2	0.1	
Petasites frigidus			0.8	1.9	0.1			0.1							
Petasites hyperboreus								0.4							
Saussurea angustifolia													+		
Senecio atropurpureus						+		0.1	+	0.1		+	1.5	0.1	
Senecio cymbalaria						0.1			+	0.1		0.1	0.2	+	
Senecio lugens	+		+	0.1				0.2		0.2					
Senecio triangularis								0.1		+					
Senecio yukonensis								0.1							
Senecio sp.			1.1												
Solidago multiradiata	+									+	0.8				
Taraxacum alaskanum								+							
Crassulaceae															
Rhodiola integrifolia				0.1	+			0.7							

Cruciferae
- Braya purpurascens
- Braya richardsonii
- Cardamine bellidifolia
- Cardamine pratensis
- Draba albertina
- Draba corymbosa
- Draba longipes
- Eutrema edwardsii
- Lesquerella arctica
- Parrya nudicaulis
- Smelowskia borealis

Cyperaceae
- Carex sp.
- Carex aquatilis aquatilis
- Carex aquatilis stans
- Carex atrofusca
- Carex aurea
- Carex brunnescens
- Carex capillaris
- Carex glacialis
- Carex membranacea
- Carex microchaeta
- Carex microglochin
- Carex misandra
- Carex nardina
- Carex petricosa
- Carex podocarpa
- Carex rupestris
- Carex scirpoidea
- Carex vaginata

Taxon									
Braya purpurascens					+				
Braya richardsonii								0.1	0.1
Cardamine bellidifolia			+						
Cardamine pratensis				0.1					
Draba albertina					+			0.1	0.2
Draba corymbosa			0.1			+			
Draba longipes			+					+	
Eutrema edwardsii							+		
Lesquerella arctica						0.1	0.1	+	0.2
Smelowskia borealis				+					+
Carex sp.			34.5		15.0				0.2
Carex aquatilis aquatilis									
Carex aquatilis stans					9.0				
Carex atrofusca		+			+		0.4	+	+
Carex aurea	+								
Carex brunnescens			0.1						
Carex capillaris	0.1				0.5		+ 1.3 +		0.3
Carex glacialis	+	0.1						2.8	0.3
Carex membranacea				0.1	25.5	11.5			0.4 0.2
Carex microchaeta					+				
Carex microglochin					+ 2.0	9.6	0.5		
Carex misandra							1.1 1.3 1.8	+	0.6 1.2
Carex nardina									
Carex petricosa			0.1	0.2		6.3	9.1	3.2	+ 4.8
Carex podocarpa					0.7		14.5		
Carex rupestris					7.0		8.6	2.0	0.5 1.8
Carex scirpoidea	+	0.1	0.1		0.7		3.1	1.5	2.0 2.4 2.6
Carex vaginata							+		0.1

Table 2 Continued

Plant species or group	1	2	3	4	5	6	7	8	9	10	11	12	13	14	15
Eriophorum angustifolium						0.5	1.4	+	2.0	4.7			1.1	+	
Eriophorum callitrix	+								0.2	0.2			0.8		
Eriophorum scheuchzeri													0.1		
Eriophorum vaginatum							8.9				0.1				
Kobresia myosuroides						2.3				+			1.7		
Kobresia simpliciuscula						1.6			3.2	0.8		+	0.1	+	
Scirpus caespitosus												+			
Eleagnaceae															
Shepherdia canadensis	0.1														
Empetraceae															
Empetrum nigrum		1	1.2					+	+	+	+				
Equisetaceae															
Equisetum arvense	0.1		1.2	4.1				+							
Equisetum scirpoides								+			0.2				
Equisetum variegatum	0.2	0.8				+				0.9		0.2	0.1		
Ericaceae															
Andromeda polifolia							2.7			0.2		0.1			
Arctostaphylos alpina	0.2		4.3			3.5				1.5	0.8	0.1	+		
Arctostaphylos uva-ursi			2.8												
Cassiope tetragona			+			+		1.8	+	9.4	5.7	0.8			
Kalmia polifolia								0.2							
Ledum decumbens							0.6	+		+	0.2		+		
Ledum groenlandicum			5.0				+							+	
Rhododendron lapponicum						0.3				1.0	0.1	1.1	+		

Species	1	2	3	4	5	6	7	8	9
Vaccinium uliginosum		3.5			+		2.4	0.7	8.5
Vaccinium vitis-idaea		0.3			+	0.8		0.8	+
Fumariaceae									
Corydalis pauciflora					+				
Gentianaceae									
Gentiana glauca		0.1		+		0.2	+	+	
Gentiana propinqua	0.2		0.1						
Gramineae									
Agropyron violaceum	+								
Arctagrostis latifolia	0.1		2.1			4.8	0.1		0.1
Calamagrostis canadensis		3.5							
Calamagrostis lapponica								+	
Deschampsia caespitosa	+	0.5	1.9						
Elymus innovatus		2.1						7.3	
Festuca altaica	+					4.5	+	4.5	
Hierochloe alpina	+						+	2.6	
Poa sp.				0.1					
Poa alpina	+								
Poa arctica				0.8					
Poa porsildii						6.5			
Trisetum spicatum	+					+	+		
Juncaceae									
Juncus sp.				0.1					
Juncus albescens	+			0.2			0.5		0.3
Juncus biglumis				0.1				+	0.1
Juncus castaneus	+								
Luzula arcuata					0.1	0.6			
Luzula confusa								+	
Luzula nivalis							0.3		+
Luzula parviflora						3.0	+		
Luzula spicata		0.6			0.1			0.1	

Table 2 Continued

Plant species or group	1	2	3	4	5	6	7	8	9	10	11	12	13	14	15
Juncaginaceae															
Triglochin maritimum										+					
Leguminosae															
Astragalus umbellatus	34.5	2.1	+	1.0						+		+			
Hedysarum alpinum						0.6				0.2	+	0.3	0.3		
Hedysarum mackenzii										1.3		0.2			
Oxytropis sp.										0.1					
Oxytropis deflexa	0.1	0.1													
Oxytropis jordalii										+		0.1			
Oxytropis nigrescens										0.1					
Oxytropis sheldonensis										+					
Lentibulariaceae															
Pinguicula vulgaris						0.8				+		0.1	+		
Liliaceae															
Lloydia serotina						+		+		+		0.1	0.1	+	
Tofieldia coccinea						1.0					0.1	0.1			
Tofieldia pusilla	0.1					0.8				0.3		0.1	0.5	0.2	
Zygadenus elegans	+										+	+			
Lycopodiaceae															
Lycopodium sp.						0.5									
Lycopodium alpinum			+					0.1							
Lycopodium selago								+		+	0.1				
Onagraceae															
Epilobium angustifolium	+		0.6			+									
Epilobium latifolium	+	0.1									+				

Family / Species	Values
Ophioglossaceae	
Botrychium lunaria	+
Papaveraceae	
Papaver keelei	0.1 + 0.1 0.1
Papaver radicatum	0.1 0.1
Pinaceae	
Juniperus communis	+ +
Picea glauca	+
Plumbaginaceae	
Armeria maritima	+ +
Polemoniaceae	
Polemonium acutiflorum	0.3 0.9 7.5 + +
Polygonaceae	
Oxyria digyna	0.3 0.1
Polygonum bistorta	0.1 0.5 0.2 + 0.1 0.5 0.1
Polygonum viviparum	0.4 0.5 0.8 0.3
Rumex arcticus	0.1
Polypodiaceae	
Woodsia glabella	+ + +
Portulacaceae	
Claytonia tuberosa	0.1 +
Primulaceae	
Androsace chamaejasme	0.5 0.3 0.3 0.1 0.1
Pyrolaceae	
Pyrola sp.	+ +
Pyrola asarifolia	+ +
Pyrola secunda	0.1

Table 2 Continued

Plant species or group	1	2	3	4	5	6	7	8	9	10	11	12	13	14	15
Ranunculaceae															
Aconitum delphinifolium				2.9											
Anemone narcissiflora			0.2		0.2			0.4							
Anemone parviflora	0.2	0.2	+	0.6		0.6		1.5		0.1	+	+			
Anemone richardsonii								0.5							
Ranunculus eschscholtzii					+			+							
Ranunculus nivalis				+				0.1							
Ranunculus sulphureus				+				0.4							
Thalictrum alpinum			0.1	2.6		0.5			0.1	0.3	+	0.3	0.1	0.2	
Rosaceae															
Dryas integrifolia			0.8			27.7			0.3	36.5		12.1	11.4	3.8	
Dryas octopetala									0.1	16.4	5.8				
Dryas sylvatica											4.2				
Geum rossii				4.5											
Potentilla biflora										4.5					
Potentilla elegans											+	+	+		
Potentilla fruticosa		+	+	0.8		3.2				+	+	+			
Potentilla uniflora													+		
Rosa acicularis								0.1							
Rubus acaulis			+												
Rubus chamaemorus			+			+									
Sibbaldia procumbens			+		+			0.1							
Spiraea beauverdiana			+												
Salicaceae															
Salix alaxensis	6.1	+								+	+			+	
Salix arbusculoides														+	

Species													
Salix arctica	0.9	10.3			0.1	7.5	0.1	1.6		0.2	0.8	0.1	
Salix barrattiana		+											
Salix commutata		30.1				2.0		4.1					
Salix dodgeana				0.7	0.2			0.3		0.2	0.2	0.6	
Salix glauca			+						+				
Salix lanata		+		+									
Salix myrtillifolia		2.9											
Salix planifolia			29.0			2.0			4.7				
Salix polaris						1.3		+			+	+	
Salix polaris x *arctica*				+		+		+					
Salix reticulata	9.5	58.0	0.8	2.9	+	0.1	0.7	0.1	1.2	4.5	0.1	2.5	+
Saxifragaceae													
Chrysosplenium tetrandrum													
Parnassia fimbriata				+		0.1							
Parnassia kotzbuei	+	0.1				+							
Parnassia palustris	+												
Saxifraga aizoides				1.0						+			
Saxifraga foliosa						0.1		+		+			
Saxifraga hieracifolia						0.3							
Saxifraga hirculus								+			+		
Saxifraga oppositifolia				1.8				1.0		0.5	0.2	1.3	
Saxifraga punctata						0.1							
Scrophulariaceae													
Lagotis stelleri						0.2							
Pedicularis arctica								0.1		0.1	+	+	
Pedicularis capitata				0.8			0.1	1.3	0.1	+	0.1		
Pedicularis labradorica		+		+					+				
Pedicularis lanata				0.1				0.1		0.1	0.9	0.3	
Pedicularis sudetica		+					0.1	0.1	0.2	0.1	0.6		
Pedicularis sp.	0.1	1.1		+		+						+	

Table 2 Continued

Plant species or group	1	2	3	4	5	6	7	8	9	10	11	12	13	14	15
Selaginellaceae															
Selaginella selaginoides						+									
Valerianaceae															
Valeriana capitata			1.2	+											
Violaceae															
Viola epipsila								+							
Viola sp.				1.9											

a   "0.1" includes all cover values of 0.14 or less recorded from the sample quadrats
+   Present in sample stand but outside of the sample quadrat
*   Cover value included in a broader category such as "moss species," "crustose lichens," "saxicolous lichens," and "soil lichens."

Table 3 Plants with significant cover value differences (2%) between control and CANOL Project disturbance plant communities, N.W.T.

Plant species or group	1 Hedysarum alpinum-Moss (Erect deciduous shrub tundra)	2 Salix reticulata-Salix lanata (Erect deciduous shrub tundra)	3 Betula glandulosa-Cladonia stellaria-Moss (Erect deciduous shrub tundra)	4 Salix barrattiana-Moss (Decumbent shrub tundra)	5 Salix polaris-Dactylina beringica (Decumbent shrub tundra)	6 Dryas integrifolia-Carex spp. (Decumbent shrub tundra)	7 Carex membranacea-Sphagnum spp. (Sedge meadow tundra)	8 Misc. Mosses-Carex spp. (Sedge meadow tundra)	9 Carex spp.-Kobresia simpliciuscula (Sedge meadow tundra)	10 Dryas integrifolia-Cassiope tetragona (Lichen heath tundra)	11 Cladonia stellaris-Alectoria ochroleuca (Fruticose lichen tundra)	12 Dryas integrifolia-Cetraria tilesii (Fruticose lichen tundra)	13 Dryas integrifolia-Rhizocarpon umbilicatum (Cushion plant tundra)	14 Saxicolous Lichens-Lecanora epibryon (Cushion plant tundra)	15 Rhizocarpon inarense-Umbilicaria proboscidea (Crustose lichen tundra)
LICHENS															
Rinodina roscida			-RP									+0			
Cladonia arbuscula			-A												
C. coccifera															
C. crispata			-A					-RB							
C. gracilis			-B					+T							
C. macrophylla			-RBC								-FP				
C. mitis			TPO								-FP				

Table 3 Continued

Plant species or group	1	2	3	4	5	6	7	8	9	10	11	12	13	14	15
C. pocillum	+O														
C. pyxidata	+O														
C. rangiferina			-BC PAO					-RB TPA			-RB PAO				
C. stellaris			-RBT PAO					-RB TPA +T			-RFB TPAO				
C. subulata															
Collema sp.			+P									+P -RP			
Lecanora epibyron													-BP	-RBP	+B -RF BP
Lecidella stigmatea											+A -FBT PAR				
Rhizocarpon sp.															+B -RB
R. eupetraeoides															
R. inarense													-RBP		
R. umbilicatum			+P												
Toninia lobulata			+P								+P	-RBP			
Pannaria pezizoides			+P												
Psoroma hypnorum															
Asahinea chrysantha						-RBT PAO					-FBP				
Cetraria cucullata						-TAO		-O		-RB PO	-RP AO	-T			
C. islandica			-RB +A							-RP AO	-RFB PAO				
Cetraria laevigata											-RFBP				
C. nivalis			-BT PA							-FP	-RFB TPAO				
C. tilesii										-FP	-RTB PAO				

Species	Codes
Peltigera aphthosa	+C +B +RT
P. canina	+B +R
P. polydactyla	+T
P. pulverulenta	+R
P. rufescens	+B +BP
Solarina crocea	+A −A
Stereocaulon alpinum	+R
S. pachale	−RBT PAO
S. saxatile	+B +B
S. tomentosum	+RBP
Umbilicaria hyperborea	−RB PAO −RB
U. proboscidea	−RFB −BT TPAOAO −RB
Alectoria ochroleuca	−PA
Dactylina beringica	−FP
Thamnolia subuliformis	
Polyblastia sendtneri	−BP
Thelidium aenovinosum	+O +O +B
Miscellaneous lichens	+O +B +RBP
Crustose lichens	−RBP
Saxicolous lichens	−RB PO, −RFB TPA +BP PO, −RF BP, −RF +BT PO, −RBP −RF +BP
Soil lichens	+BP, +O +RF BA, +BP +RF BP, +BA +BP +P, −RFP −RP, +P −RP

Table 3 Continued

Plant species or group	1	2	3	4	5	6	7	8	9	10	11	12	13	14	15
MOSSES															
Aulacomnium palustre			-PO												
A. turgidum			-B												
Polytrichum spp.			+P					+B			+BPA				+R
P. commune			-BTPO												+P
			+A												
P. hyperboreum			-BPA								+B				
P. juniperinum											-F				
											+RP				
P. piliferum			+BPA								+R				
P. strictum			+T												+P
Sphagnum spp.							-O								
Miscellaneous mosses	-RBT	-R	-RP	-B	+R	+R	-O	-RB		-RBT			-BTP		-o
	PO				TO	TA		TP		PAO					
	+PA		+BTO								+FT	+AO			+R
VASCULAR PLANTS															
Betula glandulosa			-RBC							-F	-RF	-RBT			
			TPAO								PAO	PAO			
											+BT				
Mertensia paniculata				-RB											
				TPO											
Linnaea borealis											+B				
Minuartia arctica									+R		+R				
M. macrocarpa											+RFA				
M. rossii															
M. rubella															+R

Species										
Silene acaulis									-FP	-RFT PAO
Antennaria densifolia									+B	
A. stolonifera	+A -RPT									
Artemisia arctica					+R +R					
A. tilesii										
Chrysanthemum integrifolium									+F	
Petasites hyperboreus					+P					
Solidago multiradiata	+O									
Carex aquatilis aquatilis									+O	
C. aquatilis stans	+RB					-RBT				
C. atrosquama	+P									
C. capillaris	+B									
C. lugens										+F
C. membranacea	+B									
C. microchaeta	+T	+BP	+T	+B	+B	-RP	+OB	+RO	+B	+P
C. misandra				-R			-P -RB PAO			
C. petricosa								+T	-RBP	+R
C. podocarpa					-RTP -RP +B		-RBP		+BP	
C. rupestris				-R PA			-RFB PAO -RF PO	-RBT PAO	-BP	
C. scirpoidea	+O			+B				-RBT -R PAO	-RBP	

Table 3 Continued

Plant species or group	1	2	3	4	5	6	7	8	9	10	11	12	13	14	15
Eriophorum angustifolium							+O -O	+BP	-RP	-RPA +B			-P		
E. vaginatum													+P		
Kobresia simpliciuscula					-RP +R										
Shepherdia canadensis		-R +C													
Empetrum nigrum				-RB TPO					-R						
Equisetum arvense								+R			+A				
E. variegatum	+BT +P		+R +P												
Andromeda polifolia		-R -RP			-RBP										
Arctostaphylos alpina			+R							-FP	-A				
A. uva-ursi						-O									
Cassiope tetragona								-RB TPA		-RF BTP	-RB PO +AT				
Ledum decumbens		-R													
L. groenlandicum															
Rhododendron lapponicum		-R								-PA -RBP	+T -F +BT				
Vaccinium uliginosum															
V. vitis-idaea															
Arctagrostis latifolia				-BT PO				-RTP +B							-RTB
Calamagrostis canadensis			-TPO												

Deschampsia caespitosa		+RB -P	+B		-RT -RB TPA	-P
Festuca altaica				+TA		+F
Hierochloe alpina	+O	+RB CA				-PA +F
Poa alpina			+O		+B +R	
P. arctica				+R		+R
P. glauca						
P. porsildii					-PA +R +RBP	
P. pratensis			+O	+RTA		
Trisetum spicatum						+O
Juncus albescens		+R				
J. balticus						
J. castaneus					+B	
Luzula arcuata						+F
L. confusa						
L. parviflora				+T		
Hedysarum alpinum	-RB PO +R					+F
H. mackenzii						
Epilobium anagal-lidifolium		+C	+BO +T		+R	
E. angustifolium					+A	+P
E. latifolium			-RB TPO			
Polemonium acutifolium						
Androsace chamaejasme						+T

Table 3 Continued

Plant species or group	1	2	3	4	5	6	7	8	9	10	11	12	13	14	15
Aconitum delphinifolium			-RB TPO												
Anemone narcissiflora							-A								
A. parviflora							-B								
Thalictrum alpinum				-RB TPO											
Dryas integrifolia				-RB TPO		-RB PA				-RFB TPAO		-RB PAO +FT	-RBT PAO	-RP	
D. octopetala	+A		+P						+P		-FB PA -FP				
D. sylvatica								-BPA			+A				
Geum rossii															
Potentilla biflora									+P	-RB PAO					
P. fruticosa						+R									
Sibbaldia procumbens					+T						-A				
Salix alaxensis	-BTP +R	+R BP	+RB						+RP BPA	+RF	+RP	+RF PA	+RA	+R	
S. arbusculoides					+T										+F
S. arctica									-O				+T		
S. barrattiana				-RB TPO			-T				-FP				
S. commutata							-RBP								
S. glauca		+RP		+R						+F					

S. lanata	-RB PO				+P	-A +RB
S. planifolia		-TP +AO	+T -RT PA	+B	-RPA +P -PA	
S. polaris						
S. reticulata	-RB PA +T	-RB PO	-FP			-A -BT
Saxifraga oppositifolia			+RTP		+P	
Pedicularis capitata				-B	+A	
P. lanata				-BP	-B	

Key

+ Cover values at least 2% higher on the disturbances than in the control plant community
- Cover values at least 2% lower on the disturbance than in the control plant community

A Gravel pit access road
B Bladed trail
C Camp yard
F False start road
O Oil spill
P Gravel pit
R Road
T Bulldozer track

Table 4 Floristic similarity coefficients comparing control plant communities with those of CANOL Project disturbances, Northwest Territories.

Plant communities	Road	False start road	Bladed trail	Camp yard	Bulldozer track	Gravel pit	Gravel pit access road	Oil spill	x̄
Disturbance type									
Erect deciduous shrub tundra *Hedysarum alpinum-*Moss spp.	58		53		44	38	54		49
Salix reticulata-Salix lanata	56		39			18		30	36
*Betula glandulosa-Cladonia stellaris-*Moss spp.	33	51	30	75	51	44	48		47
Decumbent shrub tundra *Salix barrattiana-*Moss spp.	23		33		54	24		24	32
Salix polaris-Dactylina beringica	23				48	40	28		35
Dryas integrifolia-Carex spp.	54		46			61	42		51
Sedge meadow tundra *Carex membranacea-Sphagnum* spp.								19	19
Miscellaneous mosses-*Carex* spp.	42		49		74	47	42		51
Carex spp.-*Kobresia simpliciuscula*	33					54			44

Table 4 *Continued*

Disturbance type

Plant communities	Road	False start road	Bladed trail	Camp yard	Bull-dozer track	Gravel pit	Gravel pit access road	Oil spill	x̄
Lichen heath tundra *Dryas integrifolia-Cassiope tetragona*	50	47	63		53	42	52		51
Fruticose lichen tundra *Cladonia stellaris-Alectoria ochroleuca*	38	52	48		39	52	57	22	44
Cushion plant tundra *Dryas integrifolia-Cetraria tilesii*	53	46	66		52	58	44	52	53
Dryas integrifolia-Rhizocarpon umbilicatum	54		58		52	50	58	21	49
Saxicolous lichens-*Lecanora epibyron*	50		54			55			53
Crustose lichen tundra *Rhizocarpon inarense-Umbilicaria proboscidea*	06	00	25			00			08
Mean similarity coefficient (all communities)	41	36	49	30	55	42	44	34	41

Key

x̄ Mean similarity coefficient for all types of disturbances in that community

Table 5 Plant species richness of control and CANOL Project disturbance plant communities, Northwest Territories.

Plant communities	Road D/C	Road S	False Start Road D/C	False Start Road S	Bladed Trail D/C	Bladed Trail S	Camp Yard D/C	Camp Yard S	Bulldozer Track D/C	Bulldozer Track S	Gravel Pit D/C	Gravel Pit S	Gravel Pit Access Road D/C	Gravel Pit Access Road S	Oil Spill D/C	Oil Spill S
Erect deciduous shrub tundra																
Hedysarum alpinum-Moss spp.	6/25	21	21/22	24					15/25	16	47/24	22	18/22	24		
Salix reticulata-Salix lanata	11/6	11	29/6	11					23/13	4			20/11	6		
Betula glandulosa-Cladonia stellaris-Moss spp.	49/54	25	90/22	59	33/13	10	27/13	59	120/30	78	28/18	18	29/33	28		
Decumbent shrub tundra																
Salix barrattiana-Moss spp.	21/22	8			48/26	18			25/20	26	58/32	14			19/37	9
Salix polaris-Dactylina beringica	21/27	7							33/13	21	22/20	14	9/27	7		
Dryas integrifolia-Carex spp.	15/38	31			13/18	13					15/32	37	3/25	10		
Sedge meadow tundra																
Carex membranacea-Sphagnum spp.													4/22	3		
Miscellaneous mosses-Carex spp.	64/61	45			62/29	43			27/30	83	61/56	51	19/47	24		
Carex spp.-Kobresia simpliciuscula	9/20	7									16/11	16				

Table 5 *Continued*

Plant communities	Road		False start road		Bladed trail		Camp yard		Bull-dozer track		Gravel pit		Gravel pit access road		Oil spill	
	D/C	S	D/C	S	D/C	S	D/C	S	D/C	S	D/C	S	D/C	S	D/C	S
Lichen heath tundra *Dryas integrifolia-Cassiope tetragona*	49/85	66	25/44	31	34/54	75			37/81	67	25/52	28	20/56	41		
Fruticose lichen tundra *Cladonia stellaris-Alectoria ochroleuca*	95/28	38	56/36	49	71/17	40			18/13	10	62/47	59	20/15	23	7/29	5
Cushion plant tundra *Dryas integrifolia-Cetraria tilesii*	41/50	52	19/30	21	14/35	47			16/32	26	32/46	53	19/36	22	14/33	25
Dryas integrifolia-Rhizocarpon umbilicatum	11/46	33			30/45	52			12/30	23	31/42	37	8/17	17	7/30	5
Saxicolous lichens-*Lecanora epibyron*	11/36	23			1/26	16					9/28	23				
Crustose lichen tundra *Rhizocarpon inarense-Umbilicaria proboscidea*	73/22	3	39/16	0	28/13	7					30/16	0				

Key
D No. of species found in disturbed plant communities only
C No. of species found in control plant communities only
S No. of species common to both control and disturbance plant communities

on disturbances, had less cover on controls than on all disturbances with the exception of the road (table 3). Moss cover was less on the road, bladed trail, and bulldozer tracks, but greater on the gravel pit and gravel pit access road sites than in the adjacent undisturbed areas. This was not the result of variation in any one factor, but rather of several different components important at each site.

Similarity coefficients indicate that the road flora is most like that of the control, while the gravel pit plant community was least similar (table 4). Species richness values indicate that the number of species was greater than that of the controls in gravel pits only (table 5).

The overall decline in species richness and low floristic similarity to control plant communities may indicate that environmental conditions were more restrictive for plants on disturbances than for those in undisturbed areas. Furthermore, the "new" disturbance environments were not suitable for local species. Alternatively, 34 to 36 years may not be a sufficient length of time for natural recover to establish new plant communities. Further colonization and cover changes may yet occur. However, in comparison with other disturbance plant communities, those studied here have been slow in their recovery. A significant difference between the tall shrub control plant communities and those containing only one synusia reflected the effects of shrub removal. The moderating effects of the shrub canopy were eliminated or at least reduced for a number of years following the initial disturbance. Similar results were noted by Gill (1973) with the removal of tree and shrub canopies in the Mackenzie Delta region. In most cases, the present disturbance shrub canopy still remains open. Local plant species would be well adapted to the protected, shaded environment below a continuous shrub canopy, and as is reflected by the low similarity coefficients, its removal has precluded colonization by these species. Until the closed canopy is replaced (probably by the shrub species currently on the disturbed sites), the disturbances will continue to be substantially different from control plant communities.

SALIX RETICULATA–SALIX LANATA–MOSS COMMUNITY This plant community was dominant in seepage areas and sites where soils were moist. *Salix lanata* formed a 1.5-m-tall overstory, but the ground cover was similar to that of the *Hedysarum alpinum*–Moss plant community (table 2), with some differences in the ranking of dominants.

Salix lanata, S. reticulata, and *Hedysarum alpinum* had less cover than controls on all disturbance types than in the control plant community (tables 2 and 3). Cover values for *S. reticulata* were 57% lower than controls on the roads in this plant community. However, *S. lanata* was restricted to undisturbed areas. With the exception of oil spill sites, *S. alaxensis* had greater cover on disturbances than in control stands.

All disturbance plant communities had greater species richness than did their associated controls (table 5). However, on the basis of similarity coefficients, species composition of the road was most similar to that of the controls, while that of the gravel pit was least similar to the undisturbed flora (table 4).

Disturbances must provide radically different environmental conditions when they occur in this community, with the result that they provided habitat for a greater number of species. However, it appears to be unsuitable for local species, since the flora on disturbances was very different from that of the controls.

BETULA GLANDULOSA-CLADONIA STELLARIS-MOSS COMMUNITY This plant community occupied one of the greatest areas and was found throughout the study area, but was most common at lower elevations in the western section. Whereas the other plant communities in the erect deciduous shrub tundra category were confined to floodplains and seepage areas, the *Betula*-dominated systems covered large areas in an almost unbroken cover. *Betula glandulosa–Cladonia stellaris*–Moss plant communities mantled the areas of till, near-surface bedrock, alluvium, and lacustrine deposits. These sites could be imperfectly drained to well-drained. Shrub stature varied considerably from the extremely dense stands, which exceed 2 m in height, to those that would most properly be classified as prostrate. Differences in heights are due primarily to exposure and mean snow depth. In the category "Gramineae species" (table 2), *Deschampsia caespitosa, Trisetum spicatum,* and *Festuca altaica* had equal cover. The most common moss species included *Hylocomnium splendens, Polytrichum juniperinum, P. piliferum,* and *P. strictum.*

All of these disturbances had less *Betula glandulosa* cover than was found in control areas. In addition to this, *Cladonia mitis, C. rangiferina, C. stellaris, Cetraria nivalis, Stereocaulon paschale,* and *Polytrichum commune* all had notably less cover on at least four of the disturbance types sampled (table 3), but none of these species was restricted to undisturbed areas. *Festuca altaica* was the only species with higher cover values on four or more disturbance types than in control stands.

Species richness was lower than that of the control stands on roads, gravel pits, and oil spills, but greater on bladed trails, bulldozer tracks, camp yards, and gravel pit access roads in this community (table 5). Similarity coefficients indicate that the flora of the bulldozer track plant community was most similar to that of the control, while that of the camp yard was least similar (table 4).

The removal of the dominant shrub, *Betula glandulosa*, and the continuous lichen cover opened the area to colonizing species that formed new plant communities on disturbances. These new assemblages of plants had many traits in common with their control stands. However, the dominant species had changed. This created greater environmental diversity in this portion of the study area, where large tracts of *Betula glandulosa–Cladonia stellaris*–Moss plant communities dominated the landscape. The alterations resulting from the CANOL disturbances provide habitat for species such as *Salix alaxensis* and *S. planifolia* that otherwise have more restricted ranges than in this plant community or are found only on natural disturbances.

Decumbent Shrub Tundra

In decumbent shrub tundra plant communities, shrub layers were less than 30 cm in height, covering at least 50% of the area under plant cover (fig. 3). In the study area this type of tundra was dominated by low-growing or prostrate *Salix* and *Dryas* species. Floristically it can be divided into *Salix*

Fig. 3. Decumbent shrub tundra occurs throughout the Mackenzie Mountains and occupies high-elevation exposed sites where soil moisture is abundant (50 cm rule).

barrattiana–Moss, *Salix polaris–Dactylina beringica,* and *Dryas integrifolia–Carex* spp. plant communities.

This type of tundra was present in every ecosection. The *Salix barrattiana*–Moss community was common where soil moisture was high along small stream bottoms or in protected seepage areas at lower elevations. This contrasts sharply with other plant communities, which were common at higher elevations and in exposed sites where winter snow cover was thin and discontinuous. In these cases soils were generally well-drained.

SALIX BARRATTIANA–MOSS COMMUNITY   The 20 to 30 cm *Salix barrattiana* shrub canopy was generally dense, over a plant assemblage dominated by broad-leaved forbs and *Salix reticulata* (table 2). Moss species in this plant community included *Tortula norvegica, Hypnum lindbergii, Brachythecium starkei, Bryum pseudotriquetrum,* and *Drepanocladus uncinatus.*

Species richness was greater than that of the controls on all disturbances with the exception of roads and oil spills (table 5). The species composition of the road plant community was most different and that of the bulldozer tracks most similar to that of the controls (table 4). Floristic similarity to control stands was generally low. If the shrub canopy acts to moderate conditions beneath it in a similar fashion to that postulated for the *Salix reticulata–S. lanata* plant community, then this would explain the low similarity coefficients. This seems probable and is supported indirectly by the fact that ground cover species such as *S. reticulata* are abundant in each plant community. However, conditions vary, as is evidenced by the fact that *S. reticulata* had greater cover on all disturbances than in the controls, while in erect deciduous shrub tundra its cover was lower on all but bulldozer tracks (table 3). Perhaps the structurally reduced shrub cover in decumbent shrub tundra provided less shelter than did the taller shrubs, so that ground cover ecotypes in this community were better adapted to the exposed conditions on most disturbances.

Salix barrattiana, Mertensia paniculata, Thalictrum alpinum, Arctagrostis latifolia, Polemonium acutiflorum, Aconitum delphinifolium, and *Equisetum arvense* had less cover than controls on all of the disturbance types in this community (tables 2 and 3). Of these, the last three species were restricted to undisturbed sites only and therefore must have been particularly sensitive to disturbance or incapable of colonizing the types of disturbances here. Even after 34 to 36 years, no one species consistently recolonized disturbed sites, indicating that conditions on disturbances were severely limiting to plant colonization.

The diversity of species in disturbance plant communities has produced low similarity coefficients, indicating that disturbed sites were substantially different from their controls. Disturbances that resulted in the lowering of surfaces have greater species richness than do their controls, while elevated roads and oil spills support fewer species (table 5). Roads can be substantially drier as a result of increased runoff and reduced infiltration, and the oil seal on the surface of oil spills would also create very dry substrates. Considerable time will probably be required before these conditions change sufficiently to allow these disturbance plant communities to become more similar to those of the controls, if indeed this is possible.

SALIX POLARIS–DACTYLINA BERINGICA COMMUNITY *Salix polaris* formed almost pure stands in this plant community (table 2). A number of lichens were also present, but these were much less abundant than the willow.

Salix polaris was the only plant species to have a greater cover on all of the disturbance types than in the controls. No species had greater cover than the controls on all site types, but *Trisetum spicatum* and miscellaneous moss species were never collected from undisturbed areas and composed a significant portion of the cover on road, bulldozer track, and gravel pit access road sites (table 3).

Species richness was lower than the controls on roads and gravel pit access roads in this plant community, but higher on the bulldozer track and gravel pit sites. Similarity coefficients indicate that the species composition of the bulldozer track plant community was most similar to that of the controls, while that of the road was most different (table 4). With a decline in species richness (table 5) and low floristic similarity to control stands, it appears that disturbances here have long-term consequences that are biologically undesirable. This indicates that this plant community does not respond well after disturbances. However, it is noteworthy that *Salix polaris* successfully recovered, and that this species was also the dominant plant species in control stands.

DRYAS INTEGRIFOLIA–CAREX SPP. COMMUNITY This plant community had a diversity of species, but *Dryas integrifolia* formed a carpet under all. There was little moss cover, in response to the general dryness and excessive exposure at these sites. *Carex* spp. were present and collectively occupied enough area to be classified as subdominants (table 2). The most common sedges were *Carex rupestris*, *C. misandra*, *C. scirpoidea*, and *C. membranacea*.

Fig. 4. Sedge meadow tundra is most common on the Mackenzie Mountain Barrens where drainage is impeded (5 cm interval on range pole).

Dryas integrifolia and miscellaneous lichen species had much lower cover values on all disturbance types than in the undisturbed areas (tables 2 and 3). Several species had notably higher cover on disturbed sites, but none of these were common to all disturbance types. With the removal of the dominant control species and recolonization by fewer species, it is apparent that the disturbance plant community is a simplification of the undisturbed area community.

Species richness was less on all disturbances sampled in this plant community than in control stands (table 5). The average similarity coefficient was not low. The species composition of the bulldozer tracks was most similar to that of the controls, while the gravel pit plant community had the least in common with the undisturbed area (table 4).

Sedge Meadow Tundra

Sedge meadow tundra was dominated by herbaceous plants, primarily *Carex* and *Eriophorum* species. In areas where soil moisture was excessive as a result of topographic depressions or impeded drainage, this type of tundra was common. The most extensive areas of sedge meadow tundra occurred over tills in the western portion of the study area (fig. 4).

CAREX MEMBRANACEA-SPHAGNUM SPP. COMMUNITY This wetland was found in the Bolstead Creek–Devil's Pass ecosection. The oil spill in this plant community affected the broader area only where it extended down a slope into the wetland. Large tussocks and small mounds surrounded by standing water were common. These contained frozen material in late August of 1978 and 1979. *Carex membranacea* and *Eriophorum vaginatum* were the most abundant vascular plants with *Sphagnum* forming a high percentage of the ground cover (table 2).

The areas affected by the oil spill in this palsa fen varied greatly in their responses. Consequently two stands representing two extremes were sampled: (1) a denuded site, and (2) a site with relatively high plant cover. The site with the greater plant cover was located on a raised area and was probably less affected by the oil than were the low-lying drainage areas where the denuded site occurred.

Eriophorum vaginatum had much less cover on the oil spills than in the undisturbed areas, and *Carex membranacea* and *Andromeda polifolia* were both present in the control sites but absent from the oil spill areas. *Eriophorum angustifolium* was the only vascular plant with greater cover on these sites than in the undisturbed areas. This has also been noted by Walker et al. (1978) on recent spills in the Prudhoe Bay area. *Cetraria cucullata*, *Cladonia rangiferina*, *Sphagnum* spp., and miscellaneous mosses were all important components of the undisturbed area, but were absent from the oil spills. These findings do not agree with those of Hutchinson and Hellebust (1978), where *Polytrichum* spp., *Cetraria* spp., and *Cladonia cornuta* were found to be unaffected or were recolonizing after 3 years. Terrestrial algae (*Nostoc* sp.) were absent from the control site, but accounted for approximately 8.5% of the total cover on one spill. The flora contained only 28% as many species as were found on the control site, and the species composition of the disturbed and undisturbed sites differed greatly (table 4). Thirty-four years following the initial oil spill, the plant community on the site remains considerably simplified and reduced in cover (table 5).

MISCELLANEOUS MOSSES–CAREX PODOCARPA COMMUNITY This plant community was dominant on the Mackenzie Mountain Barrens in relatively flat, poorly drained areas. Extensive areas of palsas and peat plateaus were common in this wetland-dominated terrain. However, the plant communities on these features were not always similar to those of the surrounding area.

Carex podocarpa was the dominant vascular plant. However, *Carex aquatilis* was also common (table 2). *Artemisia arctica, Anemone narcissifolia, Petasites frigida, Gentiana glauca, Rhodiola integrifolia, Polemonium acutiflorum,* and other broad-leaved forbs were present in most of the sampled stands.

Common mosses in this plant community included *Aulacomnium palustre, Dicranum* sp., *Drepanocladus uncinatus, Hylocomnium splendens, Pleurozium schreberi, Polytrichum commune, P. juniperinum, P. strictum,* and *Rhacomitrium canescens*. Miscellaneous lichens included *Cetraria cucullata, C. islandica, C. richardsonii, Cladonia arbuscula, C. cenotea, C. ecmocyna, C. gracilis dilatata, C. gracilis gracilis, C. pleurota, Dactylina arctica, Nephroma arctica, Solarina crocea,* and *Stereocaulon tomentosum*.

When present, the following species consistently had lower cover on the various disturbances than in the controls: *Cassiope tetragona, Carex aquatilis stans, Festuca altaica, Geum rossii, Cladonia crispata, C. rangiferina, C. stellaris, Dactylina beringica,* and miscellaneous moss species (table 3). Many of these taxa were among the most abundant plants on control sites. Cover was consistently greater for *Trisetum spicatum, Epilobium* spp., *Stereocaulon alpinum, S. saxatile,* and *S. tomentosum*. The species in this latter group were rarely found in undisturbed areas and therefore were much more common in this area as a consequence of the disturbances.

Similarity coefficients showed the species composition of the bulldozer track plant community to be least different from that of the control, while those on the roads and gravel pit access roads were least similar (table 4). Bulldozer tracks and gravel pit access roads had lower species richness, while roads, bladed trails, and gravel pits had greater numbers of species in comparison with the control sites (table 5). No general statement can be made regarding the recovery of this plant community, since responses appear to be very specific to each disturbance type.

CAREX SPP.-KOBRESIA SIMPLICIUSCULA COMMUNITY This plant community occupied snow melt and seepage areas only and was never extensive. These low-angled slopes had little microtopographic variation, and plant cover was seldom continuous. During late July and August, following the melting of late-lying snow patches, the ground surface was generally dry, except during periods of frequent rain. *Carex membranacea* was the most abundant plant, while *C. misandra* and *C. podocarpa* had only slightly less cover (table 2). *Kobresia simpliciuscula* and *Eriophorum angustifolium* were also common.

In the gravel pit, cover was notably less than that of the control for *Carex membranacea, C. misandra, C. podocarpa,* and *Eriophorum angustifolium* (table 3). None of these species was found growing on the road. *Salix alaxensis* was absent from the control, but had cover values of 8% and 12% on the road and gravel pit sites respectively. Furthermore, this erect shrub layer was not present in control stands.

Species composition on the road was most like that of the control (table 4). Both the road and gravel pit had greater species richness than the control (table 5). This reflected a substantial increase in cover by woody species on the disturbed sites in comparison with the predominantly herbaceous species of the control sites.

With greater species richness and structural complexity enhanced, it appears that these disturbances have produced a diversity of habitats previously not available. These alterations in plant community characteristics are often considered to be ecologically desirable, since the more species that are present, the greater the stability of that community. Furthermore, increased structural complexity ensures that a more efficient use of the area's resources is achieved. As available resources (e.g., radiation, moisture) can be utilized by more than one layer, the production of greater biomass is theoretically possible on sites where several layers occur. Also, animals relying on these plants will have a greater choice when selecting forage and will receive greater shelter and cover from the disturbance communities that contain erect shrubs.

Lichen Heath Tundra

Lichen heath tundra was dominated by *Dryas integrifolia* and fruticose lichens with an important ericaceous shrub component less than 30 cm in height (fig. 5). This community was most frequently found over tills and colluvium, but was also noted covering bedrock where sufficient moisture was available. It was most common on north-facing slopes at lower elevations in the eastern portion of the study area. The Blue Mountain and Bolstead Creek–Devil's Pass ecosections contained the most extensive areas of lichen heath tundra, although it was also common in the Macmillan Pass area in the Ekwi-Intga-Tsichu river valleys ecosection.

DRYAS INTEGRIFOLIA–CASSIOPE TETRAGONA COMMUNITY *Dryas integrifolia* was the most abundant plant in this type of tundra, with a mean cover value of 36%. However, due to the complete plant cover and significant ericaceous shrub and fruticose lichen components, this community was classified as lichen heath tundra. Common ericaceous shrubs included

Fig. 5. Lichen heath tundra is found throughout the Mackenzie Mountains where moisture is abundant (5 cm interval on range pole).

Cassiope tetragona, Vaccinium uliginosum, Rhododendron lapponicum, and *Andromeda polifolia.* Carices were also common. Common lichens included several species of *Cetraria* as well as *Alectoria ochroleuca.* Moss species included *Tomenthypnum nitens, Rhacomitrium lanuginosum, Encalypta procera, Tortula latifolia,* and *Hypnum bambergii.* The saxicolous lichens collected were *Evernia perfragilis, Lecanora epibyron,* and *Lecidea stigmatea.* The *Dryas integrifolia–Cassiope tetragona* plant community supported a larger number of species than did any other plant community.

Dryas integrifolia was the only plant species to have notably less cover on the disturbances sampled than on the control sites (tables 2 and 3). However, several of the dominant control species, including *Alectoria ochroleuca, Cassiope tetragona, Carex petricosa, C. rupestris,* and *Potentilla biflora* had notably less cover on five or more disturbance types. The first two species in this group tended to be restricted to undisturbed areas. *Salix alaxensis* was the only species with notably higher cover values on five or more disturbance types than in the control stands.

Species richness was less than that of the controls on all disturbance types in this plant community (table 5). Insufficient data were collected to determine the species richness of the bulldozer track plant community.

Similarity coefficients indicate that the bladed trail flora is most similar to that of the undisturbed area, while the gravel pit access road plant community differed most from that of the control (table 4). Generally, floristic similarity with control stands was not low. The reduction in species richness and similarity coefficient data show that long-term recovery is proceeding. However, it is difficult to describe the rate of this process, since little comparative data are available from other studies. In human terms, 34 to 36 years is a long time; however, until data on the duration of succession in this environment are available it is difficult to discuss this in a relative sense. If the age of control stand shrubs can be used as an indication (Kershaw 1983), this time span is short and recovery rates should not be considered slow. In addition, the presence of *Salix alaxensis* has increased structural diversity.

Fruticose Lichen Tundra

Fruticose lichen tundra was similar to lichen heath tundra in many respects. However, it lacked a shrub layer, and fruticose lichens were the most abundant plants (fig. 6). Generally this community occupied well-drained areas on alluvium and colluvium. When large-scale patterned ground was present, this community generally occupied the site, with lichen-encrusted cobbles protruding through the fruticose lichen cover. In protected areas fruticose lichen tundra was found only on excessively drained sites, where coarse-grained deposits predominated. Generally it was more typical of high, exposed sites in the Blue Mountain, Bolstead Creek–Devil's Pass, and Ekwi-Intga-Tsichu river valleys ecosections.

CLADONIA STELLARIS–ALECTORIA OCHROLEUCA COMMUNITY This plant community was dominated by several species of fruticose lichens (table 2). The most common of these were *Cladonia stellaris, C. rangiferina, Alectoria ochroleuca, Cetraria cucullata, C. nivalis,* and *C. islandica.* Several vascular plant species such as *Betula glandulosa* (in a prostrate form), *Carex microchaeta,* and *Cassiope tetragona* were also noted frequently. The saxicolous lichens on the cobbles included *Umbilicaria proboscidea, U. hyperborea, Omphalodiscus virginis, O. decussatus,* and *O. krascheninnikovii.*

   Cladonia stellaris, Cetraria nivalis, Alectoria ochroleuca, and miscellaneous moss species cover on all disturbances was less than that of the controls. *Cladonia rangiferina, Rhizocarpon* spp., *Cetraria cucullata, C. islandica, C. laevigata, Cassiope tetragona,* and *Dryas octopetala* cover was less on four types of disturbances (table 3). All of these species were

Fig. 6. Fruticose lichen tundra occurs over well-drained soils. Near the Tsichu River extensive outwash deposits have numerous abandoned channels in which the coarse lag gravels are too dry to support a shrub cover.

found frequently on both disturbed and undisturbed sites. *Betula glandulosa* had greater cover on bladed trails and bulldozer tracks, but had markedly less than the controls on all other disturbances. The species with higher cover values than those recorded from the control stands varied from one site to the next.

This lichen-dominated community contained disturbances on which vascular plants dominated. These disturbances, with the exception of oil spills, provided habitats capable of supporting a larger number of species than were found in the undisturbed areas (table 5). However, similarity coefficients were lower than those of several other plant communities. The flora of the gravel pit access road most closely resembled that of the undisturbed stands, while that of the oil spills had the least similarity (table 4).

Vascular plants have been most successful at colonizing the newly available substrates on several disturbances, despite the fact that lichens dominate in many control stands. After 34 to 36 years it is difficult to determine whether sufficient time has elapsed for lichens to re-establish their dominance or whether the environmental alterations favour the growth of vascular plants. It seems probable that the length of time is most important, since many of the common control stand species were present but had little cover on disturbances. Nevertheless, disturbances have promoted diversity in species composition and in structural characteristics.

Cushion Plant Tundra

Cushion plant tundra was characterized by small patches of vegetation, usually from 30 to 70 cm in diameter, on sites where often more than 50% of the surface area was bare ground (fig. 7). These "islands" of vegetation were dominated by cushion-forming shrubs such as *Dryas* spp. and *Silene acaulis*. Cushion plant tundra commonly occurred in areas of patterned ground in the following ecosections: Joker Ridge, Blue Mountain, Plains of Abraham Plateau, and Bolstead Creek–Devil's Pass. This portion of the study area lies within the unglaciated section of the Mackenzie Mountains.

Cushion plant tundra was found most commonly at higher elevations on exposed sites, where winter snow cover was thin and discontinuous. It included three distinct plant communities: *Dryas integrifolia–Cetraria tilesii*, *Dryas integrifolia–Rhizocarpon umbilicatum*, and saxicolous lichens–*Lecanora epibyron*.

DRYAS INTEGRIFOLIA–CETRARIA TILESII COMMUNITY This plant community occurred on well-drained to excessively drained sites in exposed areas on Joker Ridge, Blue Mountain, and the Plains of Abraham. Plants were frequently observed without snow cover during the winter or were among the first sites exposed in the spring. The cushion growth form is an adaptation to these conditions (Savile 1972), and species such as *Dryas integrifolia*, *Silene acaulis*, and *Saxifraga oppositifolia* were frequently found in these areas (table 2). Other species such as *Carex petricosa*, *C. rupestris*, and *C. scirpoidea* were found within the *Dryas* mats and cushions. There was also a diverse lichen flora including such species as *Alectoria ochroleuca*, *Cetraria tilesii*, *C. cucullata*, and *Thamnolia subuliformis*. Soil lichens included *Cladonia pyxidata*, *C. pocillum*, *Pertusaria dactylina*, *Lecanora epibyron*, and *Ochrolechia androgyna*. *Toninia lobulata* and *Asahinea chrysantha* were important crustose lichens and the saxicolous lichens included *Polyblastia sendtneri*, *P. integrascens*, and *Verrucaria muralis*.

Miscellaneous lichen cover was less than that of the control sites on the road, false start road, and gravel pit, but greater on all other disturbances (tables 2 and 3). *Dryas integrifolia* cover was greater on the false start road and bulldozer tracks, but lower on the other types of disturbances than in undisturbed areas. *Betula glandulosa*, *Silene acaulis*, *Carex rupestris*, *C. scirpoidea*, *Cetraria cucullata*, and *C. tilesii* had less cover on all distur-

Fig. 7. Cushion plant tundra is most common on the Plains of Abraham, where patterned ground is also extensive. This coarse limestone regolith is excessively drained and in winter the snowcover is thin to discontinuous (100 cm rule).

bances with the exception of false start roads. No plant species appeared consistently as recolonizers of the various disturbances in this community, but *Salix alaxensis* had notably high cover values, relative to those of the controls, on four of the seven disturbances.

Species richness was lower than that of the control sites on all disturbances in this community (table 5). However, floristic similarities to controls were generally among the highest recorded for this study. The bladed trail plant community was most similar to that of the undisturbed sites, while the species composition on the gravel pit access road was most different from the control stands (table 4).

Indications are that the floristic changes have not produced radically different plant communities from those of undisturbed stands. Although richness declined, floristic similarity after 34 to 36 years was not unduly low. This suggests that little environmental change resulted from the initial disturbances, or that local species are well adapted to the colonization of newly available habitats.

DRYAS INTEGRIFOLIA-RHIZOCARPON UMBILICATUM COMMUNITY This plant community covered extensive areas on the Plains of Abraham and small parts of the Bolstead Creek–Devil's Pass ecosection. It was restricted to the most exposed portions of the flat-topped plateau, where extensive patterned ground occurred and bedrock was visible. These excessively drained, calcareous substrates had highly irregular surface microtopography as a result of the presence of frost-riven surface material that often measured 50 cm or more in diameter. Vascular plants were restricted to areas in the central portions of the sorted patterned ground where concentrations of fines had accumulated. Cushion-forming plants such as *Dryas integrifolia, Salix dodgeana,* and *Saxifraga oppositifolia* (table 2) were able to grow in these areas, where moisture was retained for longer periods and snow was less likely to be eroded in winter. Some species that do not normally grow in cushions had also adopted this growth form in this community; these included *Salix arctica* and *S. reticulata.* Many other species were commonly found growing solitarily or within cushions formed by other plants. These included *Draba corymbosa, Parrya nudicaulis,* several Carices, *Polygonum viviparum, Androsace chamaejasme,* and *Pedicularis lanata.* The most abundant lichens were saxicolous, and *Rhizocarpon umbilicatum* had the greatest cover. Moss species in this plant community (table 2) included *Hypnum bambergii, Tortella arctica, Tortella tortuosa, Hylocomnium splendens, Ditrichum flexicaule, Campylium stellatum,* and *Tomenthypnum nitens.* The soil lichens collected were identified as *Lecanora epibyron, Cladonia pocillum,* and *Solarina bispora.*

Cover by *Dryas integrifolia* was lower on all disturbance types, while road, bladed trail, and gravel pit sites had much less cover by *Rhizocarpon umbilicatum* than was found in the control area (tables 2 and 3). Relatively high cover values were noted for various species, but none were common to all disturbances. *Salix alaxensis* cover was significantly greater on the road and gravel-pit access road sites than in the undisturbed areas. The lower cover of the plant species that were dominant in the control stands, lower species richness, and the relatively low similarity coefficients indicate that disturbances in this community have had severe and long-lasting effects.

SAXICOLOUS LICHENS–*LECANORA EPIBYRON* COMMUNITY This plant community was restricted to the highest portion of the study area. Areas at elevations above 1,800 m were susceptible to the most rigorous climatic conditions in the study area. Snow-free during the winter and excessively drained in the thaw season, these sites experience severe water shortages and desiccating conditions. On the Plains of Abraham, blockfields and sorted patterned ground provide many microsites, which was particularly important in affording shelter to vascular plants. Dominant vascular plant species include *Dryas integrifolia, Carex misandra, C. scirpoidea,* and *C. rupestris,* as well as the cushion-forming *Salix dodgeana* and *Saxifraga oppositifolia.* However, crustose lichens have the greatest cover (table 2). *Lecanora epibyron* was the most abundant lichen identified, but most other specimens collected were sterile and therefore unidentifiable. These sterile crustose lichens were most commonly found on the soil surfaces, but also covered many rock substrates. Such plants have been noted in other studies and are extremely difficult to identify (Raup 1969, Woo and Zoltai 1977). Identified saxicolous lichens in this plant community (table 2) included *Lecanora intricata, Rhizocarpon chioneum, Verrucaria muralis,* and *Polyblastia hyperborea.* The soil lichens included *Solarina bispora, Baeomyces rufus,* and *Evernia perfragilis.*

Carex petricosa, C. scirpoidea, Lecanora epibyron, and miscellaneous lichen species had markedly less cover on all of the disturbances sampled in this community than on control sites (tables 2 and 3), and all tended to be restricted to undisturbed areas. Cover by *Dryas integrifolia* was less on the road and gravel pit sites, but remained similar to that of the undisturbed areas on the bladed trail. *Salix alaxensis* was the only species to have a significantly higher cover value on a disturbance than on the control sites, and this was recorded on the road site only.

Species richness was lower than that of the controls on all disturbances in this community, with the fewest species recorded for oil spills (table 5). Similarity coefficients showed the flora of the bladed trails and gravel pit access roads to be most similar to that of the undisturbed sites, while the flora of the oil spills differed to the greatest extent (table 4).

It must be concluded that disturbances in this community have resulted in a simplification of the existing plant community and a reduction in plant species diversity. Both of these changes indicate that disturbances in this plant community create long-term disruptions, and that the existing system seems incapable of recovering to a predisturbance state.

Crustose Lichen Tundra

Crustose lichen tundra occurred in a variety of sites with rock or fine-grained mineral substrates. Lichen cover varied from 100% on some blockfields to 60% on fine-grained mineral substrates. Blockfields and blockslopes in the Blue Mountain and Bolstead Creek–Devil's Pass ecosections supported this type of tundra. Individual stands varied from a few hundred square metres to 10's of hectares. These areas were within the unglaciated section of the study area and were not common on south-facing

Fig. 8. Crustose lichen tundra occurs on blockfields and is most common in the drier eastern portion of the Mackenzie Mountains. This site is on granite blocks of 50 to 100 cm diameter whereas at other sites limestones and dolomites composed the substrates.

slopes. Surfaces were composed primarily of granitic blocks that varied in size from 5 cm to 2 m, with an average diameter of 50 cm (fig. 8).

RHIZOCARPON INARENSE–UMBILICARIA PROBOSCIDEA COMMUNITY Field identification of crustose and saxicolous lichens was not attempted, and therefore it was not possible to subdivide some of the groups included in table 2. However, sufficient information was collected that, with the identification of voucher specimens, it was possible to describe the plant community in some detail.

Umbilicaria proboscidea and *Rhizocarpon* spp. were dominant. Species of several other genera of crustose lichens were also found, including *Lecidea, Parmelia, Polyblastia, Physcia,* and *Omphalodiscus.* No vascular plants were found, and mosses were restricted to sheltered sites on the sides of and in cracks on the blocks.

Lecidella stigmatea was the only plant species to have less cover on all disturbances than it did in the control plant community (tables 2 and 3). Most of the dominant species in control stands had low cover or were absent from the disturbances. The plants colonizing disturbances varied from one site to the next, and most of the species were not found in the undisturbed areas.

It is apparent that plant communities on disturbances were substantially different from those of the controls. This indicates that environmental conditions have been fundamentally changed by the disturbance. Crustose lichens have been noted for their slow growth rates (Hale 1974), and it is expected that these species will take a longer time to recover, if indeed this plant community can ever re-establish itself on the disturbed areas.

Disturbances here have provided environmental conditions less restrictive to vascular plant growth. Furthermore, the relatively simple control plant community has been replaced on disturbances by one that has greater structural and floristic diversity, particularly in its vascular plant component.

Due to the difficulty in identifying crustose and saxicolous lichen species, these sites may appear to support fewer species than was actually the case. It is interesting to note, however, that vascular plants, while essentially absent from the undisturbed plant community, were an important component of that on the disturbances.

Summary

Table 3 includes all species, genera, or plant groups with cover values differing by 2% or more from the associated control values. A number of taxa that represented major components of the undisturbed plant communities (table 2) had significantly less cover on disturbed sites (table 3). These included the fruticose *Cladonia* spp., *Lecanora epibyron*, *Rhizocarpon* spp., *Cetraria* spp., *Stereocaulon paschale*, *Alectoria ochroleuca*, *Betula glandulosa*, *Silene acaulis*, *Carex misandra*, *C. petricosa*, *C. rupestris*, *C. scipoidea*, *Arctostaphylos* spp., *Cassiope tetragona*, *Vaccinium uilginosum*, *Hedysarum alpinum*, *Dryas integrifolia*, *Salix barrattiana*, *S. polaris*, and *S. reticulata*. Conversely, several taxa or plant groups had greater cover on many disturbed site types and/or in several areas. These included *Collema* spp., *Peltigera* spp., *Stereocaulon alpinum*, *S. saxatile*, *S. tomentosum*, most *Polytrichum* spp., *Minuartia* spp., many Composites, *Carex microchaeta*, *C. membranacea*, *Trisetum spicatum*, *Juncus balticus*, *Epilobium* spp., *Salix alaxensis*, and *S. glauca*. Many plant species and groups were inconsistent in their responses to disturbance or were not common enough to show trends. In general, taxa that were abundant on control sites had lower cover values on disturbances, while introduced or locally rare plants often became dominant or had greater cover on the disturbed sites than in controls. No non-native species were found colonizing disturbances except where horses had recently used the area. There were no weedy annuals present on disturbances, only native perennials. However, species such as *Salix alaxensis* are common colonizers on natural disturbances and therefore have many of the characteristics of weedy plants. The flora of the control plant communities was substantially different from that of the disturbances, and in some cases, an entirely new assemblage of plants was present.

Bliss (1979) noted that several native vascular plant species were potentially useful for the revegetation of disturbances in the Low and High Arctic. Of these, *Arctagrostis latifolia, Calamagrostis canadensis,* and *Deschampsia brevifolia* were also noted in this study. *Arctagrostis latifolia* and *Deschampsia brevifolia* had significantly greater cover on some disturbances. However, all three species on disturbances also had significantly less cover than was found in control stands (tables 2 and 3). Hernandez (1973) noted a number of species capable of naturally colonizing disturbances. Three of these also were found to be important components of CANOL disturbance floras. *Poa arctica, Luzula confusa,* and *Epilobium angustifolium* all had greater cover on disturbances than in controls (table 3). All of these species were observed in fruit, and it seems probable that local seed sources could be harvested or crop areas established to provide seed for revegetation programs.

Deschampsia brevifolia was an important colonizer on man-induced disturbances in the High Arctic (Babb and Bliss 1974). *Luzula confusa, Dryas integrifolia, Salix arctica,* and *Saxifraga oppositifolia* had less cover than on controls. CANOL disturbance data show little consistency among plant communities and disturbance types for these species (table 3), despite the fact that the floristic and physiognomic characteristics in certain plant communities affected by CANOL disturbances have similarities to High Arctic areas (Brassard 1972). However, the disturbances discussed by Babb and Bliss (1974) had less recovery time than did CANOL disturbances.

The "organic crusts" (Raup 1969) on many disturbance substrates often covered substantial areas. It was apparent at the time of sampling that a significant area was occupied by these plants. However, these sterile crusts were not identifiable (Marsh personal communication 1980). Woo and Zoltai (1977) have also noted the presence of these crusts in the High Arctic and Canadian Rockies (Zoltai personal communication 1982). These surfaces can be composed of various species of mosses, liverworts, and lichens. However, species of *Polyblastia* (Marsh personal communication 1980) are thought to be the most important components of the organic crusts sampled on CANOL disturbances.

The highest similarity coefficients were found in association with bulldozer track floras, which had values of 74 and 75 in two cases. Bladed trails and bulldozer tracks had the smallest differences, and bulldozer tracks were the only type of disturbance with an average of more than 50% of the flora in common with control sites. These represent the only disturbances that initially would have retained, intact, large portions of the existing plant community, and in this way could still have substantial portions of the original flora. In the case of bladed trails, Bliss and Wein (1972*a*) described recovery from roots and rhizomes that were left following surface blading. Sprouting from these contributed to the 30% to 50% cover achieved in 5 years. The lowest similarity coefficients were recorded for gravel pit and false start road sites, which had no species in common with the control stands in the *Rhizocarpon inarense–Umbilicaria proboscidea* community. Disturbances in this community had similarity coefficients consistently lower than those recorded in any other community type. False start roads and gravel pits had resulted in a complete floristic change, and roads held only 3 of a possible 98 species in common with the undisturbed areas (table 5). These crustose lichen communities develop slowly and, based on their similarity coefficients, require much more than 35 years to re-establish a plant community similar to that of the control. In addition, disturbances provide rooting areas and moisture not otherwise available in these areas. Given the degree of substrate alteration, it is doubtful that disturbances other than bladed trails will support plant communities similar to those of the control stands.

The *Salix polaris–Dactylina beringica, Salix barrattiana*–Moss species, and *Salix reticulata–S. lanata* plant communities generally had low similarity coefficients on all disturbances sampled (table 4). The structure of these communities had been altered by the removal of the dominant shrub component and its subsequent replacement with *Salix alaxensis* and/or *S. planifolia,* which were generally more robust shrubs. Increases in shading and the competition for other resources that would result from these fast-growing shrubs would substantially alter the environmental conditions, even on the less disruptive disturbances such as bulldozer tracks.

The *Dryas integrifolia–Cetraria tilesii* and saxicolous lichens–*Lecanora epibyron* plant communities had the highest average similarity coefficients of all disturbances sampled (table 4). Floristically these communities tended to be least changed, implying that environmental alterations resulting from disturbances were smaller here than in other plant communities, or that the species here were better able to colonize disturbances. The combination of lower species richness, removal of the dominant control species, and low floristic similarity to control sites indicate that several plant communities have sustained severe, long-lasting, negative impacts. Plant communities affected in this way include *Carex membranacea–Sphagnum*, *Hedysarum alpinum*–Moss, *Salix polaris–Dactylina beringica*, *Dryas integrifolia–Rhizocarpon umbilicatum*, and saxicolous lichens– *Lecanora epibyron*. Conversely there are also cases where, despite cover reductions, the species richness increased and floristic similarity was high. The greater species diversity and structural complexity on disturbances in these plant communities can be considered to be beneficial and the impacts positive after 34 to 36 years. Plant communities that provide examples of this include *Carex* spp.–*Kobresia simpliciuscula*, *Betula glandulosa–Cladonia stellaris*–Moss, *Cladonia stellaris–Alectoria ochroleuca*, and *Rhizocarpon inarense–Umbilicaria proboscidea*.

Considering that recovery on CANOL disturbances has been natural, these plant communities have achieved high cover values. After 34 to 36 years, cover in cushion plant tundra is an average of 81% of that of the undisturbed plant community, whereas in crustose lichen tundra this value is only 30% (table 6). A comparison of disturbances indicates that oil spills have only 31% of the total plant cover of their controls, while bulldozer tracks average 87%. Certainly the degree of recovery is closely related to the extent of the original disturbance. However, in some cases little difference in plant cover occurs between fundamentally different types of disturbances. For example, in erect deciduous shrub tundra there was only a 2% difference between plant cover on gravel-pit access roads and on bulldozer tracks (table 6).

In all cases disturbances have greater bare ground than do control stands. After 34 to 36 years of natural recovery these differences are evident, even on the relatively lightly disturbed sites such as bulldozer tracks. It is also noteworthy that severe initial disturbances such as gravel pits can support a plant cover approaching control conditions. In the case of oil spills, they offer little opportunity for plant colonizaton and consequently support the lowest plant cover with the greatest reduction from control plant communities. It is evident that rehabilitation is necessary in order to enhance recovery on such long-lasting disturbances.

Studies on oil spills have noted plant recovery after short time intervals (e.g., 0 to 5 years) (MacKay et al. 1980, Hutchinson and Hellebust 1978, Bliss and Wein 1972b, Wein and Bliss 1973, Babb and Bliss 1974, Walker et al. 1978, McKendrick and Mitchell 1978). Results from CANOL oil spill studies confirm that recovery is occurring but more slowly than the literature suggests, perhaps as a result of the recontamination mentioned by MacKay et al. (1980) or the delay in response noted by McCown and

Table 6 Control-corrected bare ground cover (%) on CANOL Project disturbances, Northwest Territories, in tundra plant communities

Plant community	Mean control values	Road	False start road	Bladed trail	Camp yard	Bull-dozer track	Gravel pit	Gravel pit access road	Oil spill	Mean cover difference (all disturbances)
Erect deciduous shrub tundra	8	+52	+45	+44	+10	+18	+60	+20	+94	+43
Decumbent shrub tundra	17	+58	+52	+76		+28	+70	+42	+49	+47
Sedge meadow tundra	14	+46		+1		+32	+26		+58	+33
Lichen heath tundra	8	+56		+22			+72		+74	+56
Fruticose lichen tundra	3	+53	+75	+72		+3	+79	+25		+51
Cushion plant tundra	57	+29		+14			+28	+5		+19
Crustose lichen tundra	6	+75					+66		+69	+70
Mean cover difference (all communities)		+53	+57	+38	+10	+13	+57	+23		

Deneke (1972), but not evident with short-term studies. Given the low plant cover on CANOL oil spills, it seems reasonable to assume that these sites remain hostile as a result of their extreme aridity, as is suggested by the results of studies by Bliss and Wein (1972*b*) or because they remain phytotoxic, as is suggested by the lack of oil degradation on these sites (Kershaw 1983).

Conclusions

Within each of the 15 floristically defined plant communities, responses to disturbances were often unique. It was generally found that when lichens were common in the control they were absent from or much reduced on the disturbances. Several groups of plants, and most commonly the dominant species of the undisturbed stands, had notably less cover. Higher cover values were noted for approximately the same number of taxa as had lower values. The cover differences relative to controls often varied with the type of disturbance and the plant community affected (table 5). Consequently it is difficult to make generalizations regarding the suitability of a particular species for revegetation.

However, a number of plant species consistently demonstrated an ability to disperse from areas outside of those directly affected by the CANOL Project and to establish themselves successfully on most types of disturbances (table 7). Several of these species have been noted in other studies for their potential for use in revegetation programs. Few studies have reported on colonizing bryophytes. None of the bryophytes noted on disturbances by Johnson et al. (1978) were common on CANOL disturbances. *Polytrichum juniperinum* was noted recolonizing after a fire in the Tuktoyaktuk region. In Alaska, *Polytrichum* sp. was one of the earliest colonizers on natural tundra disturbances (Peterson and Billings 1978). Few of the lichens included in table 7 have been noted in the literature as disturbance colonizers. Peterson and Billings (1978) noted that *Stereocaulon* sp. and *Cetraria* spp. were important components of the plant cover on natural disturbances in the Alaskan arctic coastal plain. *Stereocaulon* spp. have also been described as increasing in cover in areas of heavy reindeer grazing in Finland (Kallio et al. 1969, Pulliainen 1971). In these areas trampling and grazing can substantially alter the plant communities. Several of the vascular plant species noted in this study have been reported as successful colonizers of natural and man-induced disturbances in other studies. These included *Carex aquatilis* (Bliss and Wein 1972*a*, Hernandez 1973,

Table 7 Plant species with demonstrated ability for long-distance dispersal and colonization success on CANOL Project disturbances, Northwest Territories.

Lichens	Mosses	Vascular plants	
Cladonia macrophylla	Polytrichum sp.	Calamagrostis canadensis*	Minuartia rubella
C. pyxidata	P. hyperboreum*	Carex aquatilis*	Poa alpigena*
C. subulata	P. juniperinum*	C. lugens*	P. pratensis*
Collema sp.	P. piliferum*	C. membranacea*	Salix alaxensis*
Lecidella stigmatea	P. polydactyla*	C. microchaeta*	S. arbusculoides*
Pannaria pezizoides	P. strictum*	C. podocarpa*	S. glauca*
Peltigera aphthosa*		Dryas octopetala*	S. lanata*
P. canina*		Empetrum nigrum	S. planifolia*
P. polydactyla*		Epilobium anagallidifolium	Trisetum spicatum*
P. rufescens*		E. angustifolium*	Veronica wormskjoldii
Rinodina roscida		E. latifolium*	
Stereocaulon alpinum*		Hedysarum mackenzii*	
S. paschale*		Hierochloe alpina*	
S. saxatile*		Juncus balticus*	
S. tomentosum*		Linnaea borealis	
Toninia lobulata		Minuartia macrocarpa	

Key

* Species consistently forming a high proportion of standing crop and cover on disturbances

Johnson et al. 1978, Lore 1977), *Empetrum nigrum* (Johnson 1980, Wein and Bliss 1973), *Epilobium* spp. (Hernandez 1973, Johnson 1980), *Calamagrostis canadensis* (Bliss and Wein 1972a, Hernandez 1973, Johnson 1980, Johnson and Van Cleve 1976, Lambert 1972, Peterson Billings 1978, Younkin 1976), *Poa* spp. (Bliss and Wein 1972a, Hill et al. 1980, Johnson 1980, Johnson et al. 1978, Peterson and Billings 1978, Younkin 1976, Yurtsev and Korobkov 1979), *Dryas* spp. (Babb 1973, Hill et al. 1980, Johnson 1980), and *Salix* spp. (Bliss and Wein 1972a, Druzhinina and Zharkova 1979, Hill et al. 1980, Johnson 1980, 1981, Johnson et al. 1966, Johnson 1978, Lambert 1972, Peterson and Billings 1978). Not all the studies have been of long-term disturbances. However, it is apparent that these species or genera are successful colonizers of disturbances on a circumpolar basis.

Long-distance dispersal is an important component of vegetation recovery in most ecosections. It would therefore be difficult to determine what native species are best adapted to disturbed environments by selecting only from promising local species on undisturbed areas. The species that consistently demonstrate the capability for colonization and long-distance dispersal have been identified (table 7) and should be considered as preadapted for similar northern environments. Furthermore, several of these species consistently form a high proportion of disturbance standing crop and cover. These species should therefore be carefully considered, since they appear to be ideally suited for revegetation programs in northern reclamation projects.

Species composition was simplified relative to control stands on disturbances in the *Dryas integrifolia*-dominated plant communities. Oil spills and roads most often had relatively low species richness. Definite assemblages of plants were found on each disturbance type, though these varied as the associated control plant communities changed. All CANOL disturbances were small in area, and therefore the transport distance was not significantly different among the eight types of disturbances sampled. Assuming that rates of migration into disturbed areas at a given site were equal, those disturbances with high species diversity would appear to have created environments conducive to the growth and reproduction of a wide variety of plants. Low species diversity could reflect a relatively severe environment where only the hardiest species are able to survive, or a habitat especially suited to only a few species.

Considering all of the parameters discussed above, oil spill, road, and gravel-pit plant communities showed the greatest degree of difference after 34 to 36 years, while bulldozer-track plant communities were most similar to the undisturbed stands. However, all disturbances have produced changes, and in some cases it is difficult to conclude that a predisturbance state can be re-established, even within an extremely long time frame. The environmental conditions that led to the formation of the control plant communities have been changed by the disturbances, and it is reasonable to assume that successional processes will lead to the establishment of stable plant communities that are substantially different from those that were undisturbed. It is also evident in cases where the initial disturbance was slight (i.e., bulldozer tracks) or where disturbances produced environmental alterations not substantially different from control conditions (i.e., in cushion plant tundra) that the product of long-term evolution will result in plant communities that are more similar to controls than at present.

Disturbances with their own distinctive plant communities are generally linear disruptions in an otherwise gradually changing landscape composed of a mosaic of plant communities. In extensive, homogeneous plant communities, the abrupt change in species composition, plant cover, standing crop, and structure provide variability that otherwise occurs only along natural disturbances. In the region containing the stable, homogeneous plant community the relatively rapidly changing disturbance plant community can be a positive ecosystem asset. It is here that new species can become established and aid in diversifying the region's ecological characteristics. Although examples of ecosystem simplification occur (e.g., CANOL oil spills), there are also cases of ecological diversification. The removal or partial destruction of a plant community will result in long-term changes that may be positive or negative ecosystem alterations, varying in degree, but essentially permanent.

Finally, there is no evidence to suggest that the plant communities on disturbances are currently stable. It appears that the rate of change varies with the type of disturbance and the plant community in which it occurred. In any case, it is difficult to predict the point at which structural and floristic stability will be reached and the time required to achieve this result.

Acknowledgments

This research was supported by grants from the Boreal Institute for Northern Studies, the Arctic Institute of North America, and the National Wildlife Federation. In addition, logistical assistance was provided by AMAX

Northwest Mining Co. Ltd. and Imperial Oil Ltd. The Department of Geography reprographics section provided fig. 1. I gratefully acknowledge the professional and personal contribution that Linda Kershaw has made to all phases of this research. This work has benefited from valuable contributions by many people, and I thank them all.

References

Argus, G. W. 1973. The genus *Salix* in Alaska and the Yukon. *In* Publications in botany, No. 2. Nat. Mus. Can., Ottawa.

Babb, T. A. 1973. High Arctic disturbance studies. *In* L. C. Bliss and R. W. Wein, eds. Botanical studies of man-modified habitats in the Mackenzie Valley, eastern Mackenzie Delta region and the Arctic Islands. ALUR 72-73-14: 150-162.

———, and L. C. Bliss. 1974. Effects of physical disturbance on arctic vegetation in the Queen Elizabeth Islands. J. Appl. Ecol. 11: 549-562.

Bliss, L. C. 1979. Vegetation and revegetation within permafrost terrain. pp. 31-50. *In* Third International Permafrost Conference on Permafrost, Proceedings Volume 2. Nat. Res. Council of Can., Ottawa.

———, and R. W. Wein. 1972*a*. Plant community responses to disturbances in the western Canadian Arctic. Can. J. Bot. 50: 1097-1109.

———, and R. W. Wein, eds. 1972*b*. Botanical studies of natural and man-modified habitats in the eastern Mackenzie Delta region and the Arctic Islands. ALUR 71-72-14. Dep. Indian Affairs and Northern Dev., Ottawa.

Brassard, G. R. 1972. Mosses from the Mackenzie Mountains, Northwest Territories. Arctic 25(4): 308.

Crum, H. A., W. C. Steere, and L. E. Anderson. 1973. A new list of mosses of North America north of Mexico. Bryologist 76(1): 85-130.

Druzhinina, O. A., and Y. G. Zharkova. 1979. A study of plant communities in anthropogenic habitats in the area of the Vorkuta Industrial Centre. *In* The effect of disturbance on plant communities in tundra regions of the Soviet Union. Biol. Pap. Univ. Alaska 20: 30-53.

Finnie, R. S. 1945. CANOL, the sub-arctic pipeline and refinery project constructed by Bechtel Price-Callahan for the corps of engineers, U.S. Army 1942-44. San Francisco.

Flood, M. J. 1946. Arctic journal and other works. Wetzel Pub. Co., Los Angeles.

Gauch, H. G. 1979. COMPCLUS – A FORTRAN program for rapid initial clustering of large data sets. Ecol. and Syst., Cornell Univ., Ithaca.

Gill, D. 1973. Modification of northern alluvial habitats by river development. Can. Geogr. 17(2): 138-153.

Hale, M. E. 1974. The biology of lichens. 2nd ed. Edward Arnold, London.

———, and W. L. Culberson. 1970. A fourth checklist of the lichens of the continental United States and Canada. Bryologist 73: 499-543.

Hemstock, R. A. 1945. Report of operations on the Canol no. 1 pipeline, Northwest Territories and Yukon. Imperial Oil Ltd.-Norman Exploration, Hemstock collection, Calgary.

Hernandez, H. 1973. Natural plant recolonization of surficial disturbances, Tuktoyaktuk Peninsula region, Northwest Territories. Can. J. Bot. 51(11): 2177–2196.

Hill, N., B. Freedman, and J. Svoboda. 1980. Seed banks and seedling floras in a variety of plant communities. pp. 79–82. *In* J. Svoboda and B. Freedman, eds. Ecology of a High Arctic lowland oasis, Alexandra Fjord, Ellesmere Island, N.W.T. Canada, 1980 Progress Report. Deps. Botany, Univ. Toronto and Dalhousie Univ.

Honsaker, J. L. 1981. SPCSORT. Dep. Geogr., Univ. Alberta, Edmonton.

Hutchinson, T. C., and J. A. Hellebust. 1978. Vegetational recovery in the Canadian Arctic after crude and diesel oil spills. ALUR 75-76–83. Ottawa.

Johnson, A. W., B. M. Murray, and D. F. Murray. 1978. Floristics of the disturbances and neighbouring locales. *In* Lawson et al., eds. Tundra disturbances and recovery following the 1949 exploratory drilling, Fish Creek, Northern Alaska. CRREL Rep. 78–28: 30–40. Hanover, New Hampshire.

———, L. A. Viereck, R. E. Johnson, and H. R. Helchoir. 1966. Vegetation and flora. pp. 277–354. *In* N. J. Wilimovsky and J. N. Wolfe, eds. Environment of the Cape Thompson Region, Alaska. U.S. Atomic Energy Commission, Division of Technical Information, Washington, D.C.

Johnson, L. A. 1980. Revegetation and restoration investigations. *In* J. Brown and R. L. Berg, eds. Environmental engineering and ecological baseline investigations along the Yukon River–Prudoe Bay haul road. CRREL Rep. 80–19: 129–150. Hanover, New Hampshire.

———. 1981. Revegetation and selected terrain disturbances along the trans-Alaska pipeline, 1975–1978. CRREL Rep. 81–12. Hanover, New Hampshire.

———, S. D. Rindge, and D. A. Gaskin. 1981. Chena River Lakes Project revegetation study: Three-year study. CRREL Rep. 81–18. Hanover, New Hampshire.

———, and K. Van Cleve. 1976. Revegetation in arctic and subarctic North America: A literature review. CRREL Rep. 76–15. Hanover, New Hampshire.

Kallio, P., U. Laine, and Y. Makinen. 1969. Vascular flora of Inari Lapland. 1. Introduction and *Lycopodiaceae–Polypodiaceae*. Rep. Kevo Subarctic Res. Sta. 5: 1–108.

Kershaw, G. P. 1983. Long-term ecological consequences in tundra environments of the CANOL crude oil pipeline project, N.W.T., 1942–1945. Ph.D. thesis. Univ. Alberta, Edmonton.

Lambert, J. D. H. 1972. Botanical changes resulting from seismic and drilling operations, Mackenzie Delta area. ALUR 1971-72–12. Ottawa.

Lore, J. 1977. Tundra disturbance study: Burwash Uplands, Yukon Territory. M.Sc. thesis. Dep. Geogr., Univ. Alberta, Edmonton, Alberta.

Mackay, D., T. Wing Ng, Wan Ying Shiu, and B. Reuber. 1980. The degradation of crude oil in northern soils. ALUR Environmental Studies No. 18. Ottawa.

McCown, B. H., F. J. Deneke, W. E. Rickard, and L. L. Tieszen. 1972. The response of Alaskan terrestrial plant communities to the presence of petroleum.

pp. 34–43. *In* Proceedings of the Symposium on the Impact of Oil Resource Development on Northern Plant Communities. Occ. Pub. on Northern Life, No. 1. Institute of Arctic Biology, Univ. Alaska, Fairbanks.

McCown, D. D., and F. J. Deneke. 1972. Plant germination and seedling growth as affected by the presence of crude petroleum. pp. 44–51. *In* Proceedings of the Symposium on the Impact of Oil Resource Development on Northern Plant Communities. Occ. Pub. on Northern Life, No. 1. Institute of Arctic Biology, Univ. Alaska, Fairbanks.

McKendrick, J. D., and W. W. Mitchell. 1978. Fertilizer and seeding oil-damaged arctic tundra to effect vegetation recovery, Prudhoe Bay, Alaska. Arctic 31(3): 296–304.

Mueller-Dombois, D., and H. Ellenberg. 1974. Aims and methods of vegetation ecology. John Wiley and Sons, New York.

Peterson, K. M., and W. D. Billings. 1978. Geomorphic processes and vegetational change along the Mead River Sand Bluffs in northern Alaska. Arctic 31(1): 7–23.

Porsild, A. E., and W. J. Cody. 1980. Vascular plants of continental Northwest Territories, Canada. Nat. Mus. Can., Ottawa.

Pruitt, W. O., Jr. 1978. Boreal ecology. Edward Arnold, London.

Pulliainen, E. 1971. Nutritive values of some lichens used as food by reindeer in northeastern Lapland. Zool. Fennici 8: 385–389.

Raup, H. M. 1969. Observations on the relation of vegetation to mass-wasting processes in the Mesters Vig District, northeast Greenland. Medd. om Grønl. 176(6): 180–183.

Richardson, H. W. 1944*a*. Finishing the Alaska Highway. Eng. News-Rec. 132 (Jan.): 94–103.

———. 1944*b*. Controversial Canol. Eng. News-Rec. 132(May): 78–84.

Savile, D. B. O. 1972. Arctic adaptations in plants. Can. Dep. Agr. Monogr. No. 6. Ottawa.

Steere, W. C. 1978. The mosses of arctic Alaska. Bryoph. Bibl. 14. J. Cramer, Lehre.

Thomson, J. W. 1967. The lichen genus *Cladonia* in North America. Univ. Toronto Press, Toronto.

———. 1979. Lichens of the arctic slope. Univ. Toronto Press, Toronto.

Tranquillini, W. 1979. Physiological ecology of the alpine timberline: Tree existence in high altitudes with special reference to the European Alps. Springer Ecol. Stud., Vol. 31. Springer Verlag, New York.

Truman Committee. 1944. pp. 9287–9935. *In* United States Congress, hearings before a special committee investigating the National Defence Program, 78th congress, 1st session, part 22, the Canol Project. U.S. Gov. Printing Office, Washington, D.C.

United States Army. 1950. Historical record CANOL Project. Alaska Corps of Engineers, Anchorage.

Walker, D. A., P. J. Webber, K. R. Everett, and J. Brown. 1978. Effects of crude and diesel oil spills on plant communities at Prudhoe Bay, Alaska, and the derivation of oil spill sensitivity maps. Arctic 31(3): 242–259.

Wein, R. W., and L. C. Bliss. 1973. Experimental crude oil spills on arctic plant communities. J. Appl. Ecol. 10: 671–682.

Woo, V., and S. C. Zoltai. 1977. Reconnaissance of the soil and vegetation of Somerset and Prince of Wales islands, N.W.T. Information Rep. NOR–X–186. Northern Forest Res. Centre, Edmonton.

Younkin, W. E. 1976. Revegetation studies in the northern Mackenzie Valley region. Biol. Rep. Ser., Vol. 38. Canadian Arctic Gas Study Ltd., Calgary.

Yurtsev, B. A., and A. A. Korabkov. 1979. An annotated list of plants inhabiting sites of natural and anthropogenic disturbances of tundra cover: Southernmost Chukchi Peninsula. *In* The biological effect of disturbance plant communities in tundra regions of the Soviet Union. Biol. Pap. Univ. Alaska 20: 1–18.

Zoltai, S. C., and V. Woo. 1978. Sensitive soils of permafrost terrain. pp. 410–424. *In* Forest soils and land use: Proceedings of the fifth North American forest soils conference. Colorado State Univ., Fort Collins, Colorado.

Implications of Upstream Impoundment on the Natural Ecology and Environment of the Slave River Delta, Northwest Territories

M. C. English

Introduction

In Canada deltas constitute a very small percentage of the total land area north of the southern limit of discontinuous permafrost (as defined by Brown 1960). By comparison with other landforms in this region the biological and economic significance of deltas is disproportionately high. Because of the very high botanical productivity of nutrient-rich emergent macrophytes and the large diversity of plant species, these wetlands are important feeding, staging, and breeding habitat for large numbers of waterfowl, muskrat, and other wildlife. Studies of the economic importance of the Peace-Athabasca, Mackenzie, and Slave River deltas indicate significant use and heavy dependence on these landforms by local native populations (Peace-Athabasca Delta Project Group 1972, Berger 1977, Bodden 1981).

The impetus for studying the geomorphological process and biological productivity of deltaic environments in Canada has largely been instigated by past or proposed alterations of natural river regimes by hydroelectric power developments (Reinelt et al. 1971, Day 1971, Fraser 1972, Taylor et al. 1972, Glooschenko 1972, Richardson 1972, Armstrong 1973, Gill 1973, Kellerhals and Gill 1973, Kerr 1973, Gill and Cooke 1974, Maddock 1976, English 1979). Development of northern rivers will, to varying degrees, alter their natural flow regimes and consequently disrupt the natural environment of their deltas.

This paper will discuss the major processes governing the progradation of the Slave River Delta into Great Slave Lake (fig. 1), the relationship between landforms and plant assemblages on the delta, and potential environmental and ecological effects of a proposed upstream impoundment of the Slave River.

Origins of the Slave River Delta

Fifteen thousand years before the present (BP) the Keewatin ice sheet extended into the southwestern corner of the present-day Northwest Territories. Tongues of the main glacier extended up the Peace and Athabasca valleys and westward into the Selwyn Mountains. Glacial retreat began in this region approximately 10,000 years BP (Bryson et al. 1969, Prest 1969), resulting in the formation of glacial Lake McConnell. Drainage of Lake McConnell through the present-day Churchill River system into Hudson Bay was impeded by isostatic rebound. Northward drainage

Fig. 1. Location of Slave River Delta, Northwest Territories.

down the Mackenzie Valley began after the Selwyn ice tongue retreated (Cameron 1922). Lowering of the glacial lake water levels resulted in the differentiation of this large body of water into two separate lakes: present-day Great Slave Lake and Lake Athabasca. Alluvial material carried by the Peace and Athabasca rivers entered the southern arm of Great Slave Lake via the Slave River, and formation of the Slave Delta began. As lake levels dropped, the southern arm of Great Slave Lake slowly filled in with alluvium to its present position, where the active delta still progrades into the lake.

The Active Delta

The active delta of the Slave River is approximately 300 km², which is roughly 3% of the area of deltaic deposits occupying the old southern arm of Great Slave Lake. The semi-aquatic nature of the active delta is illustrated in fig 2. Fifty-one percent of the delta is submergent, either at or below the summer low-water level of Great Slave Lake. The remaining 49% is emergent, above the summer low-water level.

Botanically and geomorphologically the Slave River Delta can be di-

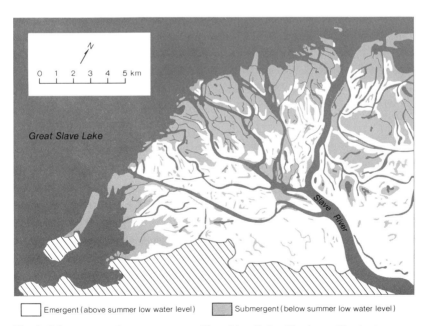

Fig. 2. Submergent and emergent areas—Slave River Delta, Northwest Territories.

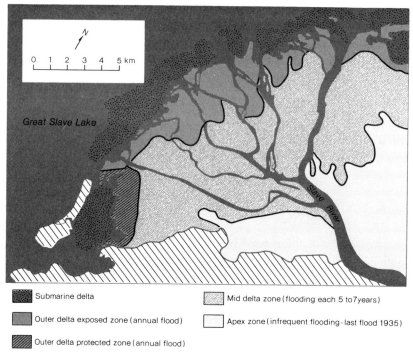

Fig. 3. Flood frequency zones of the Slave River Delta, Northwest Territories.

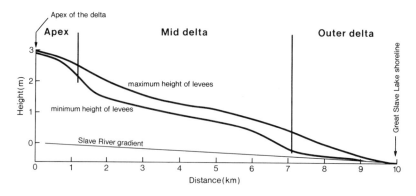

Fig. 4. Levee elevation on the three flood frequency zones of the Slave River Delta, Northwest Territories.

vided into three distinct areas: the outer delta, the mid-delta, and the apex (fig. 3). Further, the outer delta consists of three zones: the submarine delta, and the exposed and protected areas of the subaerial delta. The geomorphological and botanical differences between the three areas relate primarily to their relative elevations above either the lake or river summer low-water levels and thus to the frequency of spring flooding. The subaerial levees in the outer delta rarely exceed 0.5 m above river/lake summer low-water levels (fig. 4). Generally the vegetation in the outer delta consists of semi-aquatic emergent plants and *Salix* spp. (fig. 5). In the mid-delta, levees range between 1.0 m and 2.0 m above summer low-river levels. Plant assemblages in this portion of the delta are diverse, ranging from *Equisetum* to *Populus* assemblages. The *Alnus-Salix* assemblage is by far the most representative assemblage in the mid-delta (fig. 6). Levees along active distributaries in the apex project 2.0 m to 3.0 m above summer low-river levels. The *Picea* assemblage dominates the apex (fig. 7).

Ninety-five percent of the landform types comprising the outer delta are submergent. As a consequence of annual spring flooding, nutrient-rich sediment is deposited on the large expanses of *Equisetum* assemblages occupying the interlevee depressions in the outer delta. The annual influx of sediment is essential for the maintenance of high productivity in these assemblages. *Equisetum fluviatile*, the dominant plant on the outer delta,

Fig. 5. *Equisetum* assemblage in interlevee depression of a cleavage bar island.

Fig. 6. An *Alnus-Salix* assemblage inhabiting a cut-bank levee.

actively accumulates potassium and sodium from the sediments and water column (Shepherd and Bowling 1973). In turn, *E. fluviatile* is an important source of sodium for herbivorous animals (Hutchinson 1975) such as the large and economically important muskrat population that inhabits the Slave Delta (Geddes 1981).

The mid-delta area is transitional between the water-dominated landscape of the outer delta and the elevated, relatively dry apex area. Twenty-five percent of the mid-delta is aquatic. Approximately 50% of the aquatic area is comprised of former channels now elevated above the summer low-water levels by sedimentation during occasional spring flooding. The remaining portion of the aquatic area of the mid-delta consists of channels that remain partially open to the river's seasonal fluctuations. Soil profile records and verbal accounts from local residents indicate that flooding of the mid-delta occurs every 5 to 7 years. The relatively poor bryophyte development in the mid-delta zone attests to the occurrence of occasional flooding. Gill (1978) reports that areas in the Mackenzie Delta that experience sediment-rich floods support few bryophytes.

The apex is the oldest portion of the active Slave Delta. Increment boring of mature *Picea glauca* indicates that the *Picea* assemblage invaded these landforms approximately 250 years BP. The aquatic areas, comprising 6% of the apex zone, are elevated above the water level of the surrounding distributaries by as much as 2.5 m. Their water supply is mainly

Fig. 7. *Picea* assemblage in apex zone.

snowmelt runoff; occasionally it is river water, when ice damming in the distributing channels instigates unusually high flood levels.

Processes

Essential to an understanding of the processes responsible for the maintenance of high biological productivity on a delta is an understanding of the intersection of a river and the large body of less-turbulent water into which it drains. The formation of the submarine sedimentary platform upon which the emergent landforms of the delta are constructed is a product of the degree of density differential between the river and the body of water into which it is draining.

According to Thakur and Mackay (1973) arcuate deltas, such as that of the Slave River, are formed when the densities of the river and the body of water into which it is flowing are equal. The buildup of the submarine platform upon which the cleavage and wave-built bars will eventually form occurs primarily at or near the mouths of the largest distributary channels, which transport the bulk of suspended sediment and bedload. The primary factor modifying and eroding the shallow submarine platform on the Slave Delta is wave action during the ice-free months. A counterclockwise current in the western arm of Great Slave Lake generated primarily by the Slave River discharge and secondarily by the prevailing northwesterly winds is responsible for removing quantities of sediment from the distributary mouths and transporting the sediment into the lake (fig. 8).

The direction of the Slave River Delta's growth and the rate with which it prograde into Great Slave Lake are partly due to the reduction of the river's energy as it diverges from one channel into several. Such energy loss directly affects the ability of the river to transport bedload and suspended sediment. Hence, the deposition of sediment and formation of river bars occur near the diverging point of the main river channel. On the delta, sandbar formation due to energy reduction occurs at the confluence of the

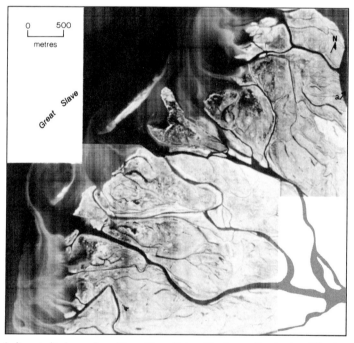

Fig. 8. Sediment discharge from Slave River Delta, Northwest Territories. The thermal imagery flown 7 September 1977 illustrates the flow of the discharging sediment. The discharge from Resdelta Channel (a) creates a current sufficiently strong to draw the discharging water from the other channels to the south along the outer perimeter of the delta. Courtesy of the Canadian Wildlife Service.

main distributaries. Significant growth of these river bars will likely occur when ice damming inhibits flow down major distributaries during the spring melt. Sand bar formation at this critical juncture in the river plays an important role in reducing discharge in some distributaries, concomitantly increasing it in others. The direction and pattern of the growth of the submarine platform, and therefore the subaerial delta, is determined by the dispersion of river energy at the distributary concourse. Deltaic growth will occur mainly at the mouths of channels where deposition of sediment exceeds the erosive capacity of the lake.

Cleavage Bar Development

The major geomorphic process operating on the outer delta is cleavage bar development. As the delta expands into the lake, cleavage bars are formed at the mouths of active distributaries. Coarse fractions of the suspended sediment and bedload are deposited along submarine levees that run parallel to the concourse of the discharging channel. Dahlskog et al. (1972) state that where lake action dominates over river discharge, the development of bars will dominate along one side of the channel mouth. Although

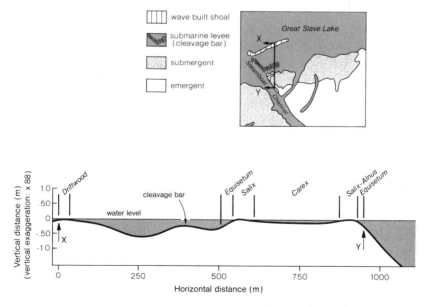

Fig. 9. Transect 14, illustrating plant assemblages and submarine cleavage bar genesis at the mouth of Steamboat Channel (sampled and surveyed 18 July 1977).

Fig. 10. Driftwood accumulation along the outer levees of the cleavage bar islands protects the developing islands from wave action. The levees are further stabilized by pioneer species of *Salix interior* and *Equisetum fluviatile* (right centre of photograph).

wave action is an important force in the geomorphological development of the Slave River Delta, the discharge through Resdelta channel (the main distributary) offsets the effects of wave action. The lake current (fig. 8) created by the discharge of Slave River into Great Slave Lake creates a counterclockwise vortex within the western arm of the lake and is a factor in the development of submarine and subaerial landforms on the outer delta.

Fig. 9 illustrates the genesis of a cleavage bar at the mouth of a discharging channel. The submerged levee is forming in a V-shape; hence the term cleavage. The buildup of the submerged levee is largely due to the deposition of bedload during the spring flood (Dahlskog et al. 1972). When the levee builds up to a sufficient elevation, driftwood lodges on it (fig. 10) and stabilization occurs, followed by an invasion of pioneer species of emergent vegetation such as *Equisetum fluviatile*. The upstream portion of the bar matures first, and *Salix* spp. invade the cleavage bar levee, further stabilizing it. Continued development of this deposit into an enclosed cleavage bar island may be aided by the construction of a wave-built shoal along the open end of the V-shaped cleavage bar, as shown in fig. 9.

In some cases protection from the erosive action of the lake is provided

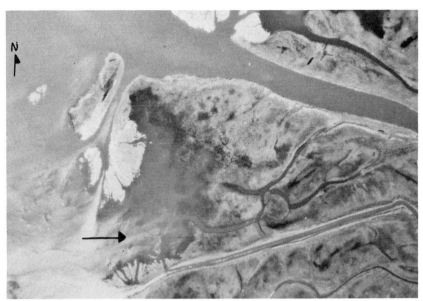

Fig. 11. Panchromatic aerial photograph of a cleavage bar island on the outer delta. This island is still partially open at the lake side (arrow), but submarine levees that are forming from each of the open ends will soon build up and close the cleavage. Courtesy of the Canadian Wildlife Service.

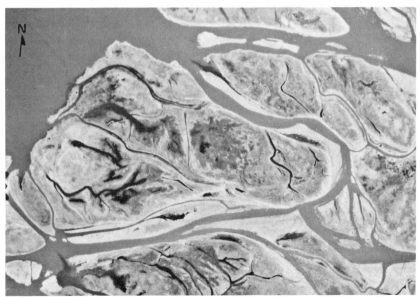

Fig. 12. Cleavage bar island formations, Slave River Delta. This complex of cleavage bar islands is in the final stage of being closed off from the lake. Levees along the outer portions of the island are inhabited by the *Salix-Equisetum* assemblage, while interlevee depressions are occupied by an *Equisetum* assemblage. Courtesy of the Canadian Wildlife Service.

by the construction of one or more levees, which may partially (fig. 11) or completely (fig. 12) seal off the open end of the cleavage bar. By using tree-ring analysis, the duration of cleavage bar island development from the submarine levee stage was estimated for an island in the outer delta (fig. 12). The elevated upstream portion of the island was supporting *Salix* spp. in 1922. The two diverging arms of the cleavage bar were invaded by *Salix* sp. 10 years later. The levees bordering Great Slave Lake began supporting *Salix* spp. in 1963. The centres of these landforms—the interlevee depressions—support very productive stands of *Equisetum fluviatile*.

Botanical Development

When classifying shoreline vegetation in the Great Slave Lake–Lake Athabasca region, Raup (1975) concluded that the concept of plant assemblages was more practical than that of plant communities. Plant communities have a similar floristic composition among sample plots, which does not occur in these northern wetlands. Instead there is a wide variation in species composition among sample plots, which necessitates a term such as assemblage because it "carries few implications of relationships that are non-existent or unknown" (Raup 1975). This variation among sample plots or assemblages was attributed to ecotypic variations within populations, which in turn have been historically conditioned to the frequent flooding that occurs in these wetland areas (Raup 1975: 138).

Twelve plant assemblages have been identified on the Slave Delta (English 1979); in this chapter the nine most significant assemblages will be discussed. In those assemblages occupying landforms frequently inundated during the spring flood, succession is governed largely by environmental factors such as the elevation of the substrate above or below summer low-water levels. Allogenic succession occurs on the outer delta and on 30% of the mid-delta. The assemblages concerned are *Equisetum*, *Carex*, *Salix–Equisetum*, and *Salix*. Species composition within the assemblages suggests that Raup's previously stated observations apply to those portions of the delta under the influence of allogenic succession. On the outer and mid-delta the wide variation of species composition within a single plant assemblage is most evident in the herb layer and to a lesser extent in the low shrub layer. The frequent flooding in large areas of the delta means that the landforms have a shorter period of stability than the life span of most of the perennial plants that make up the deltaic vegetation. Thus, plant succession within the affected landforms appears fragmented.

A factor that appears to be partially responsible for the fragmented distribution of subdominant plants in the outer delta is the disruption of rooting systems by the annual formation of ice in the sediment. Where the sediment is saturated or is in shallow water (less than 25.0 cm) the temperature gradient from the sediment to the air is sufficient to freeze the sediment to a depth that includes all of the rooting systems of most plant species. The sediment freezes in a stratigraphic pattern: alternating layers of sediment and pure ice. The thickness of the individual ice and sediment layers range between 0.5 cm and 2.0 cm. Observations indicate that in those areas where the sediments are frozen, they remain so after the ice on the standing water over the sediment has thawed. The nature of the sediment-ice stratigraphy indicated that the melting of the ice layers may be delayed in the spring, thus inhibiting the development of emergent vegetation. Further, existing rooting systems may be damaged by the ice formation. These factors contribute to the fragmented distribution of subdominant plants on the outer delta.

Equisetum Assemblage

The *Equisetum* assemblage occupies the interlevee depressions of cleavage bar islands in the outer delta and the abandoned elevated channels and narrow stretches on point bars and sand bars along distributary channels in the mid-delta.

Species composition differences exist between those assemblages found in the interlevee depressions and those on the abandoned channels. Except for a reduced occurrence of aquatic plants the species composition of *Equisetum* assemblages growing on the submerged portions of sandbars resembled that found in the outer delta. *Equisetum fluviatile*, with a cover of 95%, dominates both of these habitats. This species is prominent along the littoral boundary of point bars, usually occupying a narrow strip no wider than 2.0 m. On both point bars and sandbars the environmental gradient from the littorial zone (aquatic) to the terrestrial zone (dry) is sharp, and the transition from *Equisetum* to *Salix-Equisetum* to *Salix-Alnus* occurs over a very short distance. The close proximity of the structurally taller *Salix* spp. in the *Salix-Equisetum* assemblage affects the species composition of the *Equisetum* assemblage. The shade cast by the willow results in the exclusion of shade-intolerant *Equisetum fluviatile* and the successful growth of shade-tolerant *E. palustre* in the semi-aquatic zone closest to the *Salix* assemblage. This species gradient also occurs on sand bars and some abandoned channels.

Because of annual flooding in the outer delta the species successfully occupying the *Equisetum* assemblages are either aquatic (for example,

Rorippa islandica, Lemna minor, Hippuris vulgaris, Utricularia vulgaris) or have an adventitious rooting system (for example, *Equisetum fluviatile, Salix interior*). The species composition of the *Equisetum* assemblages on the elevated, abandoned channels in the mid-delta reflect an environment not exposed to annual flooding and sedimentation. *Carex rostrata* cover is more significant in these assemblages along the abandoned channels than on the outer delta. Thus it appears that the *Equisetum* assemblages will succeed to either the *Equisetum-Carex* or *Carex* assemblage. On the outer delta and point bar and sandbar sites on the mid-delta, succession of the *Equisetum* assemblage is directly related to the depth of water. Continued deposition of sediment on the cleavage bar formations, point bars, or sandbar sites will elevate the *Equisetum* beds such that *Salix* spp. can invade. With time these assemblages will succeed to the *Salix-Equisetum* assemblage. In the calm, clear water of the interlevee depressions on the outer delta, *Salix glauca, S. interior,* and *S. arbusculoides* grow in up to 35.0 cm of water. Although their cover is less than 5%, in water depths of less than 20.0 cm their vigour is good. The semi-aquatic nature of these *Salix* spp. may be attributable to respiratory transport of gases by aerenchyma cells on the stems (Hutchinson 1975). Similar aquatic invasion of *Equisetum* assemblages by *Salix* spp. does not occur along the active channels where suspended sediment concentrations are high. Possible reasons for this include reduced sunlight because of high turbidity and an unstable, shifting sediment bed in which the willow root.

Succession of the *Equisetum* assemblage on the outer delta and active channels of the mid-delta is different from that observed in the *Equisetum* assemblages along the abandoned channels. The reason for this successional divergence appears to be partly a function of nutrient supply and demand and the availability of direct sunlight within the different environments.

Due to the absence of tall shrubs or trees the interlevee depressions are open to incoming solar radiation. By contrast the abandoned channels are comparatively narrow (5.0 to 25.0 m wide), and the levees usually support *Alnus-Salix* or *Alnus* assemblages, with shrubs over 4.0 m high. As such, direct sunlight on the surface of the channel is reduced, especially along the shallower channel sides. Of the ten abandoned channels investigated, only two have *Salix* sp. present in the channel shallows; their areal coverage is less than 3% and their vigour is poor. *Carex rostrata* grow more abundantly and vigorously in this environment than on the outer delta or active channels of the mid-delta. As well, *C. rostrata* is a successful subdominant species in the shallow portions of abandoned channels and receives a significant number of hours of direct sunlight per day. The insignificant growth of *C. rostrata* in the interlevee depressions may be due to a combination of the substrate's pH and the calcium demand on this species. The reason for this is as follows. The aquatic nature of the interlevee depressions appears not to be a restricting factor, at least in the shallower perimeter of the landform. Hutchinson (1975) reports that *C. rostrata* may be as aquatic as *Equisetum fluviatile,* the dominant plant in the interlevee depressions. The substrate of the interlevee depressions is composed largely of decaying *E. fluviatile* and sediment. Because *E. fluviatile* actively accumulates potassium and sodium and not calcium or magnesium (Shepherd and Bowling 1973), the amount of Ca in the substrate will be low. Samples taken from several interlevee depressions had Ca concentrations ranging from 2.0 to 20.0 ppm and a pH ranging from 8.0 to 8.3. At these pH levels, *C. rostrata* requires Ca concentrations of 35.0 to 50.0 ppm to support vigorous growth (Lohammar 1938, reported in Hutchinson 1975). Generally speaking, the higher the pH, the higher the Ca demand of *C. rostrata.* The low Ca concentration in the substrate of the interlevee depressions appears to be one reason for the absence of *C. rostrata.* The *C. rostrata* in the abandoned channels of the mid-delta must obtain Ca from the alder and willow leaves that accumulate in the channels each fall. In some of the abandoned channels the standing water, derived from snowmelt runoff, rainfall, and occasional flooding is brown in colour. This indicates the addition of organic acids from decaying vegetation. In some *Equisetum* assemblages this addition may lower the pH sufficiently to reduce the Ca requirements of *C. rostrata,* allowing for vigorous growth in these habitats.

Carex Assemblage

The *Carex* assemblage is relatively unimportant on the Slave Delta. The largest associations are found in the protected zone of the outer delta, on a small bird's foot delta. Smaller associations are found in thin strips along the littoral zones of most channels within 0.5 km of Great Slave Lake, in small associations on the distal portion of the exposed outer delta (it is a pioneer species on some small shoals elevated above water level), and along some of the elevated abandoned channels in the mid-delta and apex zones. The occurrence of this assemblage does not appear to be governed by any specific set of environmental factors, as it occurs in a variety of habitats subject to a wide range of environmental conditions.

Carex aquatilis is the dominant species in the assemblages of the outer delta, while *C. rostrata* dominates the *Carex* assemblages in the older portions of the delta. *Equisetum fluviatile* and *Typha latifolia* occur in most *Carex* assemblages; the latter species is present only in those associations that have direct contact with the river or lake.

In the *Carex* assemblages in the outer, low-lying delta, *Salix interior, S. lasiandra,* and *S. arbusculoides* are present in significant numbers and have excellent vigour, indicating a successional trend towards a *Salix* assemblage. The *Carex* assemblages found on the mid-delta and apex do not appear to succeed any other assemblage. *Salix* spp. found in these locations are low in numbers and show no signs of reproducing.

Salix-Equisetum Assemblage

The *Salix-Equisetum* assemblage occurs in a wide range of habitats. Along the outer delta this assemblage occupies cleavage bar levees and the shallow perimeters of interlevee depressions. In the mid-delta, portions of interlevee depressions of older cleavage bar formations have been elevated through sedimentation and provide habitat conducive to the establishment and success of the *Salix-Equisetum* assemblage.

This assemblage is transitional between the *Equisetum* and *Salix* assemblages. As the willow canopy becomes more dense, the shade-intolerant, subdominant *Equisetum fluviatile* will die out. On the littoral portions of levees on the outer delta, *E. palustre* may replace *E. fluviatile* as the subdominant herb, as the former species is more shade tolerant. With further sedimentation and elevation of the levees to at least 0.3 m above summer low-water levels, *E. arvense* replaces *E. palustre*. However, in these situations, *E. arvense* always has poor vigour.

The successional trend in the maturing interlevee depressions appears to be somewhat different. The dominant shrub, *Salix arbusculoides,* grows in small clumps thus enabling *Equisetum fluviatile* and other herbs to grow in the open spaces. The topography of the maturing interlevee depressions is varied, ranging from swales to sloughs. The vegetation reflects landform patterns, with *Salix arbusculoides* occupying the swales and *Equisetum fluviatile, Calamgrostis canadensis, Potentilla palustris,* and *Glyceria maxima grandis* inhabiting the level, saturated plain of the depression in a pattern that is best described as fragmented. The sloughs are shallow and occupied by *Equisetum fluviatile* and a few aquatic species such as *Utricularia vulgaris* and *Lemna minor*. Although there are no *Alnus tenuifolia* present in the sampled plots, the species is beginning to invade the perimeters of these elevated interlevee depressions, where soil moisture conditions are dryer and warmer. Successional direction on this landform depends on the frequency of flooding. Flooding in this portion of the delta, the mid-delta, is infrequent. The likely progression of plant assemblages occupying this landform is from a *Salix-Equisetum* to *Salix* to a prolonged occurrence of *Alnus-Salix*. Infrequent flooding of this landform will probably result in a prolonged dominance of the *Salix-Equisetum* or *Salix* assemblage. The saturated condition of the sediment in the maturing interlevee depressions is maintained by the high water table, snowmelt runoff, and rainfall. Without frequent flooding and the accompanying sedimentation that increases the elevation of this landform, succession to the *Alnus-Salix* assemblage will not occur. Lowering of the water table either naturally, or as a result of manipulation of the Slave River's natural flow regime by man, could result in the invasion of these maturing interlevee depressions by *Alnus tenuifolia*.

Salix Assemblage

The *Salix* assemblage is found along elevated levees in the distal portions of the delta. In the mid-delta, *Salix* assemblages occupy elevated interlevee depressions and elevated portions of point bars and sand bars. The *Salix* assemblage habitats are considerably dryer than the *Equisetum, Carex,* or *Salix-Equisetum* habitats. The sampled sites in the *Salix* stands have 35% soil-moisture content by weight, compared with saturation conditions or standing water found in the other assemblages.

Salix interior and *S. arbusculoides* dominate the *Salix* assemblages in the frequently flooded areas of the delta, while *S. planifolia* is dominant in the more elevated, less frequently flooded assemblages. *Alnus tenuifolia* was recorded in small numbers but with vigorous growth in all sampled plots, indicating a successional trend from *Salix* to *Salix-Alnus* assemblage. Gill (1975a) reports a similar successional sequence in the Mackenzie Delta. Small numbers of healthy *Populus tremuloides* occur in some of the sampled plots in the mid-delta, indicating that the successional sequence of *Salix* to *Salix-Alnus* may be short-lived. Cordes and Strong (1976) report that the *Salix* community in the Peace-Athabasca Delta usually succeeds to a *Populus* community.

Autogenic Succession

The assemblages discussed in this section occupy mainly the older delta landforms: the apex zone and the islands of the mid-delta. The successional sequence of the plant assemblages inhabiting the older, more elevated landforms of the delta is: *Alnus-Salix, Alnus, Populus,* decadent

Populus, and *Picea. Picea* is the climax stage of plant assemblage succession on the rarely flooded portions of the delta. Autogenic succession occurs in these assemblages because of the reduced frequency of spring flooding. It is mainly influenced by vegetation structure and composition, rather than by physical factors. With increasing elevation above river level resulting in a reduced frequency of spring flooding, autogenic succession becomes more pronounced.

Alnus-Salix Assemblage

Along the outer delta, *Alnus-Salix* assemblages are found only on the most elevated upstream portions of cleavage bar islands. This assemblage is widespread in the mid-delta, commonly occupying the backslopes of active levees. Levees along abandoned channels, abandoned point bars, backswamps behind cut-bank levees, and upstream portions of islands are among the varied habitat that the *Alnus-Salix* assemblages occupy in the mid-delta.

Although the sample plots are located on diverse habitats, they have a similar species composition. Even the herb layer shows a degree of uniformity that is not present in earlier successional stages. This is directly related to the decreasing influence of the river regime on this assemblage. *Alnus tenuifolia,* with a cover of 75%, is the dominant tall shrub, while *Cornus stolonifera,* with a cover of 45%, is dominant in the low-shrub layer. The herb layer is largely composed of *Equisetum arvense,* which has an average cover of 45%.

Succession from the *Alnus-Salix* assemblage appears to be directed along two paths: toward either a *Populus* assemblage or an *Alnus* assemblage. In *Alnus-Salix* assemblage plots adjacent to either *Populus* or decadent *Populus* assemblages, successional direction is toward a *Populus* assemblage. Where the *Alnus-Salix* assemblage plots are not adjacent to the *Populus* or decadent *Populus* assemblages, the *Alnus-Salix* assemblage will succeed to an *Alnus* assemblage.

Alnus Assemblage

Most *Alnus* assemblages are found on levees along abandoned channels and on large expanses in the interior of the more-elevated portions of mid-delta islands.

The dominant plant species in the shrub, low-shrub, and herb layers is identical to that found in the *Alnus-Salix* assemblage. The only difference is the absence of *Salix* spp. and the increased significance of *Alnus tenuifolia* (from 75 to 85%). Environmentally the two assemblages are similar in elevation above river level, soil moisture, and temperature. The observed differences are soil composition, exposure to wind, litter fall, tall-shrub competition, and total organic carbon content of the soil. The *Alnus-Salix* assemblage occupies more exposed sites, where litter accumulating on the ground surface may be subject to removal by strong onshore winds from Great Slave Lake. *Alnus* sites are sheltered from the wind. Thus the removal process is not as efficient. Soil profiles in the *Alnus* sites have an average litter depth of 2.0 cm, while the litter layer in the *Alnus-Salix* assemblage sites averages 0.5 cm.

The sandy loam soil of the *Alnus* stands has a total organic carbon content almost four times as great (4.8%) as the *Alnus-Salix* sites (1.3%).

Samples of *Alnus tenuifolia* from the *Alnus* and *Alnus-Salix* assemblages were compared to determine whether the assemblage differences or similarities (see above) were reflected in the growth rate of this shrub. It was hypothesized that if the assemblage differences were significant, the growth rates would be statistically dissimilar. Alternatively, if the similarities were significant, the growth rates would be statistically similar.

The *Alnus* from both assemblages are of a similar age. The *Alnus tenuifolia* in the *Alnus* assemblage have a mean age of 21.91 years ($s = 6.27$, $n = 20$); those in the *Alnus-Salix* assemblage have a mean age of 21.57 years ($s = 4.69$, $n = 20$). A student-t test (Freund 1972) performed on these two samples demonstrated no significant difference at the 99% confidence interval. Although statistically the sample can be considered to be part of the same age class, there are significant differences in growth rates between the two samples. For this comparison, shrub height and DBH (diameter at breast height) were used as indicators of growth and the student-t test performed accordingly. The heights of *A. tenuifolia* in the *Alnus* assemblage average 6.33 m ($s = 1.16$, $n = 20$), while those in the *Alnus-Salix* stands have a mean height of 4.11 m ($s = 0.58$, $n = 20$). DBH measurements in the *Alnus* stands average 6.85 cm ($s = 1.54$, $n = 20$), while those in the *Alnus-Salix* assemblage have a mean value of 4.62 cm ($s = 0.74$, $n = 20$). It seems significant that the standard deviations for height and DBH in the *Alnus-Salix* are half those of the *Alnus* assemblage. This may reflect unconformities in the environment of the *Alnus* assemblages and environmental similarities among *Alnus-Salix* assemblages. Because the growth rates are significantly dissimilar (student-t, 99%), it seems probable that the environmental and ecological differences between the assemblages, namely soil composition—both particle size and chemistry (due to the litter accumulation differences), are the factors primarily responsible for the increased growth rates in the *Alnus* assemblage. Other physical factors examined were eliminated as possible contributors to differential growth rates. There was no measured difference in soil moisture, soil temperature, or wind exposure between the *Alnus* and *Alnus-Salix* assemblage plots.

Successional direction is only obvious in a small percentage of the *Alnus* plots. Only the plots occupying the most-elevated sites appear to be succeeding toward a *Populus* assemblage. Successional trends within the *Alnus* assemblage occupying low-lying sites is not apparent. With increased elevation of the sites by sedimentation during occasional flooding, soil conditions may permit the invasion of *Populus balsamifera*.

Populus Assemblage

With few exceptions, the *Populus* assemblage occupies the mesic environment of elevated levees, particularly cut-bank levees. Some *Populus* stands also occur on older levees along ponded sloughs.

Populus balsamifera, the dominant species in this assemblage, is an important species on other northern deltas. *Populus* communities occupy the high, well-drained levees of the Saskatchewan Delta (Dirschl 1970), the Peace-Athabasca Delta (Dirschl et al. 1974), and the Colville Delta in Alaska (Bliss and Cantlon 1957). In the Mackenzie Delta, well-developed *Populus* stands occur only on coarse point-bar deposits (Gill 1972, 1975a).

In this assemblage the tree layer is composed solely of *Populus balsamifera*, with an average cover abundance of 85%. The tall- and low-shrub layers are dominated by *Alnus tenuifolia* and *Cornus stolonifera* respectively. The average cover for each is 12%.

The *Populus* assemblages sampled in the mid-delta zone do not appear to be succeeding to the *Picea* assemblage as they are on the apex zone. The primary reason for this is the inability of white spruce to produce adventitious rooting during the first few years of its life (Gill 1971). Therefore in an area such as the mid-delta, which experiences frequent flooding (every 5 to 7 years), the white spruce cannot successfully germinate, and succession to a *Picea* assemblage will not occur. Consequently succession from the *Populus* assemblage on the mid-delta is directed toward a decadent *Populus* assemblage.

Decadent Populus Assemblage

This assemblage usually occupies levees along channels that have long been cut off from active distributaries in the mid-delta. The decadent *Populus* assemblage is the edaphic climax forest of the mid-delta.

The decadent *Populus* assemblage on the Slave Delta differs slightly from those reported by Gill (1971) on the Mackenzie Delta. The *Populus* stands in the Mackenzie Delta become "decadent" when flooded during lateral migration of point bars and occupation in meander depressions. Those stands on the Slave Delta would more correctly be termed stagnant or subclimax (Kershaw 1973), as the natural succession has been terminated by frequent flooding, not by geomorphic change. The term "decadent" has been used for the Slave Delta because the stands are visibly deteriorating.

Picea Assemblage

The distribution and ecology of *Picea glauca* forests in northern alluvial habitats is well documented (Jeffrey 1961, Dirschl 1972, Dirschl et al. 1974, Gill 1975a, Cordes and Strong 1976).

Picea glauca clearly dominates the *Picea* assemblages, with an average cover of 95%. In decreasing order of significance the shrub layer consists of *Rosa acicularis, Alnus tenuifolia, Cornus stolonifera,* and *Vebernum edule. Equisetum arvense* dominates the herb layer, with *Pyrola secunda secunda* and *Linnaea borealis* as major subdominant species.

Since flooding is rare on the expanses of *Picea* forest in the apex zone, the bryophyte layer has become extremely well developed. *Hylocomnium splendens* and *Aulacomnium palustre* co-dominate in this layer. The very efficient insulating capacity of the thick (up to 0.4 m) moss carpet in this assemblage results in the soil temperatures being maintained below 0 C below the active layer. The seasonal active layer in the permafrost is a function of bryophyte thickness and proximity to the stems of *Picea glauca* (Gill 1975b). Similar observations were made on the Slave River Delta; the active layer under a bryophyte cover of 6.0 cm ($n = 25$) averaged 45.0 cm ($s = 7.0$ cm), while the active layer beneath a moss layer of 16.0 cm ($n = 25$) averaged 29.0 cm ($s = 8.0$ cm).

Potential Environmental and Ecological Impacts of River Impoundment on the Slave Delta

Generally speaking the disruption of a river system, which causes changes both in its normal discharge and in the concentration of suspended sediment and bedload, will alter the biophysical regime of downstream wetlands. This is because the wetland vegetation seres are a direct response to the frequent inundation by floodwaters and the deposition of sediment. The degree of downstream impact is a function of reservoir size, the filling schedule of the reservoir, and the hydroelectric-plant operating strategy. The short-term effects are a result of the length of time it takes to fill the reservoir and the season(s) in which this is done. The long-term effects may be initiated by the reservoir filling schedule, but will largely be a product of the retention of river sediment in the reservoir and the degree to which the natural discharge pattern of the river has been altered.

Short-Term Implications

Calculation of reservoir volume was accomplished using a head of 209.1 m asl. This was derived from a report that examined various versions of hydroelectric dams that could be built on the Slave River (Montreal Engineering 1978). The version used in the volume calculation is the most economical of those proposed in terms of power production per dollar invested. Fig. 13 illustrates the number of days required to fill the reservoir, given different conditions of closure (100%, 75% 50%, 25%). Calculations are made on a monthly basis: the volume of water contributed to the reservoir (calculated as $7.308756 \times 10^9 m^3$) in each month as a product of that month's mean discharge ($n = 23$ years). Employing these calculations it is possible to foresee and discuss some of the downstream environmental and ecological implications during the initial impoundment period. Fig. 14 summarizes some of these implications in a subjective manner.

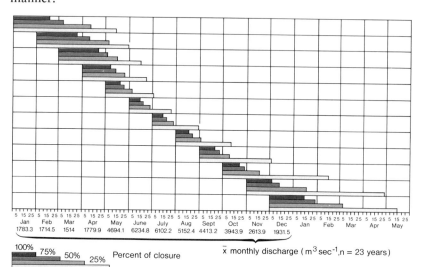

Fig. 13. Initial impoundment period required to fill the Slave River reservoir (with head 209.1 m asl) given certain conditions of closure beginning at the first of each month.

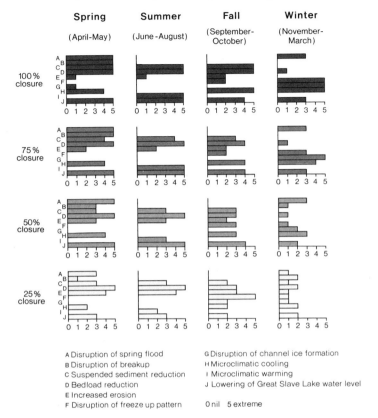

Fig. 14. Initial impoundment period and possible environmental consequences on the Slave River Delta, Northwest Territories.

A complete closure of the river during the spring period would be most disruptive to the downstream delta. The spring flood and the sediment load it carries, if not stopped completely, would at least be significantly reduced. As the spring flood is the primary initiator of ice breakup in the deltaic distributaries, the date of breakup would be delayed. This delay would affect the microclimate of the delta and therefore the phenology of its vegetation. This reduction of the length of the growing season would be most pronounced in the outer delta. The productivity of the *Equisetum* marshes occupying the interlevee depressions would decline significantly in the ensuing summer, as the annual deposit of sediment, from which these emergent plants absorb nutrients, would not occur.

By employing an equation formulated to predict Great Slave Lake water levels from the discharge of the Slave River:

$$y = .82 + .79 \times 10^{-4}x$$

where y = Great Slave Lake water level (m)
x = discharge (m³ sec⁻¹)

(English 1979), the effect of reduced river discharge on the level of Great Slave Lake can be predicted given the hydroelectric-plant operating strategy. For example, if the river discharge is completely shut off for 13.5 days in June (fig. 13) and allowed to discharge at the calculated mean monthly rate of 6234.8 m³ sec⁻¹ for the remainder of the month, the level of Great Slave Lake would be reduced 0.18 m below the August 1977 summer low-water level.

Impacts on the delta resulting from the initial impoundment period may have significant short-term effects. The seriousness of these impacts, as illustrated in fig. 14, will largely be a function of both the season during which the initial filling of the impoundment takes place and the volume of water allowed past the impoundment during this period.

Long-Term Environmental and Ecological Implications

Regulation by the dam will curtail the natural extremes of discharge in the Slave River. The ecology of the delta has evolved around these extremes.

The spring flood will become a product of the hydroelectric-plant operating stategy. The depletion of water in the reservoir due to winter energy demand will result in a portion of the spring flood being used to refill the reservoir. The reduction of high spring discharge levels will end the annual flooding that a large portion of the outer delta experiences. The intermittent flooding of the mid-delta and the rare floods inundating the apex will no longer occur.

One of the most important and immediate effects of river impoundment will be the loss of bedload and the coarse fraction of the suspended sediment that is trapped or settles out in the reservoir. The deposition of sediment in hydroelectric reservoirs can range from 95 to 100% of the available load (Day 1971). The sediments currently deposited on the Slave Delta range mostly between 2.00 ∅ and 4.25 ∅ (fine sands to coarse silts). Fine silts and clay represent less than 4% of the deposited sediment. This means that the larger sediment fractions, which are responsible for the continued progradation and aggradation of the delta, will be retained in the reservoir. The formation of the submarine sedimentary platform, submarine levees, wavebuilt shoals, and other subaerial landforms depends upon the continued supply of this sediment load; its loss would largely curtail the growth of the delta.

The increased erosive capacity of the water discharged from the dam would enable the Slave River to regain some sediment between the dam and the delta. However, the amount would be much less than the average daily sediment transport of approximately 60,000 tonnes currently transported by the Slave River above the proposed dam site at Fort Fitzgerald (Water Survey of Canada 1977).

Maddock (1976) states that elimination of the spring flood decreases the width and increases the slope of a floodplain channel, thus providing ideal conditions for the invasion of riparian vegetation along the littoral zone of the old channel. On the other hand, he states that under some conditions the elimination of sediment transported downstream promotes bank erosion, thereby increasing the width of the channel and reducing its slope. Rains (1978) adds that the channel slope immediately below the dam would be reduced. As the load contributed from this source increases downstream from the dam the erosional effects would become less significant. The zone of relatively minor erosion would include the delta area until a post-dam equilibrium is reached. The rather low frequency of flooding that a large portion of the delta currently experiences suggests that after the impoundment of the Slave River, a general widening of the channels and reduction in slope would occur. This may result in a greater erosion of levees and the elimination of much emergent littoral vegetation. Sand bar and point bar development will be retarded as the bedload is greatly reduced.

Sprague (1972) and Gill (1973, 1974) documented the warming effect of spring breakup on the bioclimate of the Mackenzie Delta. Reduction of the spring flood levels and sediment loss would delay breakup in the Slave Delta, prolonging the length of winter conditions and effectively reducing productivity.

The lowering of water levels and reduction of sediment deposition in the interlevee depressions would allow vegetation of later successional stages to invade and displace the highly productive emergent plants. *Salix* spp. would rapidly invade these marshes and displace the now-dominant, shade-intolerant *Equisetum fluviatile*. The existing successional sequence indicates that *E. palustre* would replace *E. fluviatile*. *E. palustre* would not maintain the large herbivore population currently supported by *E. fluviatile* because the former plant manufacturers two poisonous alkaloids, palustrin and palustridin, both of which would discourage consumption of the plant (Hutchinson 1975).

The elimination of flooding will allow white spruce to successfully germinate on the mid-delta, and eventually, portions of the outer delta. Invasion of bryophyte species such as *Hylocomnium splendens* will occur, and soil temperatures will drop. Organic decomposition in the soil will be reduced, the pH of the soil will become slightly acidic, and the active layer of the permafrost will restrict root growth. After a period of time, the mid-delta will resemble the present vegetation composition of the apex zone. The large-scale reduction of plant production and species change in the outer delta will have severe economic repercussions on the village of Fort Resolution, due to a reduction in economically important wildlife populations.

Acknowledgments

Special thanks are due to Ross Hastings for his outstanding contribution to the fieldwork. Jim English and Ross Hastings edited and made useful comments on this manuscript. H. Beaulieu, K. Bodden, and S. Schiff helped with the fieldwork. R. Hastings and K. England identified the vascular plants and bryophytes.

A special note of acknowledgment is due to the late Don Gill, whose tutelage and encouragement significantly aided all stages of this research. Financial assistance was provided by the Boreal Institute for Northern Studies at the University of Alberta, through a grant-in-aid for research to the author; the Inland Waters Directorate of Environment Canada, through a special projects grant to Don Gill; and the National Research Council of Canada, through a research grant (number 8700), also awarded to Don Gill.

References

Armstrong, E. L. 1973. Dam construction and the environment. Question 40 −Response 46. pp. 217–236. *In* Commission Internationale des Grands Barrages, Madrid.

Axelsson, V. 1967. The Laitaurre Delta: A study of deltaic morphology and processes. Geogr. Ann. 49-A: 1–127.

Bates, C. C. 1953. Rational theory of delta formation. Amer. Assoc. Petrol. Geol. Bull. 9: 2119–2162.

Berger, T. R. 1977. Northern frontier, northern homeland: The report of the Mackenzie Valley Pipeline Inquiry. Vol. 2: Terms and conditions. Supply and Services Can. 268 pp.

Bliss, L. C., and J. E. Cantlon. 1957. Succession on river alluvium in northern Alaska. Amer. Midland Natur. 58: 452–469.

Bodden, K. 1981. The economic use by native peoples of the resources of the Slave River Delta. M.A. thesis. Dept. Geogr., Univ. Alberta, Edmonton. 177 pp.

Brown, R. J. E. 1960. The distribution of permafrost and its relation to air temperature in Canada and the U.S.S.R. Arctic 13(3): 163–177.

Bryson, R. A., W. M. Wendland, J. D. Ives, and J. T. Andrews. 1969. Radiocarbon isochrones on the disintegration of the Laurentide ice sheet. Arctic and Alpline Res. 1: 1–14.

Cameron, A. E. 1922. Post-glacial lakes in the Mackenzie River basin, Northwest Territories, Canada. J. Geol. 30: 337–353.

Cordes, L. D., and W. L. Strong. 1976. Vegetation change in the Peace Athabasca Delta: 1974–75. Rep. to Dep. Env. 192 pp.

Dahlskog, S., A. Damberg, P. Harden, and L. E. Liljelund. 1972. The Kvikkjokk Delta: A progress report on a multidisciplinary research project on a boreal mountain-lake delta. Dep. Phys. Geogr., Univ. Uppsala, Sweden. 78 pp.

Day, J. C. 1971. The effects of dams on stream channel morphology and vegetation. Unpub. paper presented at the Can. Assoc. of Geogr. Meeting, Univ. Waterloo. 10 pp.

Dirschl, H. J. 1970. Ecology of the vegetation of the Saskatchewan River Delta. Ph.D. thesis. Univ. Saskatchewan, Saskatoon. 215 pp.

―――. 1972. Evaluation of ecological effects of recent low water levels in the Peace-Athabasca Delta. Can. Wildl. Serv. Occ. Pap. Ser. No. 13. 27 pp.

―――, D. L. Dabbs, and G. C. Gentle. 1974. Landscape classification and plant successional trends: Peace Athabasca Delta. Can. Wildl. Serv. Rep. Ser. No. 30. 34 pp.

English, M. C. 1979. Some aspects of the ecology and environment of the Slave River Delta, N.W.T. and some implications of upstream impoundment. M.Sc. thesis. Dept. Geogr., Univ. Alberta, Edmonton. 249 pp.

Fraser, J. C. 1972. Effects of river uses: Regulated discharge and the stream environment. pp. 263–286. In R. T. Oglesby, C. A. Carson, and J. A. McCann, eds. River ecology and man. Academic Press, New York. 465 pp.

Freund, J. E. 1972. Modern elementary statistics. Prentice-Hall Inc., Englewood Cliffs, New Jersey. 432 pp.

Geddes, Frank. 1981. Productivity and habitat selection of muskrats in the Slave River Delta. Mackenzie River Basin Study Report. Supplement 6. 41 pp.

Gill, D. 1971. Vegetation and environment in the Mackenzie River Delta: A study in subarctic ecology. Ph.D. thesis. Dept. Geogr., Univ. British Columbia, Vancouver. 694 pp.

―――. 1972. The point bar environment in the Mackenzie River Delta. Can. J. Earth Sci. 9(11): 1382–1393.

―――. 1973. Modification of northern alluvial habitats by river development. Can. Geogr. 17 (2): 138–153.

———. 1974. Significance of spring break-up to the bioclimate of the Mackenzie River Delta. pp. 585–588. *In* J. C. Reed and J. E. Sater, eds. The coast and shelf of the Beaufort Sea. Arctic Inst. of North Amer. 750 pp.

———. 1975a. Floristics of a plant succession sequence in the Mackenzie Delta, Northwest Territories. Polarforschung 43(1/2): 55–65.

———. 1975b. Influence of white spruce trees on permafrost-table microtopography, Mackenzie River Delta. Can. J. Earth Sci. 12(2): 263–272.

———. 1978. Some ecological and human consequences of hydro-electric projects in the Mackenzie River Delta drainage system, northwestern Canada. pp. 73–88. *In* L. Muller-Wille, P. J. Pelto, Li. Muller-Wille, and R. Darnell, eds. Consequences of economic change in circumpolar regions. Boreal Institute for Northern Studies, Occ. Pub. No. 14. 269 pp.

———, and A. D. Cooke. 1974. Controversies over hydroelectric developments in Sub-Arctic Canada. Polar Rec. 17(107): 109–127.

Glooschenko, V. 1972. The James Bay power proposal. Nature Can. 1(1): 5–10.

Hutchinson, G. E. 1975. A treatise on limnology. Volume III: Limnological botany. John Wiley and Sons, Toronto. 660 pp.

Jeffrey, W. W. 1961. Origin and structure of some white spruce stands on the lower Peace River. Can. Dept. Forestry, Forest Res. Branch, Tech. Note No. 103. 20 pp.

Kellerhals, R., and D. Gill. 1973. Observed and potential downstream effects of large storage projects of northern Canada. Question 40 — Response 46. pp. 731–754. *In* Commission Internationale des Grands Barrages, Madrid.

Kerr, J. A. 1973. Physical consequences of human interference with rivers. pp. 1–31. *In* Ninth Canadian Hydrology Symposium, Fluvial Processes and Sedimentation. Univ. Alberta, Edmonton.

Kershaw, K. A. 1973. Quantitative and dynamic plant ecology. Elsevier Pub., New York. 308 pp.

Maddock, T., Jr. 1976. A primer on floodplain dynamics. J. Soil and Water Conserv. 31(2): 44–47.

Montreal Engineering. 1978. Mountain Rapids hydroelectric project: An assessment for Calgary Power. 107 pp.

Peace-Athabasca Delta Project Group. 1972. The Peace-Athabasca Delta: A Canadian resource. A report on low water levels in Lake Athabasca and their effects on the Peace-Athabasca Delta. Summary report. Prepared jointly by the Environmental Ministers of Canada, Alberta and Saskatchewan. Env. Can., Ottawa. 144 pp.

Prest, V. K. 1969. Geological Survey of Canada Map. 1257A.

Rains, R. B. 1978. Associate Professor, Dept. Geogr., Univ. Alberta. Personal communication.

Raup, H. M. 1975. Species versatility in shore habitats. Arnold Arboretum 56: 126–163.

Reinelt, E. R., R. Kellerhals, M. A. Molot, W. M. Schultz, and W. E. Stevens. 1971. Implications, findings and recommendations. Proceedings of the Peace-Athabasca Delta Symposium. Water Resources Centre, Univ. Alberta, Edmonton. 359 pp.

Richardson, B. 1972. James Bay: The plot to drown the north woods. Clarke, Irwin and Co. Ltd., Toronto. 190 pp.

Shepherd, R. H., and D. J. F. Bowling. 1973. Active accumulation of sodium by roots of five aquatic species. New Phytol. 72: 1075–1080.

Sprague, J. B. 1972. Aquatic resources in the Canadian north: Knowledge, dangers and research needs. National Workshop on People, Resources and the Environment North of '60. Can. Arctic Resources Committee. 28 pp.

Taylor, C. H., G. J. Young, B. J. Grey, and A. F. Penn. 1972. Effects of the James Bay development scheme on flow and channel characteristics of rivers in the area. Dept. Geogr., McGill Univ., Montreal. 53 pp.

Thakur, T. K., and D. K. MacKay. 1973. Delta processes. pp. 1–21. *In* Ninth Canadian Hydrology Symposium, Fluvial Processes and Sedimentation. Univ. Alberta, Edmonton.

Water Survey of Canada. 1977. Discharge values for the Slave River.

Land Use

Planning for Land Use in the Northwest Territories

Norman M. Simmons
John Donihee
Hugh Monaghan

Abstract

There is currently no integrated public process for the management of resources north of Canada's 60th parallel. Commitments to resource use are rapidly diminishing the management options, with little influence by northern residents. In 1981 the federal Department of Indian Affairs and Northern Development, the Crown land manager in the Yukon and Northwest Territories, announced a new policy that made integrated land-use planning the precursor to land-use decisions. A short while later they released a discussion paper describing their northern land-use planning process and the free exchange of information essential to planning. In 1982 a revised draft of the discussion paper was released to the public. In both drafts the control of the proposed management structures would be highly centralized in Ottawa, and tentacles of federal control would reach into the community planning organizations. We propose principles of public participation in land-use policy setting and interpretation in the Northwest Territories, public access to the planning process, and a free exchange of information essential to planning. The Minister of Indian Affairs would still be the acknowledged land manager, but northern residents would have strong influence on land-use decisions through decentralized management structures. Native political organizations would be involved in policy interpretation in the territories, in recognition of their unique role in northern politics. Our system is far from ideal, but it is a workable interim measure that will gain public support—support that has been absent from two previous attempts to plan for land use in the North.

Introduction

> "The concept of the last frontier is no longer a play on words. We now recognize that the north is a region of the country which we have the opportunity to develop in special ways; we recognize that if it is developed carefully and wisely it could play a powerful role in the development of our society; we recognize that it could greatly alter our dependency on the culture, the markets, and the technology of other countries...." (CARC 1973a: 1)

This voice from the past decade has been echoed repeatedly (Berger 1977: 31, Rees 1978: 42), yet it remains true to this day. The critical relationship between land-use planning and resource management has not been acknowledged by the federal government until recently. Decisions that effectively diminish our options to use the land and its resources continue to be made in a reactionary manner with the imprimatur of the federal Department of Indian Affairs and Northern Development (DIAND), an agency with provincial powers in the North. These decisions are usually made without any significant collaboration of the residents of the North. This system has generated Lord Bryce's "'fatalism of the multitudes,' a sense of the insignificance of individual effort, the belief that the affairs of men are swayed by large forces whose movement may be studied but cannot be turned" (CARC 1976b: 5).

What is seriously lacking in the Northwest Territories is a system of integrated resource-management programs orchestrated through a public land-use planning process. The solution to this problem should be a long-range concern of the zoologists, geographers, botanists, and foresters who have contributed to this book. Their common goal is the conservation of northern ecosystems; their common problem is the opportunistic approach taken by government to major decisions to use northern lands and resources.

As this chapter is being written, a system of land-use planning is taking shape in the Ottawa offices of DIAND. Late in 1981 the federal Cabinet approved the DINA policy to establish land-use planning as a necessary preamble to land-management decisions on Crown lands in the Yukon

Planning for Land Use in the Northwest Territories 345

and Northwest Territories. In this chapter we suggest principles that should be honoured in any land-use planning process in the territories, and we propose structures through which the process could be implemented in a unique northern manner.

Synopsis of Development

For the past six years, the northern public has been dazzled by a kaleidoscopic variety of actual and proposed, minor and major, mineral and hydrocarbon, exploration and development projects that have affected the lifestyles and economy of residents of many of the communities in the Northwest Territories. The oil and gas production industries hold the greatest potential for the imposition of significant social and economic change in the territories. These industries will have both indirect and direct negative effects on the natural environment. The hydrocarbon industry has focused on a few potentially rich areas that, if developed, would attract large investments of manpower and money, and substantially alter the makeup of the northern economy. The most impressive example is in the Beaufort Sea area, where expenditures over the next two decades are forecast at $40 billion (Loken 1981). A workforce of more than 10,000 people may be required by 1990 (Dome Petroleum Ltd. et al. 1981) in an area now supporting a population of about 5,000 (1980).

Transportation systems developed to service the mining and hydrocarbon industries in the territories may have the greatest long-range impact on wildlife and habitat. They raise the potential for sharply increased access to critical habitat and wildlife population concentrations and for accident-related contamination of the environment. One pipeline in the Mackenzie Valley has already been approved, and two major pipelines traversing Arctic seas, tundra, and boreal forest are in the early planning stages. Daily ice-breaking tanker traffic from the United States and Canada through the Northwest Passage is a possibility in the 1900s (Dirschl 1980).

Since the federal "Roads to Resources" policy was inaugurated in the 1950s, the Mackenzie Highway to Hay River has been extended to Yellowknife, Fort Smith, Fort Resolution, and Fort Simpson. The Dempster Highway opened in 1980. The Liard Highway was completed in 1983. Several winter roads are opened annually in the western territories. Varying levels of research have been conducted on other road routes in the western Arctic and Subarctic (Donihee and Gray 1981).

Institutions and Problems

The complexity and pace of northern development is difficult for laymen and territorial government officials to understand. Comprehension is further frustrated by the obscure and constantly shifting responsibilities and interests of the many government agencies that control or influence territorial development. The co-ordination of these agencies' activities, from the most routine tasks to the highest level of policy recommendations, are handled by nearly 100 interdepartmental committees and working groups. Examples are the Interdepartmental Steering Committee on Beaufort Sea Drilling, the Eastern Arctic Marine Environmental Study Management Committee, the Interdepartmental Environmental Review Committee, the Senior Steering Committee on Lancaster Sound, the Joint Industry/ Federal Government Steering Committee on Problems of Arctic Hydrocarbon Development, and so forth. A wide variety of other interdepartmental committees operates at lower levels of the bureaucracy to co-ordinate program implementation, collect and share information, and develop regulations. Abele and Dosman (1981) discovered at least 70 of these committees.

Although the committee approach may be efficient and flexible, it has serious weaknesses. There is no easy way for federal Cabinet members and federal and territorial legislators to obtain information on which to base urgent decisions on northern matters. An irregularly timed series of committee meetings to consider individual projects is not conducive to the development of generally understood policies for the North. There is no forum for introducing major policy alternatives that do not involve industrial development projects, since the committee structure tends to feed on project proposals from private industry. Finally, in spite of the industry bias, the absence of a public policy process often results in delays and cancellations of project proposals because of strong opposition from increasingly vocal citizens (Abele and Dosman 1981).

The frustration of public assessment and of efforts to influence northern development is assured by the complexity of the decision-making process. The diffusion of responsibility for developing land-management decisions over a myriad of committees makes public access to the decision-makers difficult. Lacking organized means of gaining information, the public remains in the dark and cannot comment on planning and policy development, even through their elected representatives. (For more on public participation in decision-making, see Gamble 1982.)

In the early 1970s managers and researchers from government and nongovernment agencies both in the North and the South warned of the dangers of *ad hoc,* unco-ordinated, mineral and petroleum development-biased, land-use decisions. At public hearings, conferences, meetings, and in publications, they urged the federal government to institute a system of land-use planning that would reflect the attitudes and aspirations of northern residents (CARC 1973*b*). J. K. Naysmith, then a senior DIAND administrator, defined the need for land-use planning in the North (Naysmith 1976: 109*ff*), and made specific suggestions for land-use policy development and implementation. The appeals for land-use planning in the North continued without success throughout the late 1970s and into the 1980s. The sense of public helplessness and apathy scarcely changed; major commitments to industry were made throughout the North, commitments that appeared to be made in selective secrecy prior to the involvement of Parliament or the Northwest Territories Legislative Assembly (CARC 1976*a*, 1976*b*, Hartman 1980, Keith et al. 1981).

Though the *ad hoc* approach to land-use decisions carried over from the 1960s into the 1980s, laudable changes in land-use management appeared in the late 1970s. Formalized assessment procedures were developed by DIAND and Environment Canada (DOE), the issuance of guidelines to industry became administrative procedure, formal inquiries into major land-use proposals were initiated, and the importance of public consultation and participation was officially recognized. However, none of these changes have been backed by legislation (Lucas and Peterson 1978); none of these activities constitute land-use planning.

Considering this bleak background, the release of a discussion paper entitled "Northern Land Use Planning" (see Appendix) by DIAND in August 1981 caused quite a stir in territorial government and private circles (DIAND 1981). In the early winter of 1981, land-use planning became Cabinet-approved DIAND policy. In late 1982 a revised version of the August 1981 document was released to the public (DIAND 1982). Although this document requires further analysis and interpretation before its guidelines can be followed, it may represent a significant breakthrough toward a public process of land-use planning.

Principles

Rees (1978) called the integrated public process of land-use planning "applied common sense." The territorial government could hardly do less than endorse DIAND's proposal in principle and then begin to work with

DIAND to translate the proposal into action. Others (Nelson and Jessen 1981, Rees 1978) protested that major changes to governments and government systems are required for proper response to hydrocarbon development plans for the North, and we agree. However, in view of the pace of developments in the North, we do not feel that it is a realistic strategy to brush aside the recent DIAND initiative and hold out for a more comprehensive overhaul of governments. What we propose as an interim measure is the establishment of principles to guide northern land-use planning, and then the prompt creation of a planning structure that best fits these principles, using the legislative and policy tools at hand. Meanwhile, the debate about major changes in governments can continue, shaped in part by the pressures of the national energy policy and native land claims. The interim planning system should adapt as changes occur. However, it must be activated promptly in order to address the already critical planning needs in areas such as the Beaufort Sea, the Mackenzie Valley, and the Mackenzie Mountains.

We view land-use planning as a continuous process that should enable the public to systematically review options for land use, and then select and refine alternatives that best reflect their long-range goals. Planners and administrators should be able to adjust their strategies continually so that optimum benefits can accrue to the public. The process should adhere to the following general principles:

1. Public participation
(a) The public, through informed elected representatives, must be allowed a voice in policy setting and policy interpretation for land-use planning.
(b) The public must have access to the land-use planning process through local and regional institutions in a manner that accommodates the northern style of decision by consensus (Lucas and Peterson 1978).
(c) Information on land-use issues, hazards, benefits, constraints, and alternatives must be readily available to the public (CARC 1976*a*, Stager 1978).
(d) The informed public should be given the opportunity to affect the planning process through their comments on land-use values and standards (Swanson 1971).
2. The land
(a) Policy makers and planners must recognize alternative land uses and optimal, rather than maximal, land use and resource use (Pepperell 1976).

(b) Maintenance of the land's productivity must be encouraged by policy (Naysmith 1976).

The first set of principles, dealing with public participation in land-use planning, seems to be the most difficult for governments to define, accept, and implement. Large sums of money have been spent and continue to be spent by governments and industry on "public consultation." There were even two unsuccessful territorial/federal efforts directed at land-use planning: the Mackenzie Delta and Lancaster Sound regional planning exercises. However, these efforts, as well as the consultation process and specific land-use regulation, have all foundered on the same rock: a determined commitment to a single-purpose development process, which suffers from a lack of informed participation and does not have the confidence of many northern people (Rees 1978). We have learned through our failures that citizens must not only have early and easy access to information, but must also *believe* that they have the power to influence decisions (Stager 1978).

Tools for Implementation

What the federal and territorial governments have in place are instruments of regulation, not planning. However, we believe that these management policies, processes, and structures can be used, with little modification, to adequately serve the land-use planning regime.

A description of the legal mandates for the management or regulation of land and resource use in the Northwest Territories is a lengthy, complex topic, worthy of separate treatment. This subject is addressed comprehensively by Monaghan (1980). For our purpose, it is sufficient to say that the Department of Indian Affairs and Northern Development holds many of the provincial-type resource management powers in the territories. The Northwest Territories government, essentially an appendage of DIAND, has been delegated authority to manage lands in and near most communities and along highways, to manage wildlife, and to address the social and economic concerns of Northwest Territories residents. Parks Canada and the Canadian Wildlife Service manage lands set aside as national parks, reserves, and bird sanctuaries.

The role of the federal Department of Energy, Mines, and Resources in land management in the North has only recently become the subject of speculation. The role of the newly created Canadian Oil and Gas Lands

Administration (COGLA) will certainly influence the management and use of Crown lands north of 60°. The exact nature of COGLA's interest and participation in a land-management process remains to be defined. Present indications from senior DIAND officials are that COGLA's role north of 60° will be subject to DIAND scrutiny and control.

These management mandates and structures describe and apply government policy in the North. Although Rees (1981) and others have lamented what they perceive to be a policy vacuum in the North, there has not yet been any rejection of the policy announced by the federal government in March 1972 (DIAND 1972). Priority was given to a "higher standard of living, quality of life and equality of opportunity for northern residents by methods which are compatible with their own preferences and aspirations," and to maintaining and enhancing the northern environment with due consideration to economic and social development (McLeod 1978: 207). The observer's confusion may be generated by what appears to be an uneven application of that policy to land-use and resource-development decisions. This results from a lack of published guidelines for resource-use decisions in the North.

Public access to the land-management decision-making forum is limited. The regulatory review process is technocratically oriented. DIAND is both judge and jury in this situation.

Overshadowing all current land-use decisions are DIAND-sponsored negotiations between the federal government and the Inuit, Inuvialuit, Dene, and Métis people of the North on the recognition of aboriginal rights to land and resources. Eventually, settlement legislation will force substantial changes to acts and ordinances. The federal and territorial governments have already agreed to the principle of regional planning commissions and to the concept that residents should be able to influence planning and land-management decisions. The selection of land for ownership by settlement beneficiaries or for retention as Crown land has already been made by the Committee of Original Peoples Entitlement (COPE) and the federal government and is part of an agreement in principle. Unfortunately, COPE land selection was not part of an open land-use planning process. However, we propose that future land-claims negotiations reflect the results of such a process conducted in the public forum rather than in secrecy.

When the Minister of Indian Affairs and Northern Development, to whom the chief federal land claims negotiators report, presented his northern land-use planning policy, he stated that the federal land-claims policy is the basis for native participation in land management and planning (DIAND 1981: 4). The policy acknowledges federal responsibility for planning initiatives, and proposes the creation of Northwest Territories land-use planning commissions and committees with native organization membership.

A Practical Solution

The land-use planning structures and general process we propose here use as their foundation the DIAND (1981) proposal for northern land-use planning, and our proposed principles for land-use planning and public participation. (The 1982 version was released too recently for review in this paper. However, major differences between the two documents will be noted.) Our intention is to leave room for future land-management developments under Bill C-48 and aboriginal-rights claims settlements.

Proposed Policy and Structures for Northern Land-Use Planning
At this time, there is no quotable statement of land-use planning policy by the federal government. The policy is embodied in "Northern Resource Use and Land Use Planning" (DIAND 1982); that paper is an official statement that land-use planning will precede management decisions on Crown land in the Yukon and Northwest Territories. In our opinion, the translation of that policy should be the task of a northern land-use policy committee and the subordinate structures described below.

Several of the following structures (fig. 1) were proposed by DIAND (1981), but the responsibilities of each are our suggestions:
1. The *Northern Land-Use Policy Committee* would provide general policy direction to the organizations in the Yukon and Northwest Territories. The committee should be chaired by DIAND's Assistant Deputy Minister of Northern Affairs. The members should be drawn from Environment Canada and Energy, Mines, and Resources (EMR), and the governments of both territories. Until the settlement of claims to aboriginal rights in the Northwest Territories, the leaders of the native political organizations that are now negotiating claims (Dene Nation, Métis Associations, COPE, and the Inuit Tapirisat of Canada or a member chosen by the regional Inuit organizations) should be members of the committee. We have recommended that the strong influence of native political leaders be clearly unleashed at this policy setting level through the native associations. A weaker mandate may cause native leaders to believe that their influence is minimal and lead them to

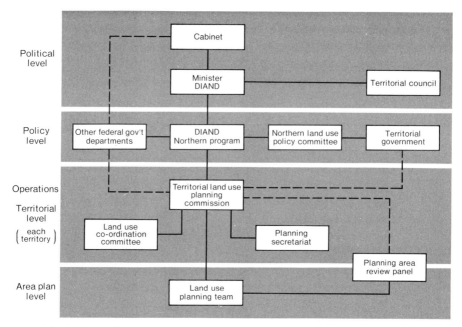

Fig. 1. Proposed organization for land-use planning in Canada's North.

refuse to support the process. This requirement may be diluted as politics evolves in the territories, but it accurately reflects the legitimate current role of these organizations. We hope that there would be an important tie to the people of Canada through the ministers to whom the committee members report.

COGLA would be conspicuously absent from the policy committee. However, that organization's interests would be well represented by the assistant deputy minister of DIAND and EMR, to which COGLA reports.

The DIAND proposals (1981, 1982) would have instead a northern land-use planning directorate of DIAND officials developing policy, guidelines, criteria, and standards in Ottawa. Involvement of Northwest Territories residents would first occur in a number of lesser policy committees, one for each planning area. The committees would be involved in the operational detail of the land-use planning process. These committees would be chaired by the Assistant Deputy Minister, Northern Program, served by a DIAND Executive Director of Land-Use Planning in Yellowknife. We consider this to be an expression of intent to centralize control of the process rather than give regional structures autonomy under policy guidelines.

2. The *Northwest Territories Land-Use Planning Commission* would coordinate land-use planning in the Northwest Territories within policy guidelines set by the Northern Land-Use Policy Committee. The chairman of the commission would be appointed by the Minister of Indian Affairs and Northern Development from nominees of the policy committee members. DIAND would provide secretariat support to the commission. Nongovernment members should be nominated by the presidents of the Dene Nation, the Northwest Territories Métis Association, by COPE, and by the Tungavik Federation of Nunavut (TFN) or a member chosen by the four regional Inuit organizations. Decisions would be made by consensus.

The principle honoured by the proposed mandate and membership is the decentralization of decision-making authority to agencies closest to the people affected by the decisions. To date, this has not been the case for DIAND and the Department of the Environment (DOE), where such authority has been held firmly by directors general and assistant deputy ministers in Ottawa.

Unlike the single Northwest Territories commission described in the

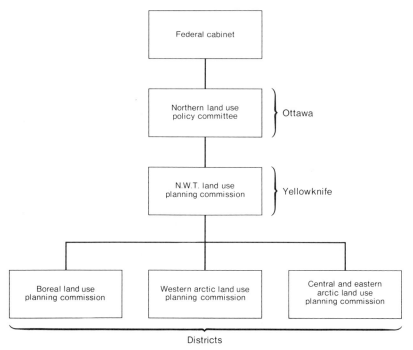

Fig. 2. Regional breakdown of planning commissions.

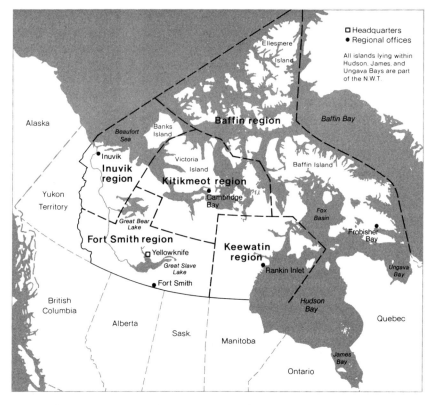

Fig. 3. Administrative regions of the Government of the Northwest Territories.

1981 DIAND proposal, the 1982 version would have a number of planning commissions established, one for each planning area. These commissions would include not only government and native association members, but also members from municipalities, "community groups," and "industry and/or other groups or individuals." Unlike our commission, these organizations would actually develop land-use plans for ministerial approval with the help of area planning teams. The minister would content himself with establishing policy guidelines, and delegate actual planning to a regional commission. Furthermore, these commissions would be disbanded after the DIAND minister has approved their area plans. We propose a single Northwest Territories commission to be a powerful decision-making organization.

We have not echoed the co-ordinating committee approach of DIAND (1981), believing the committees to be unnecessary commission appendages. We do not recommend the inclusion of specialists at

the commission level, as we see the actual planning activity occurring at the district level, where specialists will be needed.
3. Three *District Land-Use Planning Commissions* (fig. 2) would conduct land-use planning in the Northwest Territories. One would cover the central and eastern Arctic (Kitikmeot, Baffin, and Keewatin regions), another would cover COPE's Western Arctic Region (much of Inuvik Region), and the third would cover the Mackenzie Valley (roughly the Fort Smith Region). Fig. 3 locates boundaries of the Northwest Territories government's administrative regions. The boundaries of the areas of jurisdiction of the district commissions will have to be settled by the federal government, COPE, Inuit Tapirisat of Canada (ITC), the Dene Nation and the Métis Association. Even COPE's Western Arctic Region has yet to be defined to the satisfaction of governments and the other native associations.

Membership in the district commissions should be as established under agreements-in-principle signed by the federal government and the native associations. For us to suggest specific membership now would be impractical, as negotiations regarding structures in aboriginal rights claims are at an embryonic stage. As the COPE precedent indicates, the principle of regional responsibility in land-use planning will survive the negotiations. We envision strong local representation on the district commissions.

DIAND (1981) recommended that the territorial commission have three separate "components" to account for the interests of ITC, COPE, and the native people of the Mackenzie Valley. Our structure would allow greater regional autonomy in planning under the general policy direction and co-ordination of the Northwest Territories commission. It would also be more compatible with the regional structures arising from settlements of aboriginal-rights claims.

The district commissions, as we envision them, would be supported by government and nongovernment specialists, according to the demands of each planning situation. The district commissions would be empowered to hold public meetings, and they would have funds to support the attendance of special-interest organizations. We do not believe the requirement for more formal planning area review panels and land-use planning teams (DIAND 1981) or area planning teams (DIAND 1982) suggested by DIAND should be imposed on the district commissions. The district commissions should be allowed to tailor their operations as they see fit under the guidelines of aboriginal-rights agreements.

Discussion

We have listed general principles of land-use planning gleaned from the literature and from our experience with regional planning in the Mackenzie Delta and Lancaster Sound. We have then modified the structures suggested in the DIAND (1981) discussion paper to represent a more realistic approach that would accommodate those principles. The roles of each organization proposed by DIAND were substantially altered to accomplish:
1. The decentralization of land-use policy interpretation, planning, and decision-making to the Northwest Territories;
2. The involvement of native political associations at the policy-making level in Ottawa and at the policy-definition level in the territories; and
3. The accommodation of future political development in the Northwest Territories to include settlement of aboriginal-rights claims.

However, we have recognized the legitimate role of DIAND in managing Crown lands in the Northwest Territories through its proposed leadership in the policy committee and influence on district commissions.

We stop short of adopting Rees's (1978) slow evolutionary process, which begins with community-based planning for economic development. He acknowledges that most communities do not have the human resources to respond to large-scale economic development initiatives, and that many residents are at a basic literacy level. Furthermore, southern voters and taxpayers will not wait for such an ideal evolution from the grass roots, especially in the light of world market demands for scarce minerals and hydrocarbons. A planning process is needed now, even though there may be future modifications of the process, so that it better reflects federal policy shifts and aboriginal-claims settlements.

The Policy Committee would publish general land-use planning policy guidelines and would resolve questions of interpretation on request. The Northwest Territories commission would establish terms of reference for the district commissions that reflected the committee's guidelines. It would meet only when policy interpretation questions arose and when planning co-ordination between districts was required. The district commissions would actually conduct land-use planning in the North.

The proposed involvement of native association politicians representing several ethnic groups in a traditionally bureaucratic task of policy formulation, interpretation, and implementation is a reflection of the unique evolution of political decision-making in the territories. The native associations currently constitute a political force that influences the Northwest Territories Legislative Assembly. The associations do not have full confidence that the Assembly will represent their interests through the bureaucracy; in fact, the Assembly is regarded by some associations as a temporary institution. The commissions should take their ultimate direction from the Policy Committee acting on behalf of the Legislative Assembly and Parliament.

The land-use planning system must be based on solid information about natural resources, so that land-use alternatives and hazards can be defined. Many volumes of information about habitat and wildlife populations, social and economic structures and needs, and nonrenewable-resource development potentials have been generated in response to specific development proposals. Furthermore, federal projects such as the Arctic Land Use Information Series of maps produced by DOE and DIAND, Geological Survey maps, and the Arctic Ecology Map Series serve as valuable sources of information. There are also numerous publications and reports in the libraries of various government and private agencies. Although the quality of this information varies (Schindler 1976), the task of cataloguing the references in a format that makes them accessible to land-use planners and the public is an urgent one. This effort should be given highest priority, in view of the current need for the planning of projected major resource developments. Collation and cataloguing will result in the identification of gaps in the information that will be used in future proposals for resource inventories. These inventories would be conducted by responsible government agencies at the request of the district commissions.

Neither of the land-use planning systems suggested by DIAND and ourselves will adequately address the disarray of Canada's northern policy described by Abele and Dosman (1981). We recommend serious consideration of their proposed long-range institutional remedy for the lack of policy-making structures.

We have not recommended a detailed planning process. We commend to the reader Naysmith's (1976: 120*ff*) suggestions about planning areas, data collection, and a six-step approach to planning, keeping in mind our contention that plans must be flexible. He emphasizes the need to base all plans primarily on a consideration of the natural capability and limitations of the land.

We applaud the initiative taken by DIAND to institute land-use planning in the North. The proposed system could become one of the most democratic in Canada. The incentive to press ahead quickly with the implementation of a land-use planning process is the knowledge that if such planning does not become the basis for land-management decisions, the information in some of this book's excellent treatises on species and ecosystem biology may be rendered academic.

Appendix
Abstract: Northern Land-Use Planning

Discussion Paper
Minister of Indian Affairs and
Northern Development
1981

Traditional northern land-use management policy has been based on the belief that all properly regulated uses of federal Crown land could be accommodated with minimal environmental degradation. With the dramatic increase in nonrenewable-resource related activities in the 1970s, it became apparent that the traditional approach was inadequate. Conflicts between native land users and exploration companies increased, the Canadian public demanded better environmental management in the North, and industrial firms became uneasy about government policies.

There is no statutory authority for integrated land-use planning for most Crown land in the North. The paper summarizes federal pronouncements concerning land management in the Yukon and Northwest Territories, as well as federal planning and policy initiatives to date.

The paper proposes a land-use planning system that incorporates the basic characteristics of comprehensiveness, integrated management, and public participation. The system would operate within a three-level hierarchical framework: overall federal policy for northern lands; the policies of federal and territorial departments or agencies, either with land-related responsibilities or those whose activities might be affected by land-use decisions; and statements of specific project objectives. The land-use planning process would consist of five basic steps: plan initiation; resource inventory and analysis; preparation of plans to fulfil project objectives, reflecting the minister's choice of options for land use; implementation of a formal plan; and monitoring and review of results.

Planning responsibilities and expertise are contained in several departments and in two levels of government. A planning structure is proposed to cope with this complex jurisdictional and political environment (fig. 1).

"*A Northern Land Use Policy Committee* would be comprised of members from Departments having northern resource management, environmental conservation, and economic development mandates, with representation at the ADM level. Its purpose would be to:

(1) provide coordination at the policy level, including responsibility for developing and maintaining coordinated Program Objectives...;
(2) resolve policy issues that arise during planning exercises;
(3) determine priorities for planning amongst competing proposals;
(4) review and approve land use plan proposals submitted by Commissions, for recommendation to the Minister of DIAND; and
(5) monitor land use planning activities.

Committee membership would be drawn from both Federal and Territorial Governments. The Executive Director of the Land Use Planning Commission in each territory would be a member, ex officio. DIAND, through the ADM, Northern Affairs, would chair this senior policy committee.

Territorial Land Use Planning Commissions would hold responsibility for the conduct of land use planning in the Territories. They would be comprised of a Land Use Planning Coordinating Committee (an operational level reflection of the senior policy committee), but with representation from native organizations and a Planning Secretariat. The latter would contain the necessary complement of planners and other specialists to carry out planning tasks. The commission's functions would include:
(1) management of all land use planning in the Territory;
(2) coordination of operational inputs to planning projects, e.g. resource inventory by participating agencies and the obtaining of necessary cartographic support services;
(3) management of public participation, in concert with Review Panels;
(4) operation of a public information program; and
(5) functional liaison with all the various agencies involved in northern land use planning.

The Executive Director would function as chairman of the Coordinating Committee. The Commissioner would not represent any particular viewpoint in land use controversies, but operate as the neutral planning body referred to earlier.

Land Use Planning Teams would be assigned from the Commission for any particular planning projects, headed up by a Commission planner. In some circumstances, this planner would reside in the planning area for the duration of the exercise to enhance relationships with local communities and work more closely with the Review Panel.

Planning Area Review Panels would be established for each planning project, to provide a structured forum for public input during the planning process. Panel members would be appointed from amongst prominent local authorities, native associations and others (e.g. industry, conservationists, academics) who have a land use interest in the planning area.

The Department of Indian and Northern Affairs [sic], as the land manager, provides the Federal government structure through which reports, proposals and recommendations would be submitted and processed for eventual decision by the Minister. For this purpose, the Northern Affairs Program would establish a small land use planning branch in Ottawa, to assist in policy formulation and interpretation, to develop and coordinate programs, procedures and guidelines for land use planning, and to provide intra-departmental and intergovernmental coordination as well as secretariat services for the Northern Land Use Policy Committee.

DIAND would also be responsible for establishing the Planning Secretariat and providing the Executive Director for each Land Use Planning Commission.

Many Other Federal Government Departments have legislative authority and mandates in the North.... The approach to Northern Land Use Planning as outlined here will serve to provide a focus for and coordinate the activities of these departments. Departments with Northern mandates would operate as participants in the planning process, as their mandates and capabilities dictate, through representation on the Land Use Planning Commissions, and participation, at the policy level via the Northern Land Use Policy Committee.

The Territorial Governments' role would parallel that of other federal departments, with respect to their particular mandates. Matters requiring political consideration would be passed to the *Territorial Councils.*"

The Northwest Territories Land Use Commission is to "have three separate components" or "regional commissions" to accomodate the interests of ITC, COPE, and the Déné/Métis. Government membership would remain constant from one to the other.

The current reactionary process of land management can be corrected by

"adoption of a Northern Land Use Planning Policy that sets out a clear land management policy framework, within which specially designed Planning Structures can efficiently and, in a comprehensive and coordi-

nated manner, implement a Northern Land Use Planning Process with a mandate to:
— ensure orderly and planned development compatible with environmental objectives, the national interest, and the social and economic well being of native people and other residents of the Territories;
— improve coordination among departments and agencies with respect to northern land and resource use;
— establish a system for determining the allocation of lands to various uses based on the land's composite values and for guiding the application of land and related resource regulatory systems;
— provide a public consultation process and forum for assessing and proposing solutions to land use conflicts arising among the government, natives, and private sector bodies concerned with northern development and conservation activities; and
— make recommendations on changes in policy and programs necessary to achieve coordinated and comprehensive land use planning and management of territorial lands.

The operational strategy for this Northern Land Use Planning Process will include the following elements:
— the delineation of planning regions in the two territories. Delineation will be influenced by biological and physical variations, settlement, occupancy and resource utilization patterns; and consideration of the regional impacts associated with potential developments ...;
— the setting of priorities for 'planning treatment' for the planning regions. This will establish the time schedule and sequence of coverage for the planning regions. Those areas where major development or significant land use conflicts are anticipated in the short run would receive the highest priority and be the first to which planning efforts are directed. The planning schedule will also be used in guiding the timing and sequence of resource inventory and other data gathering activities in order that the information needed to conduct the planning work will be available for a given planning area at the time it is scheduled to received planning treatment;
— the identification of land and resource use options (opportunities and constraints) through assessment of biological, physical and socio-economic information, leading to the establishment of land and resource use goals, objectives and priorities and a plan incorporating these elements for each regional planning area;

— the establishment of an appropriate form of regional land zoning as an important component of each plan to guide the application of land and resource administration and regulatory systems operated by various government agencies; and
— the review and amendment of plans as required to respond to changing conditions."

Annexes to the discussion paper summarize legislation relevant to land-use planning in the North, details of the operation of the land-use planning process, examples of "planning regions," and the projected Department of Indian Affairs and Northern Development staffing and funding requirements for implementation of the proposed land-use planning structures.

Acknowledgments

We are grateful to the following individuals for reviewing and commenting constructively on drafts of the manuscript: Hiram Beaubier, Regional Director-General, Northern Affairs, Department of Indian Affairs and Northern Development, Yellowknife; Jim Bourque, Deputy Minister, Department of Renewable Resources, Government of the Northwest Territories, Yellowknife; Jan Duvekot, Inuit Tapirisat of Canada, Ottawa; Ben Hubert, Managing Principal, Boreal Ecology Ltd., Yellowknife; Ron Livingston, Planning Co-ordinator, Environmental Planning and Assessment, Department of Renewable Resources, Government of the Northwest Territories, Yellowknife; Dr. William E. Rees, Associate Professor, School of Urban and Regional Planning, University of British Columbia; and Ian Sneddon, Resource Inventory Manager, Land Management Division, Department of Indian Affairs and Northern Development, Ottawa.

We owe them much for their assistance in our research. The responsibility for the analyses and proposals presented here is entirely ours.

References

Abele, F., and E. J. Dosman. 1981. Interdepartmental coordination and northern development. Can. Public Admin. 24(3): 428–451.
Berger, T. R. 1977. Northern frontier, northern homeland: The report of the Mackenzie Valley Pipeline Inquiry. Vol. 1. Supply and Services Can., Ottawa. 213 pp.
Canadian Arctic Resources Committee (CARC). 1973a. Northern Perspectives 1(1): 1–4.
———. 1973b. Motion 81–73...carried. Northern Perspectives 1(8): 1–12.
———. 1976a. Northern Perspectives 4(3): 1–12.
———. 1976b. Northern Perspectives 4(5): 1–12.
Department of Indian and Northern Affairs (DINA). 1972. Northern Canada in the 70's. [Unpub. policy statement by the Minister, Dep. Indian and Northern Affairs.] Ottawa. 13 pp.
Department of Indian Affairs and Northern Development (DIAND). 1981. Northern land use planning. [Unpub. discussion paper, Dep. Indian and Northern Affairs.] Ottawa. 37 pp.
———. 1982. Northern resource use and land use planning in northern Canada. Land Use Planning Branch, Ottawa, Ontario. 160 pp.
Dirschl, H. J., ed. 1980. The Lancaster Sound region: 1980–2000: Perspectives and issues on resource use. Supply and Services Can., Ottawa. 113 pp.
Dome Petroleum Ltd., Esso Resources Canada Ltd., and Gulf Resources Canada Ltd. 1981. Hydrocarbon development in the Beaufort Sea–Mackenzie Delta region. [Unpub. rep.] Dome Petroleum Ltd., Calgary. 28 pp.
Donihee, J., and P. Gray. 1981. A review of road related wildlife problems and the environmental management process in the north. [Unpub. rep.] Northwest Territories Wildlife Service, Yellowknife. 20 pp.
Gamble, D. J. 1982. Northern participation in northern energy decision making: The policy and planning process. International Conference on Social Impact Assessment: State of the Art. Vancouver, British Columbia, 24–27 October 1982. 18 pp.
Hartman, G. F. 1980. Is there a future for Yukon wildlife? Northern Perspectives 8(4): 6–8.
Keith, R. F., A. Kerr, and R. Vles. 1981. Mining in the north. Northern Perspectives 9(2): 1–7.
Loken, O. H. 1981. Report of task force on Beaufort Sea developments. [Unpub. rep. to the Senior Policy Committee, Northern Development Projects, Dep. Indian and Northern Affairs.] Ottawa. 56 pp.
Lucas, A. R., and E. B. Peterson. 1978. Northern land use law and policy development: 1972–78 and the future. pp. 63–95. *In* R. F. Keith and J. B. Wright, eds. Northern transitions. Vol. 2. Can. Arctic Resources Committee, Ottawa.
McLeod, W. G. 1978. The Dempster Highway. pp. 193–250. *In* E. B. Peterson and J. B. Wright, eds. Northern transitions. Vol. 1. Can. Arctic Resources Committee, Ottawa.

Monaghan, H. J. 1980. Renewable resource management in the Northwest Territories: A proposal for change. M.Sc. thesis. Natural Resources Inst., Univ. Manitoba, Winnipeg. 249 pp.

Naysmith, J. K. 1976. Land use and public policy in northern Canada. Northern Policy and Program Planning Branch, Northern Program, Dep. Indian and Northern Affairs, Ottawa. 218 pp.

Nelson, J. G., and S. Jessen. 1981. The Scottish and Alaskan offshore oil and gas experience and the Canadian Beaufort Sea. Can. Arctic Resources Committee, Ottawa. 155 pp.

Pepperell, J. 1976. Horte—Happiness is a vague policy. Northern Perspectives 4(2): 11–12.

Rees, W. E. 1978. Development and planning north of 60°: Past and future. pp. 42–62. *In* R. F. Keith and J. B. Wright, eds. Northern transitions. Vol. 2. Can. Arctic Resources Committee, Ottawa.

——. 1981. Environmental assessment and the planning process in Canada. [Unpub. rep. to the Workshop on Environmental Assessment, Univ. Melbourne, Parkville, Victoria, Australia, 17–20 February 1981.] 31 pp.

Schindler, D. W. 1976. The impact statement boon-doggle. Science 192(4239): 1.

Stager, J. 1978. Land use planning for frontier regions. pp. 139–143. *In* R. F. Keith and J. B. Wright, eds. Northern transitions. Vol. 2. Can. Arctic Resources Committee, Ottawa.

Swanson, D. 1971. Public perceptions in resource planning. *In* W. R. D. Sewell and I. Burton, eds. Perceptions and attitudes in resource management. Policy Research and Coordination Branch, Dep. Energy, Mines, and Resources, Resource Pap. No. 2: 91–97.

Some Terrain and Land-Use Problems Associated with Exploratory Wellsites, Northern Yukon Territory

H. M. French

Introduction

Few studies describe the terrain, permafrost, and vegetation conditions of the interior northern Yukon. Even with the opening of the Dempster Highway in 1979, large areas of the territory are still inaccessible. Two of the few reports that attempt syntheses of the permafrost and terrain conditions are Zoltai and Pettapiece (1973) and Oswald and Senyk (1977). In addition, maps and reports arising from an ecological land survey (Wilken et al. 1979) are now available (Wilken et al. 1981a, 1981b). These provide a general overview of landscape types in the northern Yukon.

The present paper adds to this literature by describing terrain disturbances associated with several exploratory wellsites in the northern Yukon. The operations described here cover a period during which companies were gaining experience in working in permafrost conditions, researchers were gaining knowledge on terrain sensitivity, and regulators were learning how to set controls to minimize disturbance.

Other papers related to the environmental problems of drilling in Arctic Canada include those by Bliss and Peterson (1973) and French (1978a, 1978b, 1980). Similar studies dealing with northern Alaskan wellsites are also available (Lawson et al. 1978, Lawson and Brown 1979, Johnson 1981).

Background

Terrain problems in northern Canada frequently relate to the melt of ice-rich permafrost and subsequent ground instability. Other problems relate to the geotechnical properties of certain materials, particularly those that are unconsolidated and susceptible to either natural or man-induced failure and mass movement. In an attempt to minimize these problems, the federal government passed the Territorial Land Use Act and Regulations in 1971. Under this act, land-use permits are required for most activities and operating conditions are attached to these permits, if granted. To ensure the enforcement of these regulations, federal government land-use inspectors make periodic field visits.

With respect to oil and gas exploration, some of the early environmental concerns were related to seismic and other transportational activities (e.g., Kerfoot 1972, Bliss and Wein 1972). Recent studies in the Northwest Territories, however, suggest that the impact of modern transportational activities has been reduced to a minimum (French 1978c). Improvements in industry operating conditions (e.g., the use of vehicles equipped with low-pressure tires) and the strict application of land-use regulations (e.g., the restriction of cross-tundra vehicle movement to winter months) are important factors in the reduced impact. Viewed in this light, the potential for the most serious terrain and environmental damage is now associated with the drilling operation itself, which often extends into the critical summer months, and the disposal of waste drilling fluids.

Regional Setting

Over 50 wells have been drilled in the Yukon, most of them on either Eagle Plain or Peel Plateau (fig. 1). Eagle Plain is an extensive intermontane plateau covering an area of approximately 17,500 km². The plateau is between 650 and 800 m in elevation and is dissected by broad, shallow valleys (fig. 2), which drain northwards to the Old Crow Basin via the Eagle and Porcupine rivers. The southern edge of the plateau is delineated by a prominent escarpment that has been formed by a resistant member of the Eagle Plain Formation, a sandstone of Cretaceous age that underlies most of the plateau (Norris et al. 1963). Vegetation consists of a stunted and open boreal forest that changes to shrub tundra at higher elevations. The elevation of the alpine tree line decreases northward, from approximately 850 m in the south to 600 m in the north.

Terrain and Land-Use Problems Associated with Exploratory Wellsites 367

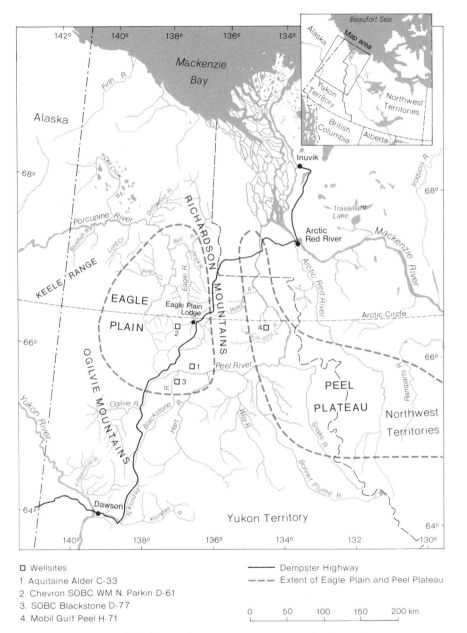

□ Wellsites
1. Aquitaine Alder C-33
2. Chevron SOBC WM N. Parkin D-61
3. SOBC Blackstone D-77
4. Mobil Gulf Peel H-71

——— Dempster Highway
- - - - Extent of Eagle Plain and Peel Plateau

0   50   100   150   200 km

Fig. 1. Location map of Northern Yukon Territory showing Eagle Plain and Peel Plateau. Wellsites discussed in this paper are also indicated.

Fig. 2. General view of the upland surface of the Eagle Plain. Vegetation consists of an open black spruce woodland and shrub tundra. Lat. 66°16′N, Long. 137°10′W.

Fig. 3. General view of the Peel Plateau, showing open spruce woodland and tussock hummock tundra. The latter is probably fire-induced. Lat. 66°22′N, Long. 134°52′W.

The Peel Plateau covers an area of approximately 20,000 km² at the eastern foot of the southern Richardson Mountains. It declines in elevation eastward from approximately 600 to 300 m. Vegetation is typical of the transition between boreal forest and forest-tundra (fig. 3). Organic material is widespread, particularly in poorly drained areas, where peat plateaus, palsas, and peat polygons occur. The plateau is underlain by shales and sandstone of Lower Cretaceous age and is dissected by a number of well-defined and relatively deeply incised valleys, such as those drained by the Caribou and Trail rivers. These constitute part of the Mackenzie River drainage system.

Although mapped by Brown (1978) as transitional from continuous to discontinuous permafrost, very little is known about the permafrost conditions of the northern Yukon. At Dawson City, discontinuous permafrost is known to extend to depths of 60 m (Brown 1978, EBA Engineering 1977), while at the Eagle River crossing of the Dempster Highway, borehole data suggest that permafrost may be as much as 90 m thick on the floodplain (Johnston 1980). Along the Eagle River, ground thermal data indicate an active layer exists between May and September and reaches a maximum thickness of almost 1 m. Only one well in the northern Yukon, Socony Mobil WM N. Cath B-62, is logged for temperature (Taylor and Judge 1974). Measurements indicate a permafrost thickness of approximately 100 m, with mean annual ground temperatures of between 0 C and − 2 C.

The northern Yukon Territory is distinctive in that much of it escaped glaciation by the Laurentide ice sheet (Bostock 1966, Hughes 1969, Vernon and Hughes 1966). Moreover, Eagle Plain may have remained unglaciated throughout the Pleistocene. As a result there is a general absence of surficial deposits from large areas of Eagle Plain, and unconsolidated, fine-grained, and high ice content sediments are restricted to lower valleyside slopes and valley bottoms. In contrast, Peel Plateau lies within the limits of the Wisconsin (Laurentide) ice sheet. Till and other fine-grained sediments occur widely. Zoltai and Pettapiece (1973: 18) report glacial till on Peel Plateau commonly having clay content in excess of 40%. High ice contents are also indicated by the ubiquitous presence of earth hummocks, areas of tilted and stunted black spruce (*Picea mariana*), and occasional mudflows and mass movements. Organic terrain in the form of peat plateaus, palsas and peat polygon areas, and fenlands are also important and complicate permafrost distributions.

Table 1 Summary of wellsite information

Wellsite	Location	Surficial materials or bedrock	Start and completion dates*	Terminal depths* (m)	Nature of drilling operation	Terrain disturbance conditions, sump-related problems, general comments
Aquitaine Alder C-33	Sloping terrain, Eagle Plain	3 m colluvium, ice-rich, overlying shale bedrock	08-03-78 04-03-79	3714	One season, winter operation	No terrain problems. No sump at site; use of borrow pit for waste fluid containment. Crushed sandstone used for rigpad. Pre-site investigations.
Chevron SOBC WM N. Parkin D-61	Upland surface, Eagle Plain	Sandstone bedrock	04-01-72 17-04-72	3352	One season, winter operation	No terrain problems. No rigpad constructed.
SOBC Blackstone D-77	Valley bottom, poorly drained	Ice-rich colluvium	10-03-62 08-01-63	4028	Two seasons, summer operation	Terrain disturbance and thermokarst. Operation conducted prior to Territorial Land Use Regulations.
Mobil Gulf Peel H-71	Upland surface, Peel Plateau	Till	05-02-77 12-06-66	3392	Two seasons, summer operation	Auxiliary sump constructed. Fluid outflow and toxicity effects.

Source: Schedule of Wells 1920-1979, Northwest Territories and Yukon Territory. Northern Non-Renewable Resources Branch, Oil and Gas Resource Evaluation Division, Exploratory Operations Section, Indian and Northern Affairs, Canada. Catalogue No. R72-21/1979.

Wellsite Terrain Disturbances

A number of abandoned wellsites, of varying ages and located in a variety of terrain and permafrost conditions, were examined in the field during the summer of 1979 (French 1981). Four are described in this paper (table 1). One (Mobil Gulf Peel H-71) was selected specifically because of known terrain and sump-related problems. A second (SOBC Blackstone D-77) was selected because it was drilled prior to the introduction of the Territorial Land Use Regulations. The two remaining wellsites (Chevron SOBC WM N. Parkin D-61 and Aquitaine Alder C-33) represent case studies of early and later wells drilled on Eagle Plain subsequent to the introduction of these regulations. At each wellsite an attempt is made to relate the nature and degree of disturbance to the site terrain sensitivity and the history of the drilling operation.

Pre–Land-Use Operations

Two factors make the Blackstone D-77 wellsite of particular interest in any assessment of wellsite disturbances in the interior Yukon Territory.

First, in contrast to most wells drilled on Eagle Plain, this well was located on gently sloping terrain near the bottom of a north-south-trending tributary valley of the Blackstone River. Poor drainage conditions together with silty ice-rich colluvial sediments produce an open forest woodland of black spruce and tamarack (*Larix* sp.), with a relatively dense shrub layer (dwarf birch, willow, ericaceous shrubs) and sedges (*Carex* spp.). In these moister environments of the interior Yukon, a fire cycle slightly different from that characteristic of well-drained sites occurs. According to Zoltai and Pettapiece (1973: 16), in moist environments, sedge and cottongrass tussocks survive fires and thrive due to the absence of competition and because of the increased moisture associated with the thickening of the active layer. This makes the re-establishment of black spruce and tamarack difficult. The result is the development of areas of fire-induced tundra that are composed of tussocks or tussock-dwarf birch. The same sequence of vegetation changes might be expected to follow wellsite disturbances in these areas.

Second, the well was drilled prior to the introduction of the Territorial Land Use Regulations. It was a two-season operation that commenced in March 1962 and ended in January 1963. In all probability there was activity around the site during the summer of 1962, but the nature of this activity is difficult to ascertain from the well records. It would appear that the well was drilled from a pile-supported platform, remnants of which still exist at the site (fig. 4).

Fig. 4. Remnants of old pile-supported rigpad are still visible at the SOBC Blackstone D-77 wellsite, drilled in 1962–63. Photo taken in July 1979.

Fig. 5. Thermokarst mounds and pools of standing water have formed in the disturbed terrain adjacent to the SOBC Blackstone D-77 wellsite. Photo taken in July 1979.

Fig. 6. Oblique air view of old seismic line traversing poorly drained ice-rich terrain, Blackstone River Valley. The seismic line was shot in summer. Note the thermokarst that has occurred. Vegetation is a fire-induced shrub tundra and tussock tundra. Lat. 65°45′N, Long. 137°15′W.

As a result of these operating procedures, it is not surprising that substantial thermokarst developed within the area of the wellsite. Thermokarst mounds that are 1 to 3 m in height, together with pools of standing water that are 2 to 4 m deep in places, surround the old drilling platform (fig. 5). Thaw depths in the disturbed terrain commonly exceed 90 cm, while those in adjacent undisturbed terrain are less than 45 cm. Revegetation in the form of sedges, mosses, dwarf birch, and cottongrass has occurred naturally, and this has helped stabilize the terrain. Also, the presence of fescue (*Festuca* sp.) and timothy (*Phleum* sp.) grasses suggests artificial revegetation at some time in the past.

Seismic lines had been shot in the summer. In the vicinity of the wellsite they reveal a common history of physical disturbance and thermokarst (fig. 6). The removal of trees and surface organic material led to the formation of water-filled trenches, 80 to 100 cm deep, colonized by aquatic plants (*Equisetum* sp., *Carex* sp., mosses). Ramparts on either side of the trench reflect the bulldozed and redeposited surface materials.

In summary, the SOBC Blackstone D-77 wellsite illustrates the inadequacies of industry operational procedures prior to the introduction of the Territorial Land Use Regulations. The resulting thermokarst topography is similar to that described elsewhere from the boreal forest and forest-tundra zones of North America (e.g., Péwé 1954, Kerfoot 1974) and the Soviet Union (e.g., French 1975: 139–143).

Post–Land-Use Operations

One of the first wells to be drilled on the upland surface of Eagle Plain following the introduction of the Territorial Land Use Regulations was the Chevron SOBC WM N. Parkin D-61 (fig. 7). Not only was it drilled completely in the winter of 1971–72, but the site was relatively well-drained and underlain by consolidated bedrock possessing low amounts of ground ice. In contrast to the Blackstone D-77 wellsite, minimal terrain disturbance occurred at this wellsite. This reflected both the one-season, winter drilling program and the low terrain sensitivity of the site. The absence of a rigpad probably reflected the lack of easily accessible surficial aggregate and the desire of the land-use regulatory bodies at that time not to excavate borrow pits in the bedrock.

Further refinements in both industry operating techniques and the application of the Territorial Land Use Regulations can be illustrated by the most recent well to be drilled on Eagle Plain. This was the Aquitaine Alder C-33 well, drilled during the late winter of 1977–78.

From a terrain viewpoint the Alder wellsite appears reasonably typical

Fig. 7. Oblique air view of the Chevron SOBC WM N. Parkin D–61 wellsite, located on the upland surface of Eagle Plain. Photo taken in July 1979.

of conditions on Eagle Plain. The well was located on a relatively well-drained, south-facing, gently sloping (2 to 5°) interfluve, several kilometres south of the main Eagle Plain escarpment. Vegetation in the area consists of an open canopy of black spruce together with a shrub layer of birch (*Betula glandulosa*), alder (*Alnus* sp.), and some Labrador tea (*Ledum groenlandicum*). A moss-lichen layer constitutes a surface organic mat exceeding 10 cm in thickness. At the wellsite the presence of white spruce (*Picea glauca*) and charred remnants of black spruce indicate a previous burn (fig. 8). In the nonburn area the black spruce average 5 to 6 m in height, while in the burn area the white spruce are rarely more than 3 to 4 m in height. Tree ring counts suggest that the trees outside of the burn are 130 to 150 years old, while those in the burn are 40 to 70 years old. These data are consistent with our understanding of the role of fire in the boreal forest ecosystem (e.g., Rowe et al. 1975, Wein et al. 1975, Zoltai and Pettapiece 1973: 15–17). Indicators of patterned ground and solifluction are absent from the site, although large hummocks, approximately 1 m in diameter and up to 15 to 30 cm in height, occur widely, and are similar to those described by Zoltai and Pettapiece (1973: 18–20). The "drunken" appearance of the boreal forest that is typical of the ice-rich, frost-susceptible soils in the Mackenzie Valley (e.g., Zoltai 1975) is noticeably absent from Eagle Plain.

Fig. 8. Oblique air view of the Aquitaine Alder C–33 wellsite prior to drilling in June 1977. The extent of the burn area is clearly visible. Lat. 65°52′N, Long. 136°54′W.

Fig. 9. Oblique air view of the Aquitaine Alder C–33 wellsite, following site abandonment and rehabilitation, July 1979. The pad consisted of crushed bedrock (sandstone) obtained from the area.

Fig. 10. Oblique air view of borrow pit area for Aquitaine Alder C–33 wellsite, Eagle Plain, following site rehabilitation. Areas stripped of soil and vegetation show meltout along joint patterns in underlying bedrock. They are *not* ice-wedge polygons. The infilled borrow pits are to the left of the photo and elongate in plan. Photo taken in July 1979.

Terrain and Land-Use Problems Associated with Exploratory Wellsites 377

During early June 1977 geotechnical and terrain investigations were undertaken at the proposed wellsite location (French 1977). The depth of the frost table and active layer temperatures were measured at several sites (table 2). As expected, thaw depths bore a close relationship to vegetation cover and type, and confirmed the important thermal role that vegetation plays in these northern boreal forest–shrub tundra environments. Shallow drilling revealed 10 to 24 cm of organic material overlying $\cong 2.4$ m of silty colluvium, which graded into highly weathered and fissile shale bedrock. Ground ice amounts were found to decrease with increasing depth (table 3), and only in the upper 2 m of the colluvial sediments were excess ice amounts of between 10 to 25% and natural water (ice) contents exceeding 20% recorded. The deeper cores retrieved from underlying shale and sandstone bedrock indicated low ground ice amounts and high structural cohesiveness of the rock.

Following these observations a land-use permit was issued for the drilling of the well during the winter of 1977–78. To prevent thaw of the ice-rich surficial material, a large gravel pad was constructed for the operation. The aggregate was obtained by crushing sandstone obtained from the main Eagle Plain escarpment to the north. A small sump was used during drilling, but most of the waste fluids were transported to the aggregate borrow area and subsequently buried. The operation was atypical, therefore, since the main rig sump was not at the site itself. Operations at the site were terminated during the summer of 1978, and site rehabilitation was accomplished the following winter.

As a result of these procedures, terrain disturbances adjacent to the rigpad were minimal (fig. 9). Moreover, no signs of permafrost thawing, as might be indicated by slumping of the pad, were visible in July 1979. It was concluded that the methods adopted to prevent thaw were sufficiently effective to offset any increase in thaw resulting from the removal of the trees and the surface vegetation layer.

The borrow pit from which the crushed bedrock aggregate had been obtained was aligned along the crest of the Eagle Plain escarpment. When examined in July 1979 a rectangular network of linear depressions had formed in those areas where only the surface soil and vegetation had been removed (fig. 10). These linear depressions undoubtedly reflect the melt-out of ice from expanded joints in the underlying bedrock. Active layer temperatures measured in the stripped borrow pit areas in 1979 were significantly higher than those measured in adjacent undisturbed terrain along the ridge crest (table 4). Similarly, as might be expected, temperatures in the rigpad itself were considerably greater than those in adjacent

Table 2 Active-layer conditions at the Aquitaine Alder C-33 wellsite, Eagle Plain, 9 June 1977 (from French 1977)

	Site 1 SEISMIC LINE	Site 2 SPRUCE FOREST; BURN AREA	Site 3 SPRUCE FOREST; BURN AREA
Depth			
Surface	15.0°C	13.0°C	15.0°C
10 cm	11.0°C	6.0°C	12.0°C
20 cm	2.0°C	—	—
Frost table (cm)			
9.6.1977	24.0	13.0	20.0
Vegetation	Sphagnum Lichen	Spruce Sphagnum Lichen	Spruce Labrador tea Lichen Cranberry
Vegetation cover (%)			
Micro-relief	Flat	Depression	Hummocky
Organic mat, cm	14.0	12.0	15.0

Table 3 Ground-ice amounts and consistency limits for surficial materials at the Aquitaine Alder C-33 wellsite, Eagle Plain (from French 1977)

Sample	Depth (cm)	Materials	Ice type*	Excess ice* (%)	Natural water content (%)	Liquid limit (%)	Plastic limit (%)	Plasticity range (%)
1	55-60	Colluvium	Vs, Vx	15-20	30.5	35.7	30.5	5.2
2	100-115	Colluvium	Vs, Vx	25	18.9	31.4	18.9	12.5
3	150-160	Colluvium	Vs, Vx, Vc	10-15	18.8	23.8	18.8	5.0
4	280-300	Weathered shale	Vc, Vx	0-5	6.2	20.8	6.2	14.6

*As defined by Pihlainen and Johnston, 1963.

Table 4 Active-layer temperatures (°C) measured at the Aquitaine Alder C-33 wellsite, 27 July 1979.

(A) Undisturbed bedrock terrain, Eagle Plain escarpment. (B) Stripped bedrock, Eagle Plain escarpment. (C) Drilling pad composed of crushed bedrock aggregate. (D) Undisturbed terrain adjacent to wellsite. Note the similarities in thermal regimes between (A) and (D) and between (B) and (C).

Site		A	B	C	D
Material		Sandstone	Sandstone	Crushed sandstone aggregate	Silty colluvium (2–4 m) overlying shale bedrock
Vegetation		Birch, poplar, white spruce, moss-lichen	Stripped of soil and vegetation	–	White spruce, moss-lichen
Depth: (cm)	Surface	17.5°C	17.5°C	18.0°C	20.0°C
	10	6.0°C	18.0°C	17.5°C	7.5°C
	20	5.0°C	18.0°C	17.0°C	–
	30	–	19.0°C	–	–
	40	–	19.0°C	17.0°C	2.0°C
	50	3.0°C	–	–	–
Frost Table:		n.d.	n.d.	n.d.	n.d.

vegetated areas. These observations highlight the high thermal conductivity of bedrock and the relatively deep thaw depths that can occur wherever bedrock is exposed.

Sump-Related Problems

The containment of waste drilling fluids in below-ground sumps has been a standard operating condition for wells drilled under the Territorial Land Use Regulations in both the Northwest Territories and the Yukon. It has also been one of the conditions that has presented the most problems. Two wells drilled on the Peel Plateau in recent years illustrate the problems posed by waste fluid containment. One, the Gulf Caribou N-25, is described elsewhere (French 1980: 801–803); the other is discussed below.

The Mobil Gulf Peel H-71 well was drilled during the summer of 1977. The wellsite is located on the flat upland surface of the Peel Plateau. Vegetation in the area consists of an open woodland of black spruce and white spruce, together with extensive areas of shrub tundra. Interest in this wellsite arose from reports that indicated that a second sump had been constructed to contain waste fluids. The conditions that led to this are not known and can only be speculated on. An inadequately sized initial sump or downhole problems leading to excessive waste fluid production are probable explanations.

Terrain disturbance around the wellsite is not extensive (fig. 11) and undoubtedly reflects the strict application of land-use procedures during the critical summer thaw period. The auxiliary sump is located on slightly lower ground adjacent to the main sump and near a shallow drainage line. The escape of toxic fluids from the downslope end of the auxiliary sump is indicated by a swath of dead or dying spruce trees, which follows the small drainage depression for approximately 150 m. The cause of this fluid escape can only be speculated on. One possibility is that the 4.0 ft (1.22 m) freeboard (as specified in the permit conditions) had not been maintained, and that seepage subsequently occurred along the top of the permafrost table. Since there is no mention of this in the various land-use inspection reports, another possibility relates to the melt-out of the ice in the silty clay that was used to construct a small retaining dyke at the lower end of the sump. This mechanism appears to be a common cause of the seepage of

Fig. 11. Air view of the Mobil Gulf Peel H–71 wellsite showing auxiliary sump and adjacent area of dead trees. Photo taken in July 1979.

sump fluids during summer, wherever retaining dykes are constructed using ice-rich and fine-grained sediments (e.g., Gulf Parsons Lake N-10, French 1978c).

Discussion

These limited observations and case histories help to assess the effectiveness of the Territorial Land Use Regulations in the boreal forest–shrub tundra environments of the interior northern Yukon. To begin with, the general sensitivity of these environments, especially in comparison with other boreal forest and tundra zones of the western Arctic, must be considered. A general conclusion is that permafrost terrain disturbances are neither as obvious nor as severe as they are in parts of the Northwest Territories, notably the Mackenzie Delta and the Arctic Islands. This is particularly true for Eagle Plain, less so for Peel Plateau.

Several factors are involved in this assessment. First, the general terrain sensitivity of much of interior Yukon is lower than it is in areas such as the Mackenzie Delta. This is because of the absence of unconsolidated ice-rich surficial materials from the unglaciated areas (e.g., Eagle Plain). There is instead a widespread occurrence of well-drained coherent bedrock surfaces of generally low ice content. The higher terrain sensitivity of Peel Plateau appears to be due to a combination of the presence of ice-rich and fine-grained glacial and lacustrine sediments, and the highly dissected nature of certain areas. Second, the revegetation and stabilization of disturbed areas in the boreal forest and shrub tundra environments of interior Yukon occur relatively quickly when compared with revegetation and stabilization further north. For example, observations from the Blackstone wellsite suggest that natural recolonization, which is similar to the changes following fire, is as successful as artificial seeding over the long term (i.e., 15 to 25 years). The wellsite also illustrates how thermokarst may occur in highly sensitive locations.

The changes effected by the introduction of the Territorial Land Use Regulations in minimizing disturbance must also be highlighted. The relative sophistication of the Aquitaine Alder C-33 operation, where an innovative waste disposal program was undertaken, and the earlier Chevron SOBC WM N. Parkin D-61 operation, where a simple application of land-use regulations was undertaken at a low-sensitivity site drilled in winter, ensured equally minimal impact. As such, these two wellsites illustrate not only the progressive evolution in the application and efficacy of the Territorial Land Use Regulations in Eagle Plain, but also the variations in wellsite permit conditions and associated operating procedures required by the permittee.

Conclusions

There appears to be no easy solution to the terrain and environmental problems of oil exploration in Arctic Canada. However, the continued application of the land-use regulations, together with the positive attitudes increasingly adopted by the operators themselves, are leading to fewer problems that cannot either be resolved or minimized. In the boreal forest–shrub tundra environments of Peel Plateau and especially Eagle Plain, recent terrain disturbances associated with well drilling appear to have been minimized. The lack of significant surficial deposits over extensive areas, together with the relative rapidity of natural revegetation, are major controlling factors in addition to the land-use regulations. In spite of this, terrain disturbance and sump-related problems may occur in some localities, as evidenced by the Blackstone D-77 and Peel H-71 experiences. They emphasize the desirability of site-specific solutions, as well as the imposition of general land-use operating conditions, to minimize the impact on the environment.

Acknowledgments

Research was funded by the Arctic Land Use Research (ALUR) Program of IAND, Ottawa. Aquitaine of Canada Ltd. provided access to the Alder C-33 wellsite in June 1977, and in July 1979 it provided helicopter support to visit other wellsites. The assistance of T. Beck and W. Sawoton in co-ordinating these logistics is gratefully acknowledged. Discussions with numerous individuals in IAND, Whitehorse, including H. Beaubier, H. Enfield, and K. Kepke, helped in the identification of those wellsites where terrain and waste drilling-fluid disposal problems had been encountered. Permission to draw upon data contained in ALUR contract report OSU-78-00366 was kindly provided by D. M. Barnett, IAND, Ottawa, who acted as Scientific Authority during the completion of this work.

References

Bliss, L. C., and E. B. Peterson. 1973. The ecological impact of northern petroleum development. In Arctic oil and gas: problems and possibilities; Proc. Fifth Int. Congress, Foundation Francaise d'Etudes Nordiques, May 2–5, 1973, Le Havre: 1–26.

———, and R. W. Wein. 1972. Plant community responses to disturbances in the Western Canadian Arctic. Can. J. Bot. 50: 1097–1109.

Bostock, H. S. 1966. Notes on glaciation in central Yukon Territory. Geol. Surv. Can. Pap. 65–36. 18 pp.

———. 1970. Physiographic subdivisions of Canada. In R. J. W. Douglas, ed. Geology and economic minerals of Canada. Geol. Surv. Can. Econ. Geol. Rep. No. 1, 10–30, Map 1254A.

Brown, R. J. E. 1978. Permafrost. Plate 32. Hydrological atlas of Canada. Dep. Fish. and Environ. Can., Ottawa.

EBA Engineering Consultants Ltd. 1977. Geotechnical investigations for utilities design, Dawson City, Yukon. Consultants' Rep., Stanley Associates Engineering Ltd. 28 pp.

French, H. M. 1975. Man-induced thermokarst, Sachs Harbour airstrip, Banks Island, N.W.T. Can. J. Earth Sci. 12: 132–144.

———. 1977. Permafrost and terrain report: Aquitaine Alder wellsite, Eagle Plain, Yukon Territory (65°52′N; 136°52′W). Consultants' Rep., Aquitaine of Can. Ltd., Calgary. 12 pp.

———. 1978a. Terrain and environmental problems of Canadian Arctic oil and gas exploration. Musk-Ox 21: 11–17.

———1978b. Why Arctic oil is harder to get than Alaska's. Can. Geogr. J. 94: 46–51.

———. 1978c. Sump Studies I: Terrain disturbances. Environmental Studies No. 6. Dep. Indian and Northern Affairs, Ottawa. 52 pp.

———. 1980. Terrain, land use and waste drilling fluid disposal problems, Arctic Canada. Arctic 33: 794–806.

———. 1981. Sump Studies IV: Terrain disturbances adjacent to exploratory wellsites, northern Yukon Territory. Contract Rep. OSU78-00366, ALUR Program. IAND, Ottawa. 41 pp.

Hughes, O. L. 1969. Surficial geology of northern Yukon Territory and northwestern District of Mackenzie, Northwest Territories. Geol. Surv. Can. Pap. 69–36, Map 1319A.

Johnson, L. A. 1981. Revegetation and selected terrain disturbances along the trans-Alaska pipeline, 1975–1978. CRREL Rep. 81–12. 115 pp.

Johnston, G. H. 1980. Polyurethane insulated road study, mile 237.2, Dempster Highway, Yukon Territory. Div. Building Res., Nat. Res. Council Can. [Unpub. manuscript.]

Kerfoot, D. E. 1972. Tundra disturbance studies in the Western Canadian Arctic. Dep. Indian Affairs and Northern Dev., Ottawa. ALUR 71–72–11. 115 pp.

———. 1974. Thermokarst features produced by man-made disturbances to the tundra terrain. In B. D. Fahey and R. D. Thompson, eds. Proc. Third Guelph Symposium on Geomorphology, 1973, GeoAbstracts Ltd.: 60–72.

Lawson, D. E., and J. Brown. 1979. Human-induced thermokarst at old drill sites in northern Alaska. Northern Eng. 10: 16–23..
———, J. Brown, K. R. Everett, A. W. Johnson, V. Komarkova, B. M. Murray, D. F. Murray, and P. J. Webber. 1978. Tundra disturbance and recovery following the 1949 exploratory drilling, Fish Creek, northern Alaska. CRREL Rep. 78-28. 81 pp.
Norris, D. K., R. A. Price, and E. W. Mountjoy. 1963. Geology, northern Yukon Territory and northwestern District of Mackenzie. Geol. Surv. Can. Map 10-1963.
Oswald, E. T., and J. P. Senyk. 1977. Ecoregions of the Yukon Territory. Can. Forest. Serv. Env. Can. Pub. BC-X-164. 155 pp.
Péwé, T. L. 1954. Effect of permafrost upon cultivated fields. U.S. Geol. Surv. Bull. 989: 315–351.
Pihlainen, J., and G. H. Johnston. 1963. Guide to the field description of permafrost. Nat. Res. Council Can. Tech. Memorandum 79. 21 pp.
Rowe, J. S., D. Spittlehouse, E. Johnson, and M. Jaccemik. 1975. Fire studies in the upper Mackenzie Valley and adjacent Precambrian uplands. Dep. Indian and Northern Affairs, Ottawa. 74-75-61. ALUR Rep. 128 pp.
Smith, M. W. 1977. Computer simulation of microclimatic and ground thermal regimes: Test results and program description. Dep. Indian and Northern Affairs, Ottawa. ALUR Program 75-76-72. 74 pp.
Taylor, A. E. and A. S. Judge. 1974. Canadian geothermal data collection: Northern wells, 1955 to February 1974. Geothermal Ser. No. 1, Geothermal Serv. Can., Earth Physics Branch, Dep. Energy, Mines and Resources, Ottawa. 171 pp.
Vernon, P., and O. L. Hughes. 1966. Surficial geology, Dawson, Larsen Creek and Nash Creek Map areas, Yukon Territory. Geol. Surv. Can. Bull. 136. 25 pp.
Wein, R. W., T. W. Sylvester, and M. G. Weber. 1975. Vegetation recovery in Arctic tundra and forest tundra after fire. Dep. Indian and Northern Affairs, Ottawa. ALUR Rep. 74-75-62. 115 pp.
Wiken, E. B., D. M. Welch, G. Ironside, and D. G. Taylor. 1979. Ecological Land Survey of the Northern Yukon. Ecol. Land Class. Ser., Lands Directorate, Env. Can. No. 7: 361–372.
———. 1981a. Ecoregions of the Northern Yukon. Map. Scale 1:1,000,000 with extended legend. Env. Conserv. Serv., Env. Can., Ottawa.
———. 1981b. Ecodistricts of the Northern Yukon. Map. Scale 1:500,000 with extended legend. Env. Conserv. Serv., Env. Can., Ottawa.
Zoltai, S. C. 1975. Tree ring record of soil movements on permafrost. Arctic and Alpine Res. 7: 331–340.
Zoltai, S. C., and W. W. Pettapiece, 1973. Terrain, vegetation and permafrost relationships in the northern part of the Mackenzie Valley and northern Yukon. Env.-Soc. Prog., Northern Pipelines, Task Force on Northern Oil Development, Rep. 73-4. 105 pp.

Energy Development, Tourism, and Nature Conservation in Iceland

Edgar L. Jackson

Introduction

The Icelandic economy developed rapidly in the twentieth century due to the success of the fishing industry. The people of Iceland now enjoy one of the highest standards of living in the world (Egertsson 1975: 176, Sigurdsson 1977: 8). To maintain this level of prosperity Icelandic policy-makers have advocated the diversification of the economy away from a dependence on fishing and toward industrial development (Egertsson 1975: 178–179, Nordal 1975: 242). The lack of domestic fossil fuel and mineral resources would appear to be a major constraint on the achievement of this goal, but Iceland has potential geothermal and hydroelectric energy resources that can be developed as the basis for industrialization (Egertsson 1975: 177–179). Also, the glaciers, lakes, rivers, waterfalls, hot-springs, volcanoes, and other elements of the landscape create a spectacular and unique tourist attraction, the revenues from which can make an additional contribution to the balance of payments position. Tourism, although not as important as energy development, plays a part in the strategy of diversification.

The exploitation of energy resources and the concentration of recreational and tourist activities in scenically attractive yet ecologically sensitive locations may have undesirable impacts on the environment. Also, tapping the energy resource potential may interfere with the environmental qualities sought by foreign tourists and Icelandic recreationists.

Other resource uses, most notably fishing and farming, are important in Iceland (Elísson 1975: 191–203, Sigthórsson 1975: 179–191), but this paper describes the development of energy resources, the pertinent aspects of recreation and tourism, and the process of nature conservation, which is directed toward alleviating the detrimental conflicts and impacts associated with these activities. Consequently it may be possible to learn about resource conflicts and management and the choices and compromises these engender not only in Iceland, but elsewhere in the North.

The Context of Resource Exploitation

Much like other countries the Icelandic government perceives its major economic objective to be the satisfaction of the demands of a growing population for a rise in material living standards (Nordal 1975: 238), but this goal can be achieved only if new strategies of economic development are followed (Egertsson 1975: 178–179, Sigurdsson 1977: 14). Continued dependence on the fishing industry, which in recent years has provided between 75 and 90% of export revenues (Sigurdsson 1977: 5), is not desirable, because economic instability has resulted from fluctuations in fish catches and product prices (Coull and Jónsson 1979: 129, Magnússon 1975: 210, OECD 1976: 22). Nor is it possible; expansion of the fishing industry will ultimately be limited by the need to conserve the fish resource (Coull and Jónsson 1979: 129–131), and increasing mechanization means that fishing and fish-processing can do little to absorb the anticipated growth of manpower in the next two decades (Björnsson 1981, Magnússon 1975: 210). This growth is due to the relatively young age-structure of the population. Furthermore, in the past decade the Icelandic economy has been plagued by unacceptably high rates of inflation, up to 50% in some years (OECD 1975: 5), and an unfavourable balance of payments position, owing partly to the escalation of world oil prices since 1973 (National Economic Institute 1979: 25, 39, OECD 1976: 5), as well as to the necessity of importing most raw materials and many manufactured goods (OECD 1975: 17, Sigurdsson 1977: 5).

Icelandic policy-makers argue that industrial diversification represents the only feasible strategy that will simultaneously tackle the problems of rising material aspirations, economic instability, inflation, and the balance of payments deficit (Sigurdsson 1977: 3, 14–15). Only industrial diversification can generate the basic jobs to absorb future manpower (table 1). Because of the lack of fossil fuels and the unacceptable burden that would

Table 1 Employment forecasts, 1980–2000

	1980	1985	1990	1995	2000
Agriculture	7,800	7,400	6,500	6,000	5,500
Fishing	5,100	5,000	5,000	5,000	5,000
Fish processing	8,500	8,500	8,500	8,500	8,500
Construction	11,300	12,000	12,500	13,000	13,000
Industry	17,900	19,400	20,900	22,900	23,900
Services	54,200	60,400	66,900	70,600	76,800
Total	104,800	112,700	120,300	126,000	132,700

Source: National Economic Institute, Reykjavík

be placed on the economy by continued imports of these materials at the necessary levels, the policy of industrial diversification calls for the large-scale development of domestic energy resources (Ingvarsson 1975: 211–212, OECD 1976: 8, Sigurdsson 1977: 1). Fortunately for Iceland, the combination of geologic, topographic, and climatic conditions means that the island has a vast hydroelectric potential (at least when viewed on a per capita basis) estimated at 35,000 Gwh per annum and a geothermal power potential of 20,000 Gwh per annum, the latter resource being exploitable not only for electrical generation, but also for space heating (Haarde 1980: 34).

The need for rapid industrialization, a process that has already begun with the Straumsvík aluminum plant (Ingvarsson 1975: 212), is recognized by all political parties (Gíslason 1973: 61). The only essential political differences concern the speed of development and the level of foreign participation (Gíslason 1973: 61, Jóhannesson 1975: 56). Given these national economic goals, regional considerations then become important (Ingvarsson 1975: 211–216). These include the presence of suitable resource opportunities for regional economic development, the availability of local manpower, and the need to allocate projects to various regions as equitably as possible, so that all Icelanders have a chance to enjoy the benefits (Haraldsson 1981 pers. com., Jónsson 1981 pers. com.).

Energy Resources

Three distinct but interrelated objectives form the basis of current Icelandic energy policy: to harness domestic resources to develop the economy, to guarantee all Icelanders access to required energy supply at minimum cost, and to reduce imports of fossil fuels by substituting domestic energy whenever possible (Björnsson 1981, Haarde 1980: 32). Strategies directed toward the first objective effectively began in 1904 with the construction of a 9 kw facility at Hafnarfjördur, near Reykjavík. This was followed by the inauguration of utilities in most towns and villages, and the construction in 1922 of a 1,032 kw facility on the Ellidáar river in Reykjavík (subsequently expanded to 3.2 Mw). The first development to serve regional as opposed to purely local needs came with the initiation of a series of power plants on the Sog river in southwestern Iceland. Between 1933 and 1960 three stations were constructed, with a total capacity of 89 Mw. More recently, and still in the southwest, the Thjorsá river has been developed. The 240 Mw Búrfell station was completed in 1972, the 150 Mw Sigalda station came on-line in 1978, and another station at Hrauneyjafoss (140 Mw) is currently under construction (Haarde 1980: 34–35, Maríusson and Björnsson 1975: 217–220).

The most rapid progress has taken place over the last two decades. Since 1962, the capacity of public power plants has increased by more than 400%, while production has increased by well over 300% (National Economic Institute 1979: 98). Hydroelectricity has accounted for most of this increase in capacity and production. The use of diesel fuel for electricity generation has always been far less important than hydro and is currently declining (National Economic Institute 1979: 98). The harnessing of geothermal energy for power production is a relatively recent phenomenon, the most notable example of the latter being the new 70 Mw geothermal power station at Krafla in north Iceland (Haarde 1980: 34).

The distribution of electricity has been uneven spatially because of the needs of the population and industrial centres (fig. 1). Other regions have not been entirely neglected, but development has been slower elsewhere.

In addition to electricity production, the use of natural hot water for space heating has played an important role in the harnessing of Iceland's domestic energy resources. The city of Reykjavík, several smaller centres, and some rural areas have been heated by natural hot water for several decades, and further expansion is both planned and under construction (Haarde 1980: 34).

Fulfilment of the second energy policy objective, guaranteed access to energy, is necessarily difficult and expensive, particularly in the case of electricity, due to the constraint of a widely dispersed and low-density population outside the southwestern part of the island. Given the spatial concentration of the major generating facilities and the desirability of

Energy Development, Tourism, and Nature Conservation in Iceland 391

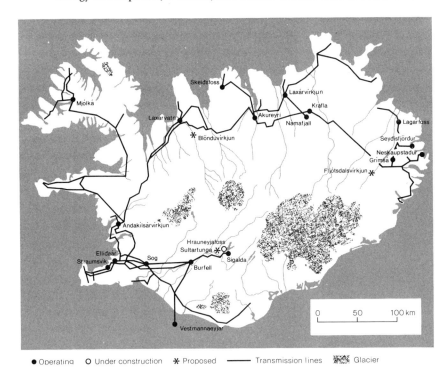

Fig. 1. Power stations and transmission lines. Source: National Economic Institute, Reykjavík.

supplying electricity to all populated parts of the country (some of which were formerly experiencing difficulties in energy availability), the need for a national electrical transmission line system was recognized in the early 1970s (Economic Development Institute 1979: 1). The first stage, a north-south line between the southern hydroelectricity plants and the Krafla station, was completed in 1977. This connected the most densely populated areas. Subsequent completion of lines to the eastern region and the Westfjords (fig. 1) has served to satisfy all present local demands throughout Iceland (Haraldsson 1981 pers. com.). The choice of future locations for power installations can now be made in the context of a national energy plan, rather than on the basis of purely local or regional considerations.

Further development of hydro and geothermal power, in conjunction with the expansion of the natural hot water space-heating system, can do much to satisfy the third objective of Icelandic energy policy, substituting domestic resources for imported fossil fuels. With the escalation of world oil prices since 1973, attaining this objective has become of pressing importance. High oil prices have had a significant negative effect on the

Icelandic economy, since the cost of oil imports rose from $27 million in 1973 to $78.5 million in 1979 (Haarde 1980: 32).

In the long run, however, high oil prices may prove beneficial to Iceland, because they have provided an incentive to develop domestic resources more rapidly than was planned a decade ago (OECD 1974: 27). For example, it was forecast in the early 1970s that natural hot water would reach only about 66% of the population by 1977, but with increased oil prices natural hot water is more economical, and by 1979 geothermal energy for space heating had reached 73% of the population. Plans are under way to increase access to a total of 77% (Haraldsson 1981 pers. com.). In Akureyri, the largest town outside the Reykjavík conurbation, the development of a local source of natural hot water has led to an absolute decline in electricity consumption. Previously much of Akureyri's power was generated in a diesel-fueled plant (Ottested 1981 pers. com.). Connection of the Eastfjords to the national grid reduced regional consumption of diesel fuel by 90% between 1977 and 1979 (Haraldsson 1981 pers. com.). In Iceland as a whole, oil sales for space heating and power generation declined by almost 46% between 1973 and 1979 (Haraldsson 1981 pers. com.). Furthermore, the use of relatively cheap domestic energy to process imported raw materials for export means that Icelandic energy is now, in effect, being exported (Haarde 1980: 32, Ingvarsson 1975: 212).

In 1979 the structure of energy use was as follows: space heating constituted the single largest consumer of energy, predominantly (76%) in the

Table 2 Energy sold to users, 1979

	Hydro (GWh)	Geothermal (GWh)	Oil (GWh)	Total (GWh)
Space heating	376	4793	1154	6323 (44.6%)
Industry	1706	591	1646	3943 (27.8%)
Fishing	–	–	1908	1908 (13.5%)
Transport	–	–	1578	1578 (11.1%)
Trade and services	163	3	3	169 (1.2%)
Other	243	4	4	251 (1.8%)
Total	2488 (17.6%)	5391 (38.0%)	6293 (44.4%)	14,172 (100.0%)

Source: National Economic Institute, Reykjavík

form of natural hot water (table 2). Industry was the second-largest energy consumer, as well as the largest consumer of both electricity and oil. Third and fourth in importance were the fishing industry and transportation, their consumption being entirely in the form of oil. As a consequence of the aforementioned changes in the composition of Icelandic energy resource consumption this picture will undoubtedly change, and Icelandic energy analysts estimate that the use of oil can be reduced from 44.4 to 34.0% of energy sold to users, while hydroelectricity could expand from 17.6 to 19.2%, and geothermal energy from 38.0 to 46.8% (Björnsson 1981). Clearly there are limits to the feasibility of further substitution, especially in transportation and the fishing industry, but opportunities still remain in other industrial sectors. In addition, partly as a result of high gasoline prices, there are signs of an emerging orientation toward energy conservation among the public (Haraldsson 1981 pers. com., National Economic Institute 1980: 32–34).

Recreation and Tourism

The recreational activities of both Icelanders and tourists constitute an important land use and source of revenue for Iceland. The island is richly endowed with recreational landscape resources whose unique features, outstanding scenery, and relatively untouched natural environments provide the opportunity for wilderness travel, sightseeing, photography, and nature study. Three national parks, Thingvellir, Skaftafell, and Jökulsárgljúfur (fig. 2), have been established on state-owned land (Náttúruverndarrad 1978). They are of interest for either their landscape, vegetation, or fauna, or because their historical importance is such that it is considered desirable to maintain and protect the landscape and natural features. In the latter category is Thingvellir National Park, established in 1928, with an area of 2,800 ha. It commemorates and preserves the site of the Althing, the world's first parliament, which was founded in 930 A.D. and held annually at the magnificent basalt canyon that is the primary feature of the area. More important for scientific and recreational reasons is Skaftafell National Park, which was established in 1968 and is the largest of the three national parks, having an area of 20,000 ha. It includes three glaciers partially or completely within its boundaries, glacial rivers and sandurs, and spectacular views of Vatnajökull. Not only is its vegetation representative of the rest of Iceland, but rare plant species occur there as well. The park also has a great diversity of insect and bird life.

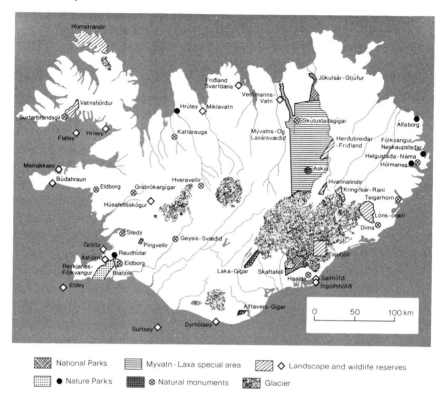

Fig. 2. National Parks and Conservation areas. Source: Nature Conservation Council, Reykjavík.

The third park, Jökulsárgljúfur, was established in 1973 and has an area of 15,000 ha. It is noted for its impressive canyon with sheer cliff walls, the waterfall Dettifoss, the huge depression of Ásbyrgi, the craters and lava flows of a volcanic landscape, an unusually luxuriant vegetation, and varied bird life (Nature Conservation Council 1980).

No statisitics are collected about Icelanders' extra-urban recreational activities, nor does the Iceland Tourist Board make any special attempts to encourage Icelanders' recreational use of the natural environment (Thórgilsson 1981 pers. com.). To a large extent this is due to the relatively informal nature of their recreational behaviour and the absence of commercialism, even in sports. Nevertheless in recent years participation has increased markedly in swimming, alpine and cross-country skiing, and horseback riding. Similarly there has been a sharp growth of rural summer cottage use (Boucher 1975: 374–380). To some extent, trends in recreation participation can be attributed to urbanization; the Reykjavík conurbation alone accounts for about 54% of the population, while the remaining towns amount to a further 24% (Hagstofa Islands 1981). This may result in a need to escape the stress of urban life from time to time in order to achieve the benefits of open space and an environment relatively untouched by man.

Considerably more data are available regarding tourism. Between 1947 and 1967, the annual number of foreign visitors increased fourfold. A marked acceleration of growth began in 1964 and culminated with 74,019 visitors in 1973. The latter figure represents an increase of 321% over the previous decade. The tourist trade has been rather more erratic since 1973; although a record of almost 77,000 visitors was reached in 1979, the following year witnessed a decline of about 14% (Kristinsson 1980: 183, Thórgilsson 1981 pers. com.). According to the Iceland Tourist Board this decrease was due to increased air fares and hotel prices, the general world economic recession, and a drop in the number of stopover passengers on Icelandair's formerly inexpensive North Atlantic route. However, advance bookings in 1981 indicated that this trend was likely to reverse and that growth in tourism will resume in the 1980s (Thórgilsson 1981 pers. com.).

Fishing the famous salmon and trout rivers of Iceland is an important tourist activity, as is visiting the historical sites characteristic of the "land of the sagas" (Kristinsson 1975: 237). The predominant reason for visiting Iceland, however, is to enjoy the unique natural environment and particularly the volcanic and glacial landscapes. According to a 1973 survey of departing tourists, one-half of the respondents cited the volcanoes and geysers, the mountains and glaciers, the geology and scenery in general, and the birdlife as reasons for visiting Iceland, and more than four-fifths of the sample described these as the "most enjoyable and interesting" aspects of Iceland (Thórgilsson 1981 pers. com.).

The basic objective of Icelandic tourist policy is to increase the number of foreign visitors (Thórgilsson 1981 pers. com.). Many Icelanders, however, react negatively to the presence of large numbers of foreigners. This is not surprising when it is recognized that the number of foreign visitors in recent years has been equal to as much as one-third of the native population. On the other hand, Icelanders do perceive the benefits of tourism. Revenues from tourism amount to about 5% of the Gross National Product (Thórgilsson 1981 pers. com.), and because tourist activity is spatially concentrated this revenue can be of great local importance. Young people especially benefit from the employment opportunities generated by the summer tourist trade (Jónsson 1981 pers. com.). At the same time, tourism-generated income helps to balance the foreign expenditures of the many Icelanders who travel abroad each year (36% of the population in 1978, and 32% in 1979). Finally, given the island's relatively isolated location, tourism is viewed as an important element in the maintenance of Iceland's cultural contact with the rest of the world, particularly Europe and North America (Thórgilsson 1981 pers. com.).

Resource-Use Conflicts and Nature Conservation

The development of Iceland's resources for energy, recreation, and other uses will continue to bring substantial material benefits to the people of Iceland. A basic principle of resource exploitation, however, suggests that economic benefits are rarely forthcoming without at least some cost or consequence. These costs are not always or necessarily directly economic in nature, but may occur as resource-use conflicts or environmental impacts. Compromise is necessary, whether it be by foregoing or modifying the actions leading to leading to impacts because the consequences are anticipated to exceed the benefits, or by accepting the impacts as a price that must be paid for the benefits that accrue from resource exploitation (Jackson and Dhanani in press).

Balancing the conflicting goals of resource exploitation and environmental conservation in Iceland is largely the responsibility of the Nature Conservation Council, instituted in 1971 under the Nature Conservation Act. According to this act, which explicitly recognizes potential conflicts and the inevitable compromises, the purpose of nature conservation is "to encourage the intercourse of man and nature in such a way that life or land be not needlessly wasted, nor sea, fresh water or air polluted; to ensure as far as possible the course of natural processes according to their own laws, and the protection of exceptional and historical aspects of Icelandic nature; and to enhance the nation's access to and familiarity with nature" (Náttúruverndarrad 1975).

Energy-Related Conflicts, Impacts, and Compromises

Despite the pressing need to exploit domestic resources, most Icelanders agree that energy development cannot proceed without respect for ecological and aesthetic considerations (Haarde 1980: 36, Maríusson and Björnsson 1975: 220). It is widely recognized that as much as one-third of the technically and economically harnessable hydroelectricity resources should be permanently set aside from development (Sigurdsson 1977: 2). In this context there are four main noneconomic and nontechnical constraints: the preservation of rivers and falls because of special scientific, aesthetic, or historical importance; the need to minimize the flooding of vegetated areas, especially grazing land; the protection of wildlife habitat; and the desirability of avoiding further concentration of hydro development in volcanic areas (Jónsson 1981 pers. com., Reynisson 1981 pers. com.). However, since only about 7% of Iceland's hydroelectric and geothermal potential has so far been harnessed, and since it has been forecast that economic conditions will require the development of only about 12% of the total energy potential by the year 2000 (Haarde 1980: 34–36), a considerably degree of flexibility is possible in energy planning and in the choice of sites for development.

The kinds of compromise involved in energy planning can be illustrated by a brief description of the case of the Laxá development in northern Iceland, and of the three proposed large-scale hydro projects that were under review in 1981 (fig.1). In 1973 work was completed on the third stage of a hydroelectric installation on the Laxá River, the capacity of which was considerably less than 25 Mw originally planned. The original project would have fulfilled electricity needs for Akureyri and other parts of the north, but the development was reduced in scale as a concession to opposition from local farmers, who were concerned about potential impacts on farmland and on the valuable salmon resource of the river. Adjustment to the anticipated power deficit took the form of construction of a 7 Mw diesel-powered station in Akureyri, the purchase of electricity from the national system, and as described above, the tapping of a local source of natural hot water for space heating uses (Ottested 1981 pers. com., Reynisson 1981 pers. com.).

As far as the proposed projects are concerned, only one will eventually be constructed. The Blönduvirkjun project in north Iceland would have a planned capacity of 177 Mw and would require a reservoir of 420 Gl. According to the Nature Conservation Council (NCC), little serious ecological impact would be involved in this project. It would flood some valuable local grazing land and is therefore opposed by local farmers, but otherwise there is general support for the project in the region. A considerably larger development, Fljótsdalsvirkjun, with a capacity of 328 Mw and a reservoir of 745 Gl in the first stage, has been proposed for eastern Iceland. In order to realize the full economic benefits of this project some form of power-intensive industry would have to be established in the east, which may not be immediately feasible. Also, some tracts of vegetated land would be flooded, to which the NCC is generally opposed. It has publicly stated its preparedness to compromise on this, however, in exchange for the permanent conservation of an important breeding area for Icelandic geese elsewhere in Iceland.

A third project, Sultartunga (124 Mw and a reservoir of 60 Gl), would be located adjacent to other sites already developed in southern Iceland. Some opposition has been raised owing to the hazards associated with locating the bulk of the generating capacity in a volcanically active region, but the NCC favours this project because the ecological impacts would likely be less serious than those of the two other proposed installations (Haraldsson 1981 pers. com., Reynisson 1981 pers. com.).

Recreation-Related Conflicts, Impacts, and Compromises

Like the development of energy resources, recreational activities cannot take place without some degree of environmental impact, especially since these activities tend to be concentrated in space and time. At many tourist attractions the predominant activities are merely sightseeing and photography, but other activities have reached such proportions as to become of major concern to the NCC. For example, the use of four-wheel drive vehicles in the interior of the island may result in the removal of vegetation and erosion of soil if these vehicles depart from marked trails. Efforts are therefore made to control movement within the interior, and visitors are discouraged from deviating from designated routes. Control is made easier to some extent by the fact that more than one-half of the visitors to the interior go there as part of guided tours, rather than independently (Reynisson 1981 pers. com.).

Other environmental concerns include water pollution and littering, which may become problematic, especially in intensively used campgrounds. A case in point is the accessible and popular but ecologically sensitive area of Landmannalaugar, where the hot pools are in danger of becoming polluted due to the large number of bathers, and the fragile vegetation is becoming excessivly trampled. Access permits have been considered but rejected thus far, and attempts are being made to divert tourists to other areas. Similar problems have occurred in Skaftafell National Park where, as a consequence of improving access upon completion of the last link in the circuminsular road in 1974, the annual number of visitors has increased from about 300 to 20,000. Problems of soil erosion, soil compaction, and impacts on vegetation and wildlife have subsequently developed. Motorized access has been prohibited, and a system of trails, guides, and regulations for behaviour has been set up so that recreational uses do not exceed carrying capacity. The problem is rather more difficult in Jökulsárgljúfur park, where vehicles have traditionally been allowed.

Attempts have been made to minimize impacts by regulating access and constructing a new road that avoids sensitive vegetated areas (Reynisson 1981 pers. com.).

These examples illustrate the central conflict involved in recreation and tourism in Iceland, which ultimately stems from the competing objectives of protecting natural areas while enabling Icelanders and foreign tourists alike to enjoy the benefits. Consequently the pre-eminent question facing policy-makers is the acceptable level of compromise in allowing free access while minimizing the environmental impacts inherently due to recreational land use.

Nature Conservation

In view of the potential problems associated with energy and recreational resource use, the NCC has been charged with the responsibility of ensuring that development proceeds with a minimum of resource-use conflict and environmental impact. This is accomplished in two ways: by setting aside areas that are protected from development, and by regulating the use of land, water, and other resources outside the protected areas (Reynisson 1981 pers. com.).

The NCC has the responsibility and power to identify and designate protection areas for aesthetic, scientific, historical, and ecological reasons (Náttúruverndarrad 1975). To date, in addition to the three national parks and the Mývatn-Laxá special area, 16 wildlife reserves, 17 natural monuments, and 8 nature parks amounting to a total area of about 80,000 ha have been instituted, as well as 10 landscape reserves with an average size of about 21,000 ha (Náttúruverndarrad 1978, see fig. 2). The NCC also develops and administers regulations relating to the use of driving trails in the interior, littering, pollution, and the location of summer cottages. The preservation of vegetated areas is also seen as an important priority (Reynisson 1981 pers. com.).

As far as the development of energy resources is concerned, the NCC must be informed at the initiation stage of all plans, and has the right to express views regarding the location and scale of projects. Indeed, in order to minimize conflicts between conservation and energy development, a standing committee of the NCC and representatives of the energy sector has been meeting regularly since 1972 to discuss present and future projects (Reynisson 1981 pers. com.). The committee has sponsored an inventory of waterfalls, lakes, rivers, and geothermal areas classified according to conservation priorities (Gardarsson 1978, Thórarinsson 1978*a*, 1978*b*). Twenty-one waterfalls have been designated in the highest priority class, which the NCC recommends should be permanently excluded from future plans to increase hydroelectric generating capacity.

It is clear that Icelanders are aware of the conflicts and impacts inherent in the large-scale use of resources; they have taken steps to ensure that the benefits of resource use are achieved at a minimum cost to the environment, while acknowledging that some degree of sacrifice is inevitable (Sigurdsson 1977: 8–9). In certain instances environmental quality must be compromised if economic objectives are to be fulfilled; for example, it may be necessary to sacrifice a valuable vegetated area if the Fljótsdalsvirkjun project goes ahead, and to sacrifice a tract of grazing land in the case of the Blönduvirkjun scheme. Similarly, allowing free access to the outstanding landscapes of the national parks will inevitably result in the degradation of these resources. Elsewhere, it may be necessary to forego the use of a particular river or waterfall for hydroelectric production, due to its outstanding ecological, aesthetic, or recreational values (Gíslason 1973: 63). This was acknowledged in the case of the Laxá hydro project in north Iceland. Similarly, it is recognized that ecological as well as economic and technical considerations define the island's hydroelectric potential. In this sense, pragmatism and compromise are the key characteristics of nature conservation in Iceland.

Conclusions

External conditions and domestic factors have required that Iceland develop a variety of economic opportunities based on the potential inherent in its physical resources. However, population growth, urbanization, and increasing material living standards, which may be traced to the benefits of resource use in the past, simultaneously create the demand for new resource uses; these in turn may lead to resource-use conflicts and environmental impacts. Icelanders, in the decisions they make about energy development, recreation and tourism, and nature conservation, are clearly seeking to achieve a balance between the competing objectives of economic growth and the preservation of the natural environment.

Some striking similarities between Iceland and other northern lands may be identified. These include the low population density, the vast and as yet largely untapped resource potential, the internal and external pressures creating a perceived need to exploit these resources, and the ecologically sensitive characteristics of the northern environment, which require that the scale and type of resource development be evaluated with great care. In many respects the problems facing Iceland resemble those currently confronting northern policy-makers and planners in Canada, the United States, Scandinavia, and the Soviet Union. This examination of the Icelandic case might prove instructive for these areas, suggesting that northern development need not imply either wholesale exploitation without regard for conservation, or that the benefits of northern resource potentials be sacrificed permanently. In the final analysis, decisions about acceptable levels of resource development in the North should be based on some form of compromise in values regarding economic development and environmental conservation. While improved information, new technologies, and better management techniques are necessary for rational resource planning, the future of Iceland and other northern lands will ultimately depend on how economic and environmental values are defined and on how the goals and objectives are formulated.

Acknowledgments

For their assistance in providing information and unpublished material in Iceland, Sigfús Jónsson, Valur Arnthorsson, Gunnar Haraldsson, Knutur Ottested, Arni Reynisson, and Birgir Thórgilsson are gratefully acknowledged. The use and interpretation of the data and other information, however, remain the responsibility of the author.

References

Boucher, A. 1975. Leisure in Iceland. pp. 374–380. *In* J. Nordal and V. Kristinsson. Iceland 874–1974. Central Bank of Iceland, Reykjavík.

Björnsson, J. 1981. On Icelandic energy resource management and energy policies. Public lecture, Akureyri, 16 January.

Coull, J.R., and S. Jónsson, 1979. Iceland and the cod war. Geography 64(2): 129–133.

Economic Development Institute. 1979. Transmission line systems in Iceland. 3 pp. mimeo.

Eggertsson, T. 1975. Structure of industry. pp. 176–179. *In* J. Nordal and V. Kristinsson. Iceland 874–1974. Central Bank of Iceland, Reykjavík.

Ellíson, M. 1975. Fisheries. pp. 191–203. *In* J. Nordal and V. Kristinsson. Iceland 874–1974. Central Bank of Iceland, Reykjavík.

Gardarsson, A. 1978. Vatnavernd: Islensk Vatnakerfi og Verndun Theirra. Fjölrit Nr. 4. Náttúruverndarrad, Reykjavík. 40 pp.
Gíslason, G. T. 1973. The problem of being an Icelander. Almenna Bókafélagid, Reykjavík. 92 pp.
Haarde, G. 1980. Domestic energy: Just a fraction tapped so far. Atlantica and Iceland Rev. (Spring-Autumn): 32–36.
Hagstofa Íslands. 1981. Bradabirgdatölur mannfjöldans I. des. 1980. 2 pp. mimeo.
Haraldsson, G. 1981. Economist. Nat. Econ. Inst., Reykjavík.
Ingvarsson, G. 1975. Power intensive industries. pp. 211–216. In J. Nordal and V. Kristinsson. Iceland 874–1974. Central Bank of Iceland, Reykjavík.
Jackson, E. L., and A. D. B. Dhanani. 1984. Resources and resource use conflict in Alberta. In P. J. Smith and B. M. Barr. Environment and economy: essays on the human geography in Alberta. Univ. Alberta Press, Edmonton.
Jóhannesson, T. 1975. An outline history. pp. 33–57. In J. Nordal and V. Kristinsson. Iceland 874–1974. Central Bank of Iceland, Reykjavík.
Jónsson, S. 1981. Economic geographer. Nat. Econ. Inst., Reykjavík.
Kristinsson, V. 1975. Communication and tourism. pp. 231–237. In J. Nordal and V. Kristinsson. Iceland 874–1974. Central Bank of Iceland, Reykjavík.
———. 1980. Samgöngur og ferdamál. Fjardamál á Tidindi 27(3): 174–189.
Magnússon, G. K. 1975. Manufacturing industries. pp. 204–211. In J. Nordal and V. Kristinsson. Iceland 874–1974. Central Bank of Iceland, Reykjavík.
Maríusson, J. M., and S. Björnsson. 1975. Power resources. pp. 217–224. In J. Nordal and V. Kristinsson. Iceland 874–1974. Central Bank of Iceland, Reykjavík.
National Economic Institute. 1979. The Icelandic economy, developments 1978–1979. Reykjavík. 111 pp.
———. 1980. The Icelandic economy, developments 1979–1980. Reykjavík. 93 pp.
Náttúruverndarrad. 1975. Nature Conservation Act, English translation. Reykjavík. 13 pp. mimeo.
———. 1978. Fridlýstir stadir á Íslandi og náttúruminjaskrá. Reykjavík. 16 pp.
Nature Conservation Council. 1980. Skaftafell National Park, Jökulsárgljúfur National Park. Reykjavík. 16 pp.
Nordal, J. 1975. Economic policy. pp. 238–242. In J. Nordal and V. Kristinsson. Iceland 874–1974. Central Bank of Iceland, Reykjavík.
OECD. 1974. Economic surveys: Iceland. Organisation for Economic Co-operation and Development, London.
———. 1975. Economic surveys: Iceland. Organisation for Economic Co-operation and Development, London.
———. 1976. Economic surveys: Iceland. Organisation for Economic Co-operation and Development, London.
Ottested, K. 1981. Managing director. Laxárvirkjun, Akureyri.
Reynisson, A. 1981. Director. Nature Conservation Council, Reykjavík.
Sigthórsson, G. 1975. Agriculture. pp. 179–181. In J. Nordal and V. Kristinsson. Iceland 874–1974. Central Bank of Iceland, Reykjavík.

Sirgurdsson, J. 1977. The Icelandic economy, riches under risk. Nat. Econ. Inst., Reykjavík. 15 pp.
Thórarinsson, S. 1978a. Fossar á Íslandi. Fjölrit Nr. 2. Náttúruverndarrad, Reykjavík. 51 pp.
———. 1978b. Hverir og Laugar: Ölkeldur og Kaldavermsl. Fjölrit Nr. 3. Náttúruverndarrad, Reykjavík. 14 pp.
Thórgilsson, B. 1981. Director. Icelandic Tourist Board, Reykjavík.

Aklavik, Northwest Territories: "The Town that did not Die"

William C. Wonders
Heather Brown

Introduction

For 50 years Aklavik was the undisputed "capital" of the Mackenzie Delta and the western Canadian Arctic. In 1953 the federal government decided to relocate the community, and in 1961 Inuvik was officially opened by Prime Minister John Diefenbaker. Aklavik, however, instead of disappearing, remained very much alive despite its displacement as regional capital by Inuvik. Although an initial out-migration of residents occurred to the new town, a reverse flow also developed, and the older community's population again increased.

This paper examines the nature of the population movement between the two centres and seeks to explain some of the main reasons for Aklavik's survival in the face of its much publicized "replacement," based on conditions 20 years after the relocation survey was launched. The experience in the Mackenzie Delta communities provides valuable insight into the complexities of the problem of relocation. It is hoped that the experience will reduce (if not eliminate) the social and economic costs of mistakes in other northern settlements.

Background

The main physical geographic features of the Mackenzie Delta have been set out by Mackay (1963) and Kerfoot (1972), and its regional setting by

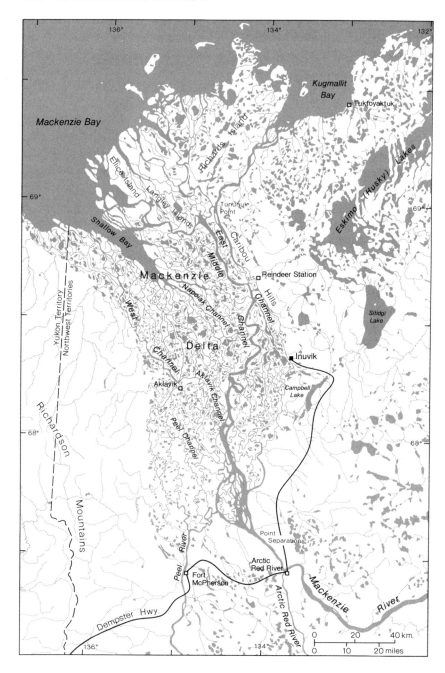

Fig. 1. The Mackenzie Delta, Northwest Territories.

Wonders (1972). It is considered the twelfth-largest river delta of the estuarine-filling type in the world (Merrill et al. 1960), extending 240 km from Point Separation to the Beaufort Sea, and has a maximum width of about 80 km. It is bounded on the west by the Richardson Mountains, which rise to over 750 m, and on the east by the 150-m-high Caribou Hills. Approximately one-half of the delta is water, including innumerable channels and myriad ponds. The upper level of the silt making up the balance lies only 3 to 4 metres above normal river level, and much of it is flooded during spring break-up. Though spruce line most of the higher channel banks in the southern half of the delta, willows, alders, and sedges dominate most of the land surface, with tundra in the north.

Aklavik was established in 1912 as a fur trade post within the concave bend of the Peel Channel, one of the main navigation routes through the Mackenzie River Delta. The local area provided excellent habitat conditions for muskrat, and trappers were attracted by rising prices for muskrat pelts. Kutchin Indians from the Peel River and coastal Inuit from both east and west moved into the delta. They were joined by white trappers who had made their way down the Mackenzie River or by other whites who had abandoned the recently defunct commercial whaling of the western Arctic. The influx created a distinctive local ethnic mosaic: Inuit, Indian, white, and Métis. Since then this assemblage has characterized delta residents, particularly the Aklavik residents.

Aklavik's importance arose from its central location in the delta (fig. 1) and the convenience this offered trappers as the principal trading post for muskrat pelts. The real base for permanent settlement came with the arrival of the missionaries—the Anglicans in 1919, the Roman Catholics in 1926 (Robinson and Robinson 1946)—and the schools and hospitals they later introduced (Wolforth 1966). Additional trading posts and stores were established, and federal government officials were stationed in the community. Native people erected permanent dwellings in Aklavik in the late 1930s and 1940s, as the town became the most important centre in the western Arctic.

The expanded government presence in the North in postwar years required large new investments in construction for the western administrative centre. Aklavik's site limitations were considered too serious to permit the necessary new construction. These constraints included a confined building area (between the eroding river and swamps and ponds), permafrost with up to 60% of the soil volume composed of ice, susceptibility to flooding, poor surface drainage, and the inability to construct an all-year airstrip close by (Merrill 1960, Robertson 1955).

In 1953 the federal government decided to move the entire settlement to a new location within the delta area. It did this without prior consultation with local residents. Nine possible sites were considered by the survey team; the ultimate choice was "East 3" on the east bank of the Mackenzie River's East Channel, 55 km east of Aklavik by air and 110 km by water. The government expected the original Aklavik to disappear as "New Aklavik" developed, but despite an initial slump in local population, old Aklavik persisted and remains. The original settlement retained its name, necessitating a new one—Inuvik—for its scheduled "replacement."

The federal government somewhat belatedly launched the Mackenzie Delta Research Project in the second half of the 1960s. This program attempted "to describe and analyze the social and economic factors relating to development in the Mackenzie Delta. Particular attention was directed toward the participation of the native people of the area, and the extent to which they are making effective adjustments to changes brought about by government and commercial expansion in the north" (A. J. Kerr in Wolforth 1966: ii).

The major recent foci of economic activity in the delta area have been the search for oil and natural gas and the critical role the region is expected to play in related pipeline systems. Native groups and others such as the Canadian Arctic Resources Committee (1976) voiced concern about the possible environmental and social impacts. Subsequently Mr. Justice Berger recommended deferment of any pipeline construction for 10 years, until native land claims are settled (Berger 1977). Acceptance of this deferment by the federal government has appreciably slowed the pace of economic activity in the delta.

The Impact of Inuvik on Aklavik During the Construction Period, 1955 to 1961

The late 1950s was a time of major economic and social change in the Mackenzie Delta. These changes were initiated almost completely by forces from outside the area. Defence requirement resulted in construction of the Distant Early Warning (DEW) Line stations. Increased government responsibilities resulted in the establishment of Inuvik. Wage employment created by construction at Inuvik came at a time when fur prices declined due to falling world demand for wild fur. Many native people abandoned their trap lines either temporarily or permanently to seek wage employment. This accelerated the process of change that had already emerged,

Aklavik, Northwest Territories: "The Town that did not Die" 409

Fig. 2. Air view north over Aklavik, Northwest Territories and down the Peel Channel of the Mackenzie River in the delta.

Fig. 3. Air view north over Inuvik, Northwest Territories, on the East Channel of the Mackenzie River in the delta (left).

Table 1 Permanent town population of Aklavik and "bush" population trading furs into Aklavik, 1955-61

Year	Permanent town population					"Bush" population			
	WHITE	METIS	INDIAN	INUIT	Total	WHITE	METIS	INDIAN	INUIT	Total
1955 a	n.a.	n.a.	n.a.	n.a.	820	n.a.	n.a.	n.a.	n.a.	n.a.
1958 b	350	n.a.	200	150	700	34	0	42	733	809
1959 c	188	187	145	148	674	n.a.	n.a.	n.a.	n.a.	n.a.
1961 b	138	53	192	222	605	0	0	0	106	106

Key
a Honigmann and Honigmann, 1970, p. 54
b Clairmont, 1962, p. 2
c Boek and Boek, 1960
n.a. not available

and led to a greater permanency of residence in the community (settlement) than out in the countryside (land). A dichotomy between "land-based" and "settlement-based" activities emerged, as did the concept of "dual allegiance" to land and town, discussed by Honigmann and Honigmann (1970), Smith (1967), Ervin (1968), and Wolforth (1971).

The trend for people to move into Aklavik involved particularly the Indian population, while the majority of the Inuit were still "bush" dwellers[1] in the mid-1950s (table 1). The construction of Inuvik resulted in both temporary and permanent migration of Aklavik residents to Inuvik. It also tended to divert the flow of some "bush" dwellers from Aklavik to Inuvik.

The initial construction phase of Inuvik (1955 to 1958) did not disrupt the normal seasonal routine of local trappers. Most trapped until the middle of June and worked on construction from then until early in September (the "off season" for trappers), when they returned to the land to obtain a supply of fish for the trapping season. Wage employment in Inuvik merely supplemented the declining income from fur trapping. Yet, increasing numbers of native people found settlement-living and its concomitant wage employment attractive, and trapping was abandoned in favour of working in the settlement.

In 1955 the total permanent population of Aklavik was 820, including a substantial number of white residents (Honigmann and Honigmann 1970). In summer the arrival of "bush" trappers almost doubled the population. During the first two summers the movement of people to Inuvik was chiefly seasonal. It involved mainly unskilled workers, who returned in winter to their permanent homes in Aklavik or elsewhere in the delta. A government campaign then began to persuade permanent Aklavik residents to move "voluntarily" to Inuvik, in hopes of emptying the older community of permanent residents. Among the inducements offered were compensation for their old buildings in Aklavik and the allocation of new lots in Inuvik. Nevertheless by spring 1955 the government conceded that many Aklavik residents would not move and if the number was large enough, a trading post, a store, and perhaps a small day school would be provided (*Edmonton Journal* 25 April 1955). Clairmont (1962) reported the permanent town population of Aklavik in 1958 to be 700, a decline of 120 people or 14.7% between 1955 and 1958.

By 1959 Inuvik began to have an even stronger impact on Aklavik. Services and permanent accommodations became available in Inuvik while they were cut back and phased out in Aklavik. In September 1959 a federal school with 30 classrooms opened in Inuvik, along with adjoining hostel accommodation for 500 children from outlying settlements. Many pupils were transferred from Aklavik to the new school as their parents relocated. The latter included representatives of several government departments, hospital personnel, and employees of commercial establishments. Between 1959 and 1961 the Aklavik federal day school population declined from 176 to 126. By September 1961 the two residential schools in Aklavik were closed.

The first Mackenzie Delta liquor store opened in Aklavik in 1959; in 1960 it was transferred to Inuvik. Many local residents viewed this as a particularly reprehensible attempt on the part of government to entice them to Inuvik! The close of the 1960 construction season saw the conclusion of all major contract work, and spring 1961 marked the completion of the relocation of functions from Aklavik.

By 1959 the permanent population of Aklavik had declined to 674, including Army Signals and RCMP personnel (Boek and Boek 1960). Of the 285 native people settled in Inuvik that year, approximately half (148) were from Aklavik or the Aklavik area. There was also an active temporary movement between the two settlements; Boek and Boek (1960) indicate that in July/August 1959, 48 of Aklavik's permanent residents were temporarily in Inuvik and 18 people listing Inuvik as their permanent residence were visiting Aklavik.

A number of "push-pull" factors influenced Aklavik residents in their decision either to move to Inuvik permanently or to remain in Aklavik when the construction phase of the new town ended. "Pull" factors in Inuvik included the federal government's offer of over 70 jobs to relatively untrained workers and another 57 positions to men who could be trained to fill them (Honigmann and Honigmann 1970). Additional employment was available from private employers. Also, housing was available in Inuvik; the government gave natives the opportunity to purchase, at a relatively low cost, some of the "512"[2] frame houses that had been used during the construction phase.

"Push" factors in Aklavik also encouraged people to move; one of them was the security of wage employment in Inuvik. The decline in local employment opportunities was obviously a vital factor. They steadily worsened as more agencies transferred to Inuvik. Moreover, trapping increasingly proved to be an inadequate source of income because of the reduced value of furs (Black 1961). By 1960 complete reliance on full-time trapping had declined to the point where only one permanent native resident in Aklavik earned his livelihood from the land (Spence 1961).

Aklavik's population sank to its nadir in 1961, the year of Inuvik's official opening. In that year the federal census listed Aklavik's population as 599 (Clairmont 1962). Forty-five percent of the population was under 15 years of age, compared with the Canadian average of 34%. Only 22.8% of the permanent residents were white, compared with 50% prior to the transfer of government officials to Inuvik (Clairmont 1962).

Declining employment possibilities in Aklavik led to greater reliance on welfare payments. In 1957–58, $13,360 was spent in social assistance; in 1958–59, $16,953; and in 1959–60, $27,066. In recognition of the need for new employment possibilities in Aklavik the government initiated two new undertakings in 1959: a fur garment project and a logging/sawmill operation. Twenty-five women (23 Inuit, 2 Indian) were enrolled in the training course for the garment project, which was designed to assist the former household help of the Aklavik-based government personnel. The business was successful and permanent. The logging project employed 25 men, including 13 seasonal workers, but was abandoned in 1966.

Social expenditures in Aklavik declined to $21,450 in 1960–61, largely because of these two employment projects, but local employment conditions remained unsatisfactory. Out of a labour force of 331 in 1961 only 69 people, or 21% of the labour force, were permanently employed (Bissett 1967). Nevertheless many Aklavik residents stubbornly refused to move, and the government was forced to recognize the fact. The *Annual Report* of the Commissioner of the Northwest Territories in 1961–62 acknowledged that "Many Aklavik residents did not take advantage of the government assistance offered to those moving and elected to remain." In 1961 the future of Aklavik remained uncertain.

The Characteristics of Aklavik-to-Inuvik Migrants, 1955 to 1961

Local interviews carried out in 1974 established certain characteristics of the migrants to Inuvik during the 1955–61 construction phase. The migrants included those who had remained in Inuvik permanently ("permanent migrants") and those who subsequently had moved back to Aklavik ("return migrants"). Out of a total of 105 interviewees[3] in Aklavik, 45 had lived in Inuvik during the construction period. In Inuvik, out of a total of 96 migrants from Aklavik who were interviewed, 60 had moved to Inuvik between 1955 and 1961. Similar numbers of males and females were involved in the two groups.

The migration involved primarily the young adult age groups both for permanent migrants to Inuvik and for return-migrants to Aklavik. Out of a total of 105 interviewed migrants, 76% were between 15 and 44 years of age in 1961. This compares with the 42.5% of Aklavik's permanent population who were in the same age group in 1961.

The small number of people in each occupational category make it difficult to determine how selective the migration to Inuvik was in terms of occupation. It is significant, however, that of the returning migrants over one-third gave their original occupation as hunter/trapper/fisherman, while none of the people who remained in Inuvik were so classified.

The move to Inuvik altered the occupations of most permanent migrants and return-migrants. By 1974 there were higher percentages of the permanent migrants to Inuvik in such occupations as professional/technical, service and recreation, and transportation and communication. At the same time, 11.7% of them were classed as labourers, compared with only 1.7% prior to their move from Aklavik.

Prior to their move, 73.6% of the return-migrants were hunters and trappers in Aklavik. In Inuvik 67.7% of the return-migrants were employed as labourers, but seem to have acquired other skills. While some return-migrants to Aklavik (24.4%) resumed hunting and trapping, a higher proportion of them were in other occupations by 1974, particularly in service and recreation (25.0%).

The most typical sequence of occupation for return-migrants was from fishing/hunting and trapping in Aklavik to either labouring or service jobs in Inuvik. On returning to Aklavik the majority of the service workers remained in this category or engaged in craft production. Although most labourers returned to fishing/hunting/trapping, they also moved into professional/technical, service/recreation, and managerial categories.

Indian and white migrants appear to have adjusted to wage employment in Inuvik more readily than did Métis and Inuit migrants; 80% of the Indian peoples and 77% of the white migrants remained there. Métis and Inuit migrants appear not to have adjusted as well, with 79% and 56% respectively of them returning to Aklavik.[4]

Employment was the major factor encouraging migration to Inuvik. It totalled 80% of the reasons given by both groups of people (permanent migrants and return-migrants). Forty-two people moved to Inuvik to work on the construction of the town: 18 were transferred there; 24 moved in search of a job. Fewer people moved because of school (6 persons), or because their parents moved to Inuvik (5 persons).

The reasons for returning to Aklavik as stated by return-migrants are set out in table 2. Nearly one-half of the respondents listed family and friends in Aklavik as the reasons for their return. Twenty percent indicated that food was more easily obtained in the Aklavik area than it was in Inuvik,

Table 2 Reasons why Aklavik return-migrants moved back (more than one reason possible)

Reason for move back	1955–61		1962–74	
	NUMBER	PERCENT	NUMBER	PERCENT
Family/friends	31	48	9	36
Food easier to obtain	13	20	5	20
Hated Inuvik	8	12	2	8
Too much drinking in Inuvik	5	8	3	12
Business	2	3	2	8
Job ended	2	3	1	4
Retirement	1	2	–	–
School	1	2	–	–
Health	1	2	–	–
Too expensive in Inuvik	–	–	1	4
Personal	–	–	2	8
	64	100	25	100

Source: Field survey, July–August 1974.

due to better hunting, trapping, and fishing in the area. A dislike of Inuvik as a place to live and of the alcohol-related habits of its townspeople were also significant factors.

Over one-half the migrants who subsequently returned to Aklavik had stayed in Inuvik for less than 2 years. Having made their choice of place of residence, both groups appeared satisfied. Eighty-five percent of Inuvik's permanent migrants and 91% of Aklavik's return-migrants wished to remain in their respective settlements.

Movements Between Aklavik and Inuvik and the Characteristics of Migrants, 1962 to 1974

Fewer permanent migrants moved between Aklavik and Inuvik during the period 1962–74 (50 families) than during the preceding construction period (105 families). From 1962 to 1974, 28 families moved back to Inuvik. Over the entire period 1955–74, Inuvik had a net gain of 96 families from Aklavik.

As was the case during the construction period, the majority of migrants (82%), both permanent and return, were aged between 15 and 44 years.

Table 3 Ethnic breakdown of Aklavik's population, 1961–78

	Indian	Inuit	Métis[1]	White[1]	Total	Sources
1961	192 (31.7%)	222 (36.7%)	53 (8.8%)	138 (22.8%)	605	Clairmont, 1962
1965	114	282	203[2]		629	Bissett, 1967
1966	n.a.	n.a.	n.a.	n.a.	611	1966 Census
1967	160	282	129	42	613	Makale et al., 1967
1970	n.a.	n.a.	n.a.	n.a.	675	Simenon et al., 1970
	450[3]		200[4]		650	Northwest Territories Government Survey, 1970
1971	n.a.	n.a.	n.a.	n.a.	665	1971 Census
1973	173	319	259[2]		708	Forth et al., 1974
1974	152 (20.4%)	321 (43.0%)	205 (27.5%)	68 (9.1%)	746	Field survey, July–August 1974
1978	352 (43.6%)	345 (45.3%)	84[4] (11.0%)		761	Government of Northwest Territories

Notes

1. Yearly discrepancies between Métis and whites indicate the differences in interpretation of a person's ethnic status.
2. Métis and whites were taken together for these years.
3. Indian and Inuit combined.
4. "Others."
n.a. not available

What might not have been expected was that 93% of the return-migrants were in that age category.

Some differences from the preceding period may be observed in the ethnic nature of the migration. Inuit made up an even larger percentage of the total number of migrants (68.0%) during this period than during the earlier one (45.7%). Once again a high percentage (57.2%) of the return-migrants were Inuit. This might have been expected, given that in 1974 they numbered 321 and represented the largest proportion (43%) of Aklavik's permanent population. The three other ethnic segments in Aklavik's population of 746 at that time were: Métis—205 (27.5%), Indian—152 (20.4%), and white—68 (9.1%). It would appear that Inuit responded mostly readily to short-term employment opportunities in the 1962-74 period, and were extremely mobile. For example, they have been actively involved in Arctic oil and gas activities (Hobart 1981).

A larger proportion of Indian residents from Aklavik appear to have adjusted to living permanently in Inuvik, with six permanent migrant families and only one return-migrant. As in the preceding period, however, three-quarters of the Métis migrants to Inuvik during this period subsequently returned to Aklavik. The comparative interchange of percentages of Métis and white populations in Aklavik in 1961 and 1974 (table 3) is due to a different interpretation of the term "Métis." Clairmont (1962) restricts its use to people who have white fathers, while we use the term to describe all people whose parents belong to different ethnic groups among those identified. The Government of the Northwest Territories community data indicate that 15% of Aklavik's population was white in 1961—a figure more in keeping with our definition. Most of the white population had been transferred to Inuvik by 1961, although a small number were transferred between 1961 and 1962. Certainly the white element in Aklavik's population has declined greatly in recent years. Between 1961 and 1974 the number of white permanent residents in Aklavik declined by 5.9% due to out-migration, some of it to places outside the delta.

While 90.8% of the migrants who had moved back to Aklavik by 1961 were in the unskilled labour or service categories, only 50.1% of those who lived in Inuvik between 1962 and 1974 were employed in these categories (35.8% labourers, 14.3% service workers). The decline in the absolute numbers of these moving to Inuvik and in the category of unskilled labourers reveals the change from Inuvik's construction phase to its administrative or functional stage. The latter demanded skills and educational qualifications not possessed by the vast majority of Aklavik's native population. As in the earlier period, however, there was a general change in return-migrants' occupational categories following their stay in Inuvik; whereas 17 had been "fishermen, etc." prior to the move, only 5 listed the same occupation after their return in 1974; only 1 was in the professional/technical category earlier, compared with 7 afterward.

Migration was also selective according to educational qualifications. Permanent migrants included larger numbers with more educational training. Twenty-four percent of Inuvik's permanent migrants had completed grades 10 to 12, compared with 18.6% of its return-migrants (and only 6.4% of the migrants who returned to Aklavik in the 1955–61 period).

The motivations of those who migrated in both directions in this later period had not changed radically from the earlier period. Seventy percent of the total reasons for migration referred to economic/employment factors as the major influences encouraging migration to Inuvik, while the motivations for return-migration were dominated by "social" reasons and by the local availability of country food (table 2).

Factors Favouring Migration to Inuvik

By far the most important factors affecting migration between Aklavik and Inuvik related to the economic conditions of the two settlements. The type and availability of employment in both Aklavik and Inuvik influenced the migrants' decisions. The availability of employment for unskilled workers in Inuvik during the construction phase acted as a major attractive force, encouraging people to move to the new town. Later, employment became available in other occupational categories, and attracted other migrants from Aklavik. This "pull" factor in Inuvik was stronger than the "push" factor in Aklavik, which was related to declining fur prices and lack of other employment.

Aklavik's wage economy was very unstable and subject to fluctuations over both the long and short term. Resource harvesting is subject to cyclical fluctuations in the number of fur-bearing animals and to fur price changes. Permanent employment opportunities in the settlement were extremely limited during the 1962–74 period and were in large part provided by government projects. In 1967 Makale indicated that only 71 persons were permanently employed out of a total population of 618 persons. Social assistance payments were drastically higher, reaching $65,636 in 1964–65 and oscillating to $62,641 in 1973–74 (Social Welfare Records, Inuvik).

Nearly one-third of the permanent migrants to Inuvik in the later period cited the relative abundance and quality of Inuvik's facilities (recreation, shopping, education, health, accommodation), as compared with Aklavik's, as factors encouraging migration to Inuvik and discouraging a return to Aklavik. The wisdom of the decision to move to Inuvik seemed to be confirmed by three particularly destructive floods in Aklavik during three successive years in the early 1960s. Many of Aklavik's facilities were left in a substandard state during the early part of this period because the government did not wish to invest in new services in a settlement that it thought would be eventually abandoned (Heinke 1974). In January 1966 the Aklavik day school was described by a senior government official as "probably one of the worst fire traps we have anywhere in the North" (*Edmonton Journal* 28 January 1966). The same year, 38 houses in Aklavik were considered to be in need of replacement (Bissett 1967), and one-half of the remaining houses were overcrowded. Only two nurses in a small frame structure provided local health care.

Factors Favouring Migration to Aklavik

As noted previously the lack of educational qualifications prevented many Aklavik residents from finding permanent employment in Inuvik, particularly after the construction phase. Though the level of formal education is now much greater among young settlement natives they still are unqualified for many jobs in Inuvik, despite the government's vigorous program to increase the percentage of natives in northern jobs.

The economic and social significance of hunting, trapping, and fishing must not be underestimated as a factor encouraging migration back to Aklavik, where natural conditions are much more favourable, and discouraging further movements to Inuvik. Despite the declining numbers of full-time trappers, a large portion of Aklavik's population continues to rely on cash and in-kind revenues from traditional resource-harvesting activities. Return-migrants to Aklavik point out that if a person is unemployed in Aklavik, he is at least able to obtain food from the land. Moreover, the traditional attachment to the land, even if only on a part-time basis, is an important part of living for many of the native residents.

Social considerations were perceived as being the most important factors encouraging migration back to Aklavik. Kinship ties are particularly strong among the native segment of the population. The friendliness of Aklavik is considered to be the major attraction of the community by both return-migrants and permanent migrants alike. Its smaller size and quieter life style are also significant attractions.

Finally, it should be noted that the physical deterioration of Aklavik was checked, and the process was reversed. The stubborn tenacity of local residents won widespread general support and the grudging admiration of government, which was symbolized by the June 1966 announcement that a new school building would be constructed (*Edmonton Journal* 3 June 1966). The new school was named after a popular former local school principal, A. J. "Moose" Kerr, who played an influential role in the resistance and gave Aklavik its motto, "Never Say Die." The era of uncertainty ended. Additional new facilities, both government and private and including modern housing, were brought into Aklavik. (In some recent cases, additional sources of seasonal income from oil and gas activities have permitted residence in Aklavik as readily as in Inuvik.) Aklavik's population has again increased to approximately that of the pre-Inuvik period. A lively interchange between Aklavik and Inuvik occurs on a temporary basis in the form of shopping and family visits. These reflect the close ties between the two communities and are not now seen as a threat by the residents of Aklavik.[5] While the community has yielded its former dominant regional role to Inuvik,[6] it is proud of its distinctive qualities and has confidence in its long-range future.

Summary

The establishment of Inuvik resulted in a significant loss of population for Aklavik. But Aklavik did not disappear, as was originally anticipated. The transference of government and commercial facilities altered the ethnic nature of Aklavik, notably by reducing the number of white permanent residents. Many native residents remained in Aklavik or returned to it despite wage employment opportunities in Inuvik. Their limited formal education reduced those opportunities after the construction phase. The greater social attraction of Aklavik as a permanent place of residence outweighed the attraction of wage income for many natives, and appreciable numbers of them returned to live in Aklavik.

Native peoples in the North have increasingly been drawn into permanent settlements over the past four decades (Wonders 1960, 1970). They have not completely readjusted to the new life style, despite better housing and expanded educational and training opportunities. Indeed, there has been renewed interest by many in a "return to the land," as demonstrated by the recent "outpost camp" program in the Northwest Territories. Most native northerners are now settlement-dwellers, however, and are directly affected by any changes in those settlements.

One might ask whether the Aklavik relocation might have been better handled and whether similar relocations could occur again. The answer to both must be in the affirmative. Though the physical aspects of the relocation of Aklavik were adequately handled, the residents were not involved before the fact. This could not happen now, given northern residents' increased awareness, political sophistication, and power over the past quarter-century. That power is recognized especially in the Northwest Territories, but less so in the provinces. There, the southern concentration of population inevitably weights political decisions.

At the same time external forces continue to pressure the North. "Aboriginal rights," "environmental protection," and other principles applying to the North are much more strongly established than they were when the Aklavik relocation was decided. Nevertheless the external forces cannot be resisted indefinitely, whether they be the demand for new mineral or fuel resources or for new sources of hydroelectric power.

The James Bay project resulted in the recent relocation of Fort George, a community of some 2,200 Indian and white residents. Though the original site had been occupied since the early eighteenth century and relocation cost $50 million, it was shifted five miles upstream. There the native residents did have input into the planning of the new settlement, but there was no doubt that they would have to relocate. The hydro power was needed not by northern residents, but by those in the South.

In northeastern Alberta the small Indian community at Fort McKay is on a much smaller scale, and has not had to relocate. It has experienced significant pressures, however, from the external forces that saw Fort McMurray expand with oil sands development. If or when the third oil sands plant is completed and a new service community constructed to the north, the prospects for the survival of the native community are in doubt. Still farther north, Fort Chipewyan suffered major setbacks in the settlement economy when the Bennett Dam was constructed far upstream on the Peace River to provide power for the South. Currently the Slave River's hydro potential is attracting attention from southern forces, with major implications for northern settlements.

In the case of the Churchill River hydroelectric project, the Saskatchewan government decided against development and disruption of the traditional northern settlements. This is an exception to the usual pattern, yet the lesson of Aklavik does seem to have been learned. Arctic Bay, the Inuit community 30 km from the new Nanisivik mine on northern Baffin Island, has been allowed to continue to function, with residents mixing wage employment with traditional hunting and trapping. It is hoped that similar compromises may be possible in future. Northern development need not totally disrupt traditional northern settlement life.

Acknowledgments

The authors are grateful to the Boreal Institute for Northern Studies, the University of Alberta, for a research grant in support of field work. The willing co-operation of many people in Aklavik and Inuvik, despite the frequency with which they so often are besieged by outside "researchers," was sincerely appreciated. We hope that the results will contribute to a greater understanding of their way of life. Thanks also go to the staff of the Inuvik Research Laboratory, and in particular to John Ostrick, who was always so ready to help in any way.

Notes

1. Local usage of the term "bush" is applied here. Native residents refer to the "bush" as being any place outside the settlements where they engage in trapping, hunting, and fishing, regardless of the vegetation in the area.
2. So called because they provided 512 ft^2 of space.
3. Usually the head or the spouse of the head of the household. In Aklavik 35 households were missed because no resident could be contacted during the summer. People who died or migrated elsewhere between 1955 and 1974 have inevitably been excluded from analysis.
4. No attempt is made precisely to define the complex question of ethnicity here; the four categories are used in Aklavik and the Mackenzie Delta, and statistics reflect the personal designations of the individuals enumerated in the field survey. Data are from interviews in Aklavik and Inuvik, July-August, 1974.
5. Space limitations do not permit commentary here on these "temporary migrations" between Aklavik and Inuvik. In 1973-74, 96 Inuvik residents visited Aklavik and 103 Aklavik residents visited Inuvik.
6. The 1979 population of Aklavik was 750 and that of Inuvik 2,892. (Government of the Northwest Territories, Information Services.)

References

Berger, T. R. 1977. Northern frontier, northern homeland: The report of the Mackenzie Valley Pipeline Inquiry. Minister of Supply and Services Can., Ottawa. 2 vols. 213 pp. and 268 pp.

Bissett, D. 1967. The lower Mackenzie region: An area economic survey. Dep. Indian Affairs and Northern Dev., Northern Administrative Branch, Industrial Division, Ottawa. 501 pp.

Black, W. A. 1961. Fur trapping in the Mackenzie Delta. Geogr. Bull. 16: 62–85.

Boek, W., and J. K. Boek. 1960. A report of field work in Aklavik and Inuvik, N.W.T. Dep. Indian Affairs and Northern Dev., Northern Coordination Res. Centre, Ottawa. [Unpub.]

Canadian Arctic Resources Committee. 1976. Mackenzie Delta: Priorities and alternatives. Ottawa.

Clairmont, D. H. J. 1962. Notes on the drinking behaviour of Eskimos and Indians in the Aklavik area, 62-4. Dep. Northern Affairs and National Resources, Northern Coordination Res. Centre, Ottawa. 13 pp.

Edmonton Journal. Various issues. Edmonton.

Ervin, A. M. 1968. New northern townsmen in Inuvik. Mackenzie Delta Research Project 5. Dep. Indian Affairs and Northern Dev., Northern Science Res. Group. 24 pp.

Forth, T. G., I. R. Brown, M. M. Feeney, and J. D. Parkins. 1974. Mackenzie Valley development: Some implications for planners. Rep. No. 73–45. Canada Task Force on Northern Oil Development, Ottawa.

Heinke, G. W. 1974. Report on municipal services in the N.W.T. 73-1. Dep. Indian Affairs and Northern Dev., Northern Science Res. Group, Ottawa. 165 pp.

Hobart, C. W. 1981. Impacts of industrial employment on hunting and trapping among Canadian Inuit. pp. 202–217. In Milton M. R. Freeman, ed. Proceedings, First International Symposium on Renewable Resources and the Economy of the Canadian North, Banff, Alberta, May 1981. Assoc. Can. Univ. for Northern Studies, Ottawa.

Honigmann, J., and I. Honigmann. 1970. Arctic townsmen. Can. Res. Centre for Anthropologie, Saint Paul Univ., Ottawa. 303 pp.

Kerfoot, D. E., ed. 1972. Mackenzie Delta area monograph. 22nd Int. Geogr. Congress, Canada. Brock Univ., St. Catharines. 174 pp.

Mackay, J. R. 1963. The Mackenzie Delta area, N.W.T. Memoir 8. Dep. Mines and Technical Surveys, Geogr. Branch, Ottawa. 202 pp.

Makale, Holloway and Associates. 1967. Aklavik, N.W.T. planning report and department plan. Part 1. Makale, Holloway and Assoc. Ltd., Edmonton. 68 pp.

Merrill, C. L., J. A. Pihlainen, and R. F. Leggett. 1960. The new Aklavik: Search for the site. Tech. Pap. No. 89, NRC 5575. Nat. Res. Council, Div. Building Res., 6 pp. Reprinted from The Engin. J. 43: 52–57.

Robertson, G. 1955. Aklavik: A problem and its solution. Can. Geogr. J. 50(6): 196–205.

Robinson, M. J., and J. L. Robinson. 1946. Exploration and settlement of Mackenzie District, N.W.T. Can. Geogr. J. 32(6): 246–255, 33(1): 42–49.

Simenon, E. T., and G. W. Heinke. 1970. An evaluation of the municipal services in Mackenzie Delta communities. Ottawa. 119 pp.

Smith, D. G. 1967. The Mackenzie Delta: Domestic economy of the native peoples. Mackenzie Delta Res. Proj. 3. Dep. Indian Affairs and Northern Dev., Northern Coordination Res. Centre, Ottawa. 59 pp.

Spence, I. 1961. Human adaptation in the Mackenzie Delta. M.A. thesis. McGill Univ., Montreal.

Wolforth, J. R. 1966. The Mackenzie Delta: Its economic base and Development. Mackenzie Delta Res. Proj. 1. Dep. Indian Affairs and Northern Dev., Northern Coordination and Res. Centre, Ottawa. 85 pp.

———. 1971. The evolution and economy of the delta community. Mackenzie Delta Res. Proj. 11. Dep. Indian Affairs and the Northern Dev., Northern Coordination and Res. Centre, Ottawa. 163 pp.

Wonders, W. C. 1960. Postwar settlement trends in the Mackenzie Valley area. Geogr. Annal. 42(4): 333–338.

———. 1970. The Canadian northwest: Some geographical perspectives. Can. Geogr. J. 80(5): 146–165.

———. 1972. Field guide, field tour Ea 1 (the Canadian northwest). 22nd Int. Geogr. Congress, Canada. Dep. Geogr., Univ. Alberta, Edmonton. 60 pp.

Acknowledgments

The editors want to acknowledge the individuals and organizations who encouraged and supported this tribute to Don Gill. Our thanks to the people at the University of Alberta Press. Mrs. Norma Gutteridge, Director, advocated the project right from the first time we approached the Press. Mary Mahoney-Robson spent many hours on the difficult task of coordinating the efforts of everyone involved in the project. Joanne Poon supplied her skills in designing the book. We would also like to thank John Eerkes, a freelance editor from Toronto, who edited the individual papers.

The Department of Geography at the University of Alberta was very helpful, especially Dr. I. A. Campbell, in the initial stages of the project. Later, the department underwrote the costs of the cartography. Our appreciation to Geoff Lester and his cartographic staff, especially Michael Fisher and Stephanie Kucharyshyn for all their fine work.

The Boreal Institute for Northern Studies assisted us in many ways. Our thanks to Robbie Jamieson, Director, and Anita Moore, Executive Assistant, for everything they have done for us.

Finally, we want to thank the University/Community Special Projects Funds Committee and the Boreal Insitute for Northern Studies for the financial support they provided to cover the costs of production of this book.

About the Contributors

Dennis Andriashek
Wildlife Technician,
Canadian Wildlife Service,
Edmonton, Alberta
B.Sc. (Alberta 1972)

Mr. Andriashek has worked on the Polar Bear Project for eleven years. He has worked on population studies of polar bears in the Beaufort Sea, High Arctic, Southeastern Baffin Island, Northern Labrador and Hudson Bay. He has also conducted ecological studies of seals in several areas of the Western and High Arctic. In 1978 he spent six weeks with a team from the University of Minnesota studying seals around the Antarctic Penninsula. Mr. Andriashek is a life member of the Arctic Institute.

William Ralph Archibald
Research Biologist, Wildlife Research Section,
British Columbia Fish and Wildlife, Victoria, British Columbia
B.Sc. (Guelph 1971)

Mr. Archibald has worked on wildlife research projects in the north since 1973. He worked for the Northwest Territories and the Canadian Wildlife Service and has published papers on seals and polar bears. Mr. Archibald worked for the Yukon Game Branch from 1976 to 1981 as fur biologist and while there conducted field investigations on the ecology of pine marten. He is currently working for British Columbia Fish and

Wildlife, studying the impacts of logging on grizzly bears in coastal British Columbia.

Heather Brown
Housewife,
Fife, Scotland
B.A. (Birmingham 1972)
M.A. (Alberta 1975)
DIP.T.P. (Heriot Watt, Edinburgh 1980)

Mrs. Brown conducted her graduate research on population movement and development strategies for communities in the Mackenzie River Delta area. In addition to private consulting she has worked for the Scottish Tourist Board Research and Planning Division and for the Regional Planning Authority in Edinburgh, Scotland. She is a member of the Royal Town Planning Institute in Great Britain. Mrs. Brown is currently at home looking after her two active toddlers.

Dr. George Calef
Wildlife Biologist,
Environmental Services North,
 Whitehorse, Yukon
B.Sc. (Chicago 1964)
Ph.D. (British Columbia 1971)

Dr. Calef conducted research into the ecology and management of caribou, wood bison, and various other wildlife species in the continental Northwest Territories while employed by the Northwest Territories Wildlife Service. His book, *Caribou and the Barren-Lands,* won The Governor General's award for non-fiction in 1981. His current work involves wildlife research and public education in the Yukon Territory.

Wendy Calvert
Biologist,
Canadian Wildlife Service,
 Edmonton, Alberta
B.Sc. (Alberta 1973)

Ms. Calvert has worked on the Polar Bear Project for ten years. Her field work in the Beaufort Sea and the High Arctic has concentrated on studies of the behaviour of polar bears; distribution, abundance, and habitat selection of the arctic seals; and underwater vocalizations of seals and walruses. She designed and manages the computer programs for coordinating the national polar bear data base. Ms. Calvert is Secretary of the Federal-Provincial Polar Bear Technical Committee.

John Donihee
Chief Environmental Planning and Assessment, Department of

Renewable Resources,
Government of the Northwest Territories, Yellowknife, Northwest Territories
B.Sc. (Carleton 1975)
M.Env. Studies (York 1977)

Mr. Donihee has worked on a variety of northern land management issues, ranging from wildlife habitat management and assessment of the impacts of major project proposals to the negotiation and development of a land-use planning process for the Northwest Territories. He has been with the Department of Renewable Resources in Yellowknife since 1979. His current work involves the implementation of Northern land-use planning policies.

Dr. M. C. English
Research Associate, National Water Research Institute,
Canada Centre for Inland Waters, Sault Ste Marie, Ontario
B.Sc. (Trent 1976)
M.Sc. (Alberta 1979)
Ph.D. (McGill 1984)

Dr. English did his M.Sc. under Don Gill on a study at the ecology and environment of the Slave River Delta, Northwest Territories. Continuing to work in subarctic systems, his Ph.D. at McGill was focussed on the role of snow in the nutrient budgets of terrestrial and aquatic ecosystems. Dr. English is currently conducting research on the interactions of acidic snowmelt runoff in Canadian Shield Ecosystems for the National Water Research Institute, Canada Centre for Inland Waters. He is a member of the American Association of Limnologists and Oceanographers.

Dr. Hugh M. French
Professor, Departments of Geography and Geology,
University of Ottawa, Ottawa, Ontario
B.A. (Southampton 1964)
Ph.D. (Southampton 1967)

Dr. French's research has focussed upon periglacial geomorphology, particularly in the western Canadian Arctic and the High Arctic islands. His recent research has included the environmental impact of resource development in permafrost regions. Dr. French is author of a widely-used textbook on periglacial geomorphology, *The Periglacial Environment*, and is currently Chairman of the International Geographical Union Commission on the significance of Periglacial Phenomena, and Chairman of the Permafrost Subcommittee, Associate Committee on Geotechnical Research, National Research Council of Canada.

Dr. Manfred Hoefs

Chief of Wildlife Management, Government of Yukon, Whitehorse, Yukon

B.Sc. (Manitoba 1967)
M.Sc. (Manitoba 1968)
Ph.D. (British Columbia 1975)

Dr. Hoefs is currently Chief of Wildlife Management for the Yukon Government's Wildlife Branch. He has worked on and supervised numerous projects on most northern game species. During 1981, he spent one year at the University of Giessen's Wildlife Institute and the Federal Government's Wildlife Research Laboratory at Bonn, West Germany studying Roe deer and Mufflon sheep. Dr. Hoefs has published 35 scientific papers in North American and European professional journals, with topics ranging from vegetation studies to range ecology of ungulates, ornithology, and wildlife management. Most of his research efforts have been concentrated on wild sheep. Dr. Hoefs is also a regular contributor to Swiss and German popular wildlife and hunting magazines, having some 40 contributions to his credit. He is a member of the Wildlife Society, Arctic Institute of North America, American Society of Mammalogists, Arbeitskreis fur Wildbiologie at the University of Giessen and the Council for Northern Wild Sheep and Goats, where he is a Board member and the organizer of the 1984 International Wild Sheep and Goat Conference.

Dr. Edgar L. Jackson

Associate Professor, Department of Geography, University of Alberta, Edmonton, Alberta

B.A. (London School of Economics 1967)
M.A. (Calgary 1970)
Ph.D. (Toronto 1974)

Dr. Jackson's current research interests include studies of environmental perceptions, values and behaviour, especially with regard to recreation and energy resource use, as well as resource use conflict and environmental conservation. However, for the past two decades, he has also maintained both a personal and an academic interest in Iceland, which he has visited many times since 1965. His M.A. thesis dealt with the historical geography of Iceland, and he has published numerous papers on his research on Iceland. Dr. Jackson is a member of the Canadian Association of Geographers, and was recently (1982–83) president of the Western Division of the CAG.

Ronald Harvey Jessup
Wildlife Technician, Wildlife Management Branch,
Yukon Department of Renewable Resources, Whitehorse, Yukon
Tech. (Lakeland 1977)

Mr. Jessup is a Wildlife Technician in the fur bearer management section of the Wildlife Management Branch of the Yukon Government. His current research includes a wolverine study, fur bearer inventory and habitant assessment, trapper and public education, and a marten transplant. Mr. Jessup is also responsible for monitoring the annual Yukon fur harvest and liaising with Yukon trappers.

Dr. G. Peter Kershaw
Assistant Professor, Department of Geography,
The University of Alberta, Edmonton, Alberta
B.E.S. (Waterloo 1973)
M.A. (Waterloo 1976)
Ph.D. (Alberta 1983)

Dr. Kershaw has conducted biogeographical research in southern Ontario, the Yukon, the Northwest Territories, Alberta and Alaska. He has studied alpine geomorphological processes in the Rocky Mountains of Alberta and the Mackenzie and Selwyn Mountains of the Yukon and the Northwest Territories. He has also examined environmental impacts of natural and man-induced disturbances in the Mackenzie Mountains, Northwest Territories. At present Dr. Kershaw is conducting research into the bioclimate of boreal mixedwood habitats, the ecology of small mammals, the long-term consequences of oil spills in alpine tundra environments and native plant revegetation of human disturbances. He is a member of The Arctic Institute of North America, The Canadian Field-Naturalists Club, The National Wildlife Federation, the Wildlife Society (USA), the Wildlife Society of Canada, and is active with the Boreal Institute for Northern Studies.

Dr. Arthur M. Martell
Senior Scientific Advisor, Wildlife Research,
Wildlife Research and Interpretation Branch, Canadian Wildlife Service, Environment Canada, Ottawa, Ontario
B.Sc. (Acadia 1967)
M.Sc. (Acadia 1969)
Ph.D. (Alberta 1975)

From 1978 to 1983 Dr. Martell worked for Canadian Wildlife Service in Whitehorse, Yukon, and conducted research on the range ecology of the Porcupine caribou herd. Dr. Martell has also studied the ecology and population dynamics of small mammals in the eastern and western Arctic and, in relation to logging, in northern Ontario. At present Dr. Martell is involved at the policy level with a number of northern conservation issues.

Hugh L. Monaghan
Assistant Deputy Minister,
 Department of Renewable
 Resources,
Government of the Northwest
 Territories, Yellowknife,
 Northwest Territories
B.Sc. (Alberta 1968)
MNRM (Manitoba 1979)

Mr. Monaghan has been involved in resource management in the Northwest Territories since the mid 1960s and is currently the Assistant Deputy Minister of the Department of Renewable Resources. His master's thesis included a recommendation to the Government of the Northwest Territories on constitutional development in the Northwest Territories.

Dr. W. Wayne Pettapiece
Officer-in-Charge, Alberta Pedology Unit,
Agriculture Canada, Edmonton, Alberta
B.Sc. (Manitoba 1961)
M.Sc. (Manitoba 1964)
Ph.D. (Alberta 1970)

Dr. Pettapiece has worked for Agriculture Canada since 1963. He has conducted soil surveys in Alberta, Manitoba and the Northwest Territories. His current interests include regional soil-climate-physiography relationships, soil genesis and mapping methodology.

Dr. William O. Pruitt
Professor, Department of Zoology,
The University of Manitoba,
 Winnipeg, Manitoba
B.Sc. (Maryland 1947)
M.A. (Michigan 1948)
Ph.D. (Michigan 1952)

Dr. Pruitt has studied the ecology of animals in the snow from Newfoundland to Alaska and from Manitoba to Devon Island; as well as in northern and eastern Finland and northern Sweden. He has studied shrews, voles and caribou but in recent years he has concentrated on comparative winter ecology of several subspecies of *Rangifer*. His Taiga Biological Station in Manitoba has become a centre for winter ecology teaching and research. His earlier book *Animals of the North* (1967) has recently been republished in paperback under the title *Wild Harmony*. He also has published a short textbook entitled *Boreal Ecology.*

Dr. J. Stan Rowe
Professor, Department of Crop Science & Plant Ecology,
University of Saskatchewan,
Saskatoon, Saskatchewan
B.Sc. (Alberta 1941)
M.Sc. (Nebraska 1948)
Ph.D. (Manitoba 1956)

Dr. Rowe has worked extensively on landscape (terrain) ecology in Manitoba, Saskatchewan, Yukon, and the Northwest Territories. As a research forester with the Canadian Forest Service he conducted studies both on the autecology of boreal trees and on forest geography, from the latter updating and rewriting the popular *Forest Regions of Canada.* At the University of Saskatchewan since 1967, he has examined vegetation-landform relationships in the arctic, subarctic, boreal and grassland zones, focussing particularly on fire ecology, northern dune systems, and ecological land classification. This research has been supported by the National Research Council, the federal departments of Environment and of Indian and Northern Affairs, and the Saskatchewan Research Council. Dr. Rowe is a member of various national and international botanical, ecological, and environmental societies.

Donald Edmund Russell
Biologist,
Canadian Wildlife Service,
Whitehorse, Yukon
B.S.F. (British Columbia 1972)
M.F. (British Columbia 1976)

Mr. Russell conducted research on the Porcupine caribou herd from 1976 to 1980 while working as a habitat biologist and until 1983 as a caribou biologist for the Yukon Wildlife Branch. The study on the winter range ecology of the caribou herds was a joint Yukon Government and Canadian Wildlife Service project. Mr. Russell is currently working for the Canadian Wildlife Service as senior management big game biologist. He has also conducted research on mountain caribou in relation to the effects of forestry practices on habitat in British Columbia.

Dr. Matti K. Seppälä
Associate Professor, Department of Geography,
University of Helsinki, Finland
B.Sc. (Turku, Finland 1965)
M.Sc. (Turku, Finland 1967)
Lic. Phil. (Turku, Finland 1969)
Ph.D. (Turku, Finland 1971)

Dr. Seppälä's investigations have included periglacial and glacial landforms and processes, loess stratigraphy in Hungary and geomorphological mapping. His field areas have been Fennoscandia,

Central Europe, Alaska, British Columbia and Spitsbergen. His is a member of the Commissions of the International Geographical Union and the International Union for Quaternary Research; correspondant of the International Glaciological Society; member of the Arctic Institute of North America; and working member and President (1984) of the Geographical Society of Finland.

Dr. Norman M. Simmons
Rancher,
Pincher Creek, Alberta
B.A. (Claremont Men's College 1956)
M.Sc. (Colorado 1961)
Ph.D. (Arizona 1969)

Dr. Simmons studied large mammals in the Mackenzie Mountains of the Northwest Territories between 1966 and 1973. From 1975 to 1980 he was the Senior Wildlife Manager for the Northwest Territories and was a member of the Mackenzie Valley Pipeline Inquiry Appraisal Team. Dr. Simmons was Assistant Deputy Minister, Department of Renewable Resources, Northwest Territories from 1980 until he retired in 1983 to ranch in southern Alberta.

Dr. Ian Stirling
Research Scientist,
Canadian Wildlife Service,
Edmonton, Alberta
B.Sc. (British Columbia 1963)
M.Sc. (British Columbia 1966)
Ph.D. (Canterbury, New Zealand 1969)

Dr. Stirling has conducted research on Antarctic seals, southern fur seals and sea lions, as well as polar bears and seals of the Canadian Arctic. Although he still maintains an active interest in Antarctic research, his current interests centre around arctic marine mammal ecology — particularly the relationships between seals, polar bears, polynyas, and pack ice. His research has been supported by the Canadian Wildlife Service, Polar Continental Shelf Project, other government departments, the World Wildlife Fund, Canada and the Natural Sciences and Engineering Research Council of Canada. He belongs to several professional societies and is the Chairman of the Federal-Provincial Polar Bear Technical Committee and is a member of the Polar Research Board.

Charles Tarnocai
Soil Correlator of British
 Columbia, Yukon and Northwest
 Territories,
Canadian Department of Agricul-
 ture, Ottawa, Ontario
B.F.S. (British Columbia 1959)
M.Sc. (Oregon State 1967)

Mr. Tarnocai has gained a wide background in soil science, having conducted soil studies and soil surveys in British Columbia, Manitoba, Yukon Territory and the Northwest Territories. He was one of the originators of the Cryosolic Order for the classification of permafrost soils. His current interests include peatlands, permafrost, soils, and the application of soil information for resource surveys. Mr. Tarnocai is chairman of both the National Wetland Working Group of the Canada Ecological Land Classification Committee and the Soil Classification Working Group of the Expert Committee of Soil Survey.

ern Region of the Canadian Wildlife Service. His current research involves studies of the impact on ungulates of vegetation manipulations on experimental watersheds as part of the Alberta Watershed Research Program. He has published extensively on ungulate winter ecology and on the impact of forest management on wildlife. He has also served as associate editor for wildlife for the Forestry Chronicle. Mr. Telfer is a member of the Wildlife Society, the Canadian Society of Wildlife and Fisheries Biologists, the Canadian Institute of Forestry and is active in the International Union of Forest Research Organizations.

Edmund S. Telfer
Research Scientist,
Canadian Wildlife Service,
 Edmonton, Alberta
B.Sc. (New Brunswick 1953)
B.Ed. (Acadia 1962)
M.Sc. (Acadia 1965)

Mr. Telfer's research has centred around forest wildlife inter-relationships in both Atlantic and Western Canada. In particular, he has studied characteristics of the habitat of big game animals and their relationships to snow. From 1978 to 1983 he served as Program Leader for National Parks Research in the Western and North-

Dr. William C. Wonders
University Professor, and
 Professor of Geography,
The University of Alberta,
 Edmonton, Alberta
B.A. (Toronto 1946)
M.A. (Syracuse 1948)
Ph.D. (Toronto 1951)
Fil.Dr.h.c. (Uppsala 1981)

For over 30 years Dr. Wonders has conducted research on a wide range of subjects in northern Canada and northern Europe. However, his major area of research has been on settlement patterns and problems, including historical aspects, in western and northern North America and Scandinavia. Amongst the northern books which he has edited and contributed to are *Canada's Changing North, The North* and *The Arctic Circle*. He is a fellow of both the Arctic Institute of North America and the Royal Society of Canada. He was the founding chairman of the Boreal Institute for Northern Studies. He is a member of many professional organizations including the Canadian Association of Geographers (President 1961–1962), the Association of American Geographers, and the Society of Sigma xi. He has served on the Canadian National Committee of the International Geographical Union, and on the National Advisory Committee on Geographical Research. He was chairman of the Royal Society of Canada's Atlas of Northern Canada Committee, and in 1983–84 was fact-finder on overlapping native land claims in the Northwest Territories for the Minister of Indian Affairs and Northern Development.

About the Editors

Rod Olson, Ross Hastings, and Frank Geddes first met Don Gill in his *Boreal Ecology* class in 1976. The leadership and enthusiasm that Don demonstrated, both in the classroom and in the field, encouraged each of them to pursue undergraduate field studies under him. At the time of Don's death in July 1979, Rod, Ross, and Frank had each initiated M.Sc. research projects under his direction.

Rod Olson
Environmental Climatologist,
Edmonton, Alberta
Dip., Envir. (Mt. Royal College 1974)
B.Sc., First Class Honors (Alberta 1978)
M.Sc. (Alberta 1983)

Mr. Olson is currently an environmental climatological consultant. He has been contracted by the federal, provincial, and private sectors on recent projects involving the Alberta Oil Sands acid rain studies, a wildlife habitat-climate winter severity classification, and the verification of non-standard Arctic climate data. In 1982 he was an invited session chairman to the National Research Council Symposium on the long-range transport of acidic materials, held in Ottawa. Rod completed a B.Sc., with first class honors at University of Alberta in 1978. His Master's research on the local climate of a mountain valley was conducted at

Don's Drysone Ranch, in the Front Ranges of the Rockies, during 1978 and 1979. Before working as a consultant, Mr. Olson had been employed as a climatologist for the Atmospheric Environment Service.

Ross Hastings
Provisional Ph.D. Candidate,
 Department of Botany, Plant Ecology,
University of Alberta, Edmonton, Alberta
B.Sc., First Class Honors (Alberta 1979)
M.Sc. (Alberta 1984)

Mr. Hastings completed his B.Sc. in biogeography under the supervision of Don Gill. He accompanied Don and his graduate students on many field expeditions in western Alberta, northern British Columbia, and the Northwest Territories, serving as a research assistant. Mr. Hastings did his M.Sc. in Botany—Plant Ecology on soil-vegetation relationships in the Mackenzie Delta area. While completing his M.Sc. he was contracted by the Canadian Forestry Service to examine wetland ecology of northern Alberta. He is currently conducting his Ph.D. research in Botany on aspects of snow-vegetation ecology in the boreal forest.

Frank E. Geddes
Wildlife Biologist,
Edmonton, Alberta
B.Sc. (Alberta 1977)
M.Sc. (Alberta 1984)

Mr. Geddes has conducted research in wildlife-related fields since graduating with a B.Sc. in zoology in 1977. In 1978, he began a M.Sc. program under Don and was conducting research on muskrats in the Slave River Delta at the time of Dr. Gill's death. In recent years, Frank has worked on biological considerations of resource developments in Alberta, the Yukon, and Northwest Territories. He is currently pursuing graduate studies in the Department of Geography at the University of Alberta. Frank is a member of the Alberta Society of Professional Biologists and is the Secretary for the Alberta Chapter of the Wildlife Society of Canada.